M000300351

WASTE IS A TERRIBLE THING TO MIND

Risk, Radiation, and Distrust of Government

JOHN WEINGART

RIVERGATE BOOKS
An Imprint of Rutgers University Press
New Brunswick, New Jersey

First published in 2001 by the Center for Analysis of Public Issues,
Princeton, New Jersey
Reprinted in 2007 by Rutgers University Press,
New Brunswick, New Jersey

Library of Congress Cataloging-in-Publication Data

Weingart, John R.
    Waste is a terrible thing to mind : risk, radiation, and distrust of
government : with a new preface / John Weingart.
        p. cm.
    Originally published: Princeton : Center for Analysis of Public
Issues, 2001.
    Includes bibliographical references and index.
    ISBN-13: 978-0-8135-4237-9 (pbk. : alk. paper)
    1. Radioactive waste disposal—Political aspects—New Jersey.
2. Radioactive waste disposal—Risk assessment—Government
policy—New Jersey. 3. Radioactive waste disposal—United States—
States. 4. Transparency in government—United States. I. Title.
    TD898.12.N5W45 2007
    363.72'8909749--dc22   2007012713

A British Cataloging-in-Publication record for this book is available
from the British Library.

Visit our Web site: http://rutgerspress.rutgers.edu

Manufactured in the United States of America

WASTE IS A TERRIBLE THING TO MIND
Risk, Radiation, and Distrust of Government

TABLE OF CONTENTS

TABLE OF ILLUSTRATIONS

Photographs, Charts and Exhibits

Maps

Maps *(continued)*

Cartoons

Preface to the 2007 Edition

The disposal of low-level radioactive waste is but one of many unresolved problems carried over from the late 20th century into the 21st. While nuclear waste is dangerous material that can cause cancer and death, most scientists and engineers believe facilities can be designed and built to isolate it with only minimal risk of any human contact or environmental harm. A solution has remained elusive due to the difficulty of finding agreement upon acceptable levels of risk in an era of general public distrust of government. This quandary has also impeded movement toward good public policy on a wide variety of other issues affecting health and safety.

From 1994 to 1998 I directed New Jersey's program to find a disposal site for the low-level radioactive waste generated within the state's borders. I wrote this book because I thought examination of the extensive federal and state efforts to manage low-level radioactive waste could offer useful perspective and lessons for better addressing many issues requiring risk assessment. The book, first published in May 2001, focuses on the period from 1980 to 2000. This preface continues the story into 2007.

The study of radioactive waste disposal intersects with a number of other important policy areas ranging from the future of nuclear power to more process-oriented matters such as the role of the public in governmental decision making and the great difficulty of making policy that depends on years of study and requires commitments to be sustained over inevitable changes in leadership and administrations. I examine these issues in the book and also try to highlight the various roles, capacities, and expectations of the local, state, and federal levels of government. I also include a journal I kept describing more than nine months of day-to-day life within a government bureaucracy.

Certainly, some aspects of radioactive waste policy are unique to its specific properties and history. But for me, both when I was a government official wrestling with this problem and later as I have tried to reflect upon, write, and speak about the experience, they are more interesting and instructive in a broader context.

For example, nuclear power has been gaining favor in recent years as trepidation about dependence on foreign oil and concern about the contribution of fossil fuels to global warming has increased. If a new national energy policy emerges with more emphasis on nuclear power, however, where will new nuclear plants be located and how will those decisions be made? It should be sobering to proponents of nuclear energy that two decades of efforts by states around the country to build new disposal facilities for the least dangerous waste resulting from nuclear power all resulted in failure.

The story of how New Jersey tried and why it failed highlights some of the obstacles that would confront any initiative to find an acceptable location for a new nuclear power plant. The lessons may be valuable for considering the feasibility of what would be the first such plant approved in the United States since the 1979 nuclear accident at Three-Mile Island in Pennsylvania.

New Jersey's search for a site for a radioactive waste disposal facility emanated from a Congressional decision in 1980 to assign states the responsibility of safely disposing of "low-level radioactive waste." Each state was to secure a location with sufficient capacity to store the low-level waste generated by its nuclear utilities, industries, medical facilities, laboratories, and research universities. The law provided incentives for states to collaborate on building regional disposal operations and allowed groups of states to form regional compacts to exclude waste from nonparticipating states.

The federal law was the result of a successful campaign by the state of South Carolina to change interstate commerce provisions so that its low-level radioactive waste disposal facility, approved back in 1969, would no longer have to accept waste from all who wanted to send it. The facility, known as Barnwell after the county in which it is located, had become an issue in the 1978 race for governor. By that time, it was one of only two operating disposal facilities in the entire country. One in Richland, Washington, served 11 states while Barnwell accommodated the low-level radioactive waste from the other 39. Noting this reality, the ultimately victorious candidate had vowed to keep South Carolina from being "the nation's dumping ground."

This book focuses on the unique voluntary siting process New Jersey employed to try to meet the mandates of the federal law. The premise was that some of the upheaval that would accompany the imposition of a controversial development on an unwilling community could be averted by seeking a municipality to volunteer to host a disposal facility. Most other states relied on more traditional approaches that first analyzed layers of data to identify sites that best met a specific set of criteria before approaching officials and other residents in the potential host community. But regardless of the methodology a state selected and regardless of whether states joined together to try to build one regional operation or, like New Jersey, worked essentially on their own, the outcome was the same—no location was approved and no new facility was constructed. That remains true in 2007 with no groundbreaking anywhere on the horizon.

Throughout the country, each time a location for a disposal facility was proposed, it was defeated by varying combinations of fierce and sustained public opposition, problems arising during the multiyear regulatory review processes, revised economic forecasts, and local, state, and federal politics. Some of the defeats occurred within weeks of a site being identified while others played out over many years.

From the passage of the federal act in 1980 to at least 1995, state governments and administrators of the entities that generate radioactive waste were forecasting an approaching crisis. On July 1, 1994, after letting several earlier deadlines pass, South Carolina finally employed the power it had been given by the federal law and stopped accepting waste from much of the country. The Nuclear Regulatory Commission and the states that were excluded responded by requiring those who generated the waste to now keep it on their site since there was no other disposal location available. Suddenly, each state had radioactive waste collecting at multiple locations. New Jersey, for example, had 100 of these temporary storage areas at sites ranging from Princeton and Rutgers Universities to community hospitals to the Oyster Creek and Salem nuclear power plants.

In retrospect, this was the moment when the efforts to gain approval for a new disposal facility had the greatest chance for success. The problem was no longer abstract. Reporters could easily find a spot in their coverage area or media market where low-level radioactive waste was piling up. They could raise concerns about the training of local staff to properly maintain these areas and the ability of the regulatory authorities to monitor so many of them. Local officials in the surrounding area had new motivation to support the establishment of a regional facility to ensure that the waste was transported away from their communities.

At the same time, state officials could gain public interest and some support by displaying maps showcasing all the temporary sites and arguing that it would be much safer to replace them with one statewide or multistate facility. All recognized that they were in this situation only because of the inability to agree on locations for new disposal facilities.

With the Barnwell, South Carolina, operation off-limits, producers and users of radioactive materials across the country warned that society would suffer if an inability to address the disposal problem forced cutbacks in the exploration of continued and new nuclear applications. Fears that the potential of nuclear medicine might be artificially limited were voiced in California as health care organizations produced studies showing how the increasing uncertainty about future radioactive waste disposal capacity could discourage researchers from pursuing otherwise promising new techniques.

But the crisis did not materialize. In 1995, South Carolina's newly-elected governor stunned those following this issue by reopening Barnwell to the nation's waste. While he added that he would annually reevaluate this policy reversal, his action nonetheless removed the urgency that had been propelling state and regional efforts to build new disposal facilities. Five years later, the next South Carolina governor led the state to yet another major change in policy. He negotiated an agreement that would permit Barnwell to accept low-level radioactive waste from only Connecticut, New Jersey, and his own state. This change was to be phased in over eight years so that generators of waste in the other states that had been relying on Barnwell would have to find a new place to put it by July 1, 2008.

That was the situation at the end of the book: A pending national crisis that had loomed since 1980 had become much more pressing in 1994 and was then abruptly downgraded to a worry in 1995. In 2000, it was resolved for three states while for 36 others it was reelevated to a problem that would have to be addressed over the next eight years.

Early in 2007, however, the South Carolina legislature considered a proposal to reverse course yet again and keep Barnwell open to all 39 states for another 15 years until 2023. Ironically, it appeared that by the 30th anniversary of the 1978 election in South Carolina that created this national issue, the problem would be resolved by agreement among all parties to effectively return to what had been in place originally. Completion of this full circle would come after the expenditure of hundreds of millions of dollars for disposal facility siting efforts around the country, the formation of hundreds of local opposition groups, and the holding of thousands of public hearings.

The bill, cosponsored by one-quarter of the South Carolina House, appeared headed for passage until the Agriculture, Natural Resources, and Environmental Affairs Committee met and, as reported in a front-page headline the next day, voted 16–0 to "Slam Door To Nuclear Waste Site." The panel's action effectively ended the latest effort to keep the Barnwell landfill open longer. One legislator who had initially supported the bill was quoted as saying, "These other states in the United States need to get up off of their backsides and start doing what's right. They want to stomp us in the ground and beat us up and say, 'You bunch of country hicks.' I'm just getting tired of it."[a]

What happens next is as uncertain now as it was six years ago when I finished this book. The major reason the operators of Barnwell were able to propose that the facility continue to accommodate much of the nation's waste is that the nuclear utilities and other generators of low-level radioactive waste implemented dramatic waste minimization practices that radically reduced the area required for disposal. As a result, Barnwell, which would have already reached its capacity if waste volumes had continued as projected in 1980, now has space to house the waste anticipated from the 39 states for many more years.

Waste minimization has been a positive outcome of this journey, propelled by the continuing uncertainty and increasing cost of available disposal options. It has not made the waste less hazardous but it has allowed the nation to continue to avoid dealing with its inability to agree on locations for new disposal facilities.

The other factors causing South Carolina to revisit this issue were largely economic. During the 1990s, a company called Envirocare began accepting what is known as "Class A" low-level radioactive waste at a facility in Clive, Utah, that was already the disposal site for a former uranium mill. Class A waste loses its radioactivity through natural decay within 100 years and constitutes most of the volume of low-level radioactive waste. Much smaller quantities are generated of Classes B and C, which require up to 300 and 500 years respectively to decay to background levels of radiation.

In recent years, states have been choosing Envirocare, now known as Energy*Solutions* Clive Operations, over Barnwell for disposal of much of their Class A waste because it is cheaper. While this has freed up even more capacity at Barnwell, it has also forced South Carolina and the operators of Barnwell to worry about the facility's economic viability. The legislation proposed in 2007 was spurred by concern that if South Carolina sticks with the policy it embraced in 2000, Barnwell will soon have to be supported almost entirely by fees from generators of B and C waste from only three states. This might result in unacceptably high disposal costs or even the shut-down of the facility.

Another future economic consideration is that both Barnwell and Envirocare were purchased in 2005 by Energy*Solutions*. This new company offers to "package, transport, process, store, and dispose of radioactive materials for nuclear utilities, universities, laboratories, and radiopharmaceutical and industrial facilities, and provide decontamination and decommissioning services."[b] As the company seeks to build a major national and international role in all stages of nuclear waste disposal, it also strives to create a positive corporate presence, most prominently through commercial sponsorship of the home of the Utah Jazz, now known as EnergySolutions Arena. Whether one dominant player for low-level radioactive waste disposal will have a positive or negative impact on disposal options, safety, and costs over the long term remains to be seen.

The problem of low-level radioactive waste disposal is not so much on the way to being solved as it is to being defined away. Generators of low-level radioactive waste as well as the affected states would be wise to wait for South Carolina to change direction yet again, or for the state of Utah to allow EnergySolutions to accept Class B and C waste. They have neither need nor incentive to do anything else. It appears there is no longer a market to support a new disposal facility and the reduction in waste volumes means that the onsite areas that nuclear utilities and the other generators designed for temporary storage have capacity for many more years' waste than origi-nally anticipated.

While the national picture on low-level radioactive waste has taken many unexpected turns over the past 25 to 30 years, it does seem that the nation's low-level radioactive waste disposal needs are going to be met without the construction of any new disposal facilities.

Those who believe that problems ignored will just go away may find false comfort in this outcome. However, if the nation has stumbled into this solution to the low-level radioactive waste problem, it has left unanswered important questions of how we can better assess relative risks and arrive at informed decisions about them. The experience of the states that grappled with this issue over several decades offers lessons and warnings for people considering other public policies involving risk and the need to find locations for facilities that society may need but no community wants to host.

All of the nation's operating nuclear power plants and radioactive waste disposal facilities received their initial approvals before the nation's first Earth Day in 1970. While state and federal agencies have issued the periodic license and permit renewals required to keep most of them open, and in some cases have allowed expansions, no major new nuclear operation has been approved and constructed in the modern era of intense regulation, litigation, interest group activity, and distrust of government. A lesson offered by this recent history is that it is much easier to add waste or other hazardous materials to a location already contaminated than to open a facility at an entirely new location.

One aspect of South Carolina's long history of policy machinations on low-level radioactive waste disposal that is surprising to some observers is that local officials and residents living closest to Barnwell have been the most outspoken supporters of keeping the facility open to as much waste as possible. This has been true from the facility's inception through the legislative debate in 2007. While their region receives greater direct economic benefit than the rest of the state through tax revenue and local jobs, one might expect them to have been the most worried about the health and environmental risks that motivated the leading opponents of the facility who lived many miles away. But most of the local area residents, who had seen the facility and perhaps knew people who worked there or had been employed there themselves, apparently concluded that Barnwell was not a bad neighbor and that any risk associated with it was acceptable.

This suggests that efforts to build future facilities that raise fears of public health or environmental impacts may be more readily embraced—or at least tolerated—in areas where there are already similar uses. As new enterprises that produce or use radioactive or other hazardous materials are needed, and if new nuclear power plants are proposed, the first sites policymakers may want to consider are those near existing nuclear or hazardous material facilities or other already contaminated areas. Not only will it be easier to earn community acceptance, but ever-evolving technologies offer the possibility that a proposal for a new facility could include a commitment to remediate the site, leading to a net reduction in the risk it poses.

A potential criticism of a policy to cluster locally unwanted land uses is that it could increase the concentration of environmental burdens, with risk disproportionately falling on people who are poorer, less powerful, and often members of racial minorities. No new facility should be built if the cumulative impact of its operation, combined with existing risk factors, is going to be unsafe, regardless of who lives in the vicinity. But if one could be designed to not only meet all applicable safety standards but to also cause a net reduction in the environmental and health risks, wouldn't it be worthy of consideration? If, for example, an approved new facility was not permitted to begin operating until preexisting hazards had been removed or remediated, it is possible that the result would be a safer neighborhood.

This leads back to the central question: Who defines and determines what is safe? Even before September 11th, threats to public safety and health presented us all with difficult choices both in our daily lives and in forming opinions on a variety of community and national issues. As parents, when should we make our children wear helmets? And as a society, what should we require of the manufacturers of toys, automobiles, and other potentially hazardous products? As individuals, what is a healthy diet? And as a society what should we require of those who grow, process, distribute, and sell food?

Since 2001, the range of such questions has only expanded. Should we consider relocating to communities we believe are less likely to be targets of terrorism? Are there seasons in which we should avoid travel on certain forms of transportation or to certain countries? What analysis, if any, will lead the Federal Aviation Administration to decide that we no longer need to remove our shoes before getting on an airplane?

We all want to be as safe as possible but there times when the likely benefit of the least dangerous option is so small that the additional cost, inconvenience, or deprivation may not be worth it. On the other hand, who will tell us or how will we know when something we want to do is so dangerous that we should give it up?

The answer to all such questions requires a risk assessment. As individuals we may assess the questions intuitively, based perhaps on conversations with friends or an article or news report, or some of us may occasionally do extensive research. As a society, however, we need a structured process to calculate the potential risks, benefits, and costs and then agree upon the best policy that will govern us all.

The need to address these problems is relatively new and it is therefore understandable that we are having such trouble doing it. Amazing scientific breakthroughs, particularly during and since World War II, have led to the creation of thousands of new processes, products, and by-products. This burgeoning knowledge has been followed by growing awareness of the potential dangers posed by some of these new substances, and also by the perfecting of tools and techniques that enable us to measure increasingly minute quantities of them.

Simultaneously, events and movements growing out of the 1960s and 1970s have spawned much greater citizen awareness, activism by individuals and groups, and a general discarding of the unquestioning trust of authority that characterized much of the 1940s and 1950s.

The combined effect of these societal changes has been the production of substances that are beneficial to society but come with by-products known to be hazardous in large quantities, the ability to detect the tiniest amounts of them, and a general inability to agree on the location or sometimes even the existence of a threshold below which exposure to them is safe.

It is the job of government to evaluate potentially hazardous substances and determine which should be banned outright and, for the others, what level is considered safe or acceptable. Findings by public agencies may be considered much less authoritative than they once were, but there simply is no other entity with sufficient research and analytical resources and the ability to enforce its conclusions. Moreover, government is by definition designed to serve the public interest in a much broader sense than can be said of any private sector or nonprofit organization that might be proposed as an alternative.

But to imply that "safe" and "acceptable" are interchangeable, as I did above, points to an immediate disconnect between the scientific analyses necessary to determine permissible levels and the public confidence needed to enforce them. The nature of scientific inquiry is such that the absence of all possible risk cannot be proved. Scientists can demonstrate that the amount of risk posed is very small but they can't make absolute assertions or promises.

Yet the public when confronted with a possible new industrial or waste facility near where they live demands, sometimes explicitly, a guarantee that no further risk will be added to their lives. A scientist who concludes that exposure to an energy-producing plant will lead to a one-in-a-million increased chance of getting cancer will consider the proposal safe and would probably be unconcerned to live nearby; but many people will visualize that data as their child plucked out of a large crowd and given an increased risk of getting cancer. Promises that any added risk will be negligible are not comforting and are often heard as bureaucratic hairsplitting that may be covering up significant danger.

The public hearings where such issues are discussed, whether one views them in person or on television, are often very emotional, even somewhat theatrical events. The fear of some participants is palpable. In these settings, it is easy to notice the concerned citizens who after saying that even the smallest amount of added risk is unacceptable, then step outside to smoke a cigarette. But, their seemingly contradictory actions dramatize a feeling shared by many that whether or not they have any plans to stop smoking, they don't want someone else to make their life feel any more dangerous than it already does.

When the public demands "safety" and a government official offers "acceptable risks," the gulf usually is unbridgeable. This situation is not likely to change unless the public comes to a different understanding of the relative risks of available policy options. That, in turn, is impossible to imagine without, at a minimum, a substantial increase in confidence in government. In February 2007 when the Gallup Poll asked, "How much trust and confidence do you have in our federal government in Washington when it comes to handling domestic problems?" 8 percent said "a great deal" and 44 percent said "a fair amount." The remaining 48 percent answered "not very much" or "none at all."[c] Similar findings come from the University of Michigan's American National Election Studies, which since 1958 has computed a "Trust in Government" index. When asked, "How much of the time do you think you can trust the government to do what is right?" with 100 being "just about always" and zero being "none of the time," the index in 2004 was at 37 percent. The last time it reached 50 percent was in 1966.[d]

There are no indications that this lack of trust in government will dissipate any time soon, making it a given to be reckoned with in formulating most public policy. Even the more supportive views formed in response to the fear, sadness, and outrage evoked on September 11th were short lived. And so, government policy in the 21st century must begin with the knowledge that there is only a limited amount of public trust available and what there is must be earned.

When I began work on New Jersey's search for a site for a low-level radioactive waste disposal facility, I expected to succeed. I knew it would not be easy but I thought that an ambitious, carefully reasoned public participation process could overcome the understandable concern and hostility that would immediately greet any proposed location. I didn't account for all the players and events that could disrupt the calculus, from new governors in South Carolina to successful waste minimization around the country. I also didn't appreciate the extent to which risks, no matter how unlikely, can capture the public imagination and determine a community's reaction to a proposal, nor did I appreciate how deeply rooted is public antipathy to anything involving radiation.

Conducting a siting process like the one I describe in this book would be even more challenging in this post-9/11 world than when I was a participant. The fact that horrific, previously unimaginable acts of terrorism did transpire in 2001 makes it much harder to completely discount virtually any fear voiced about a proposed facility. In the one paragraph in this book that now makes me cringe, the example I chose to demonstrate how farfetched were some of the objections we encountered during the siting process was about a person who asked, "What would happen if a plane crashed into a concrete bunker filled with radioactive waste and exploded?"ᵉ While I can explain why I still believe there is no reason to worry about this possibility, the question seems much more reasonable today than it did when this book first appeared in May 2001.

I wrote this book because the policy implications raised by the nation's low-level radioactive waste saga are important, but also because I believe there is a good story behind it filled with suspense, surprise, frustration, some humor, and a large cast of interesting characters. I have tried to tell the story in a way that illuminates some of the inner workings of public policy implementation and also conveys my experience that a job in government, even on an ultimately unsuccessful endeavor, can be a fascinating, challenging, and satisfying career choice.

I hope you enjoy the book.

John Weingart

WASTE IS A TERRIBLE THING TO MIND

Risk, Radiation, and Distrust of Government

Introduction

Over a 48-hour period in December 1995, I spoke from two very different stages. On a frosty Sunday afternoon, I was in a high school auditorium in Frenchtown, New Jersey introducing Pete Seeger at a concert held to benefit a local holiday food and toy fund. Then on Tuesday morning I was in Phoenix, Arizona on the dais in a hotel ballroom as a member of the opening panel discussion on "The Changing Landscape and Future of Low-Level Radioactive Waste Management" at the 17th Annual U.S. Department of Energy Low-Level Radioactive Waste Management Conference.

I was invited to serve as master of ceremonies for the concert because I have hosted a Sunday evening folk music radio show on WPRB in Princeton for more than 20 years. Pete Seeger has been my hero since I was nine, so I was thrilled to be asked. My love for his music has led me to hundreds of wonderful concerts and recordings, and to my radio hobby. In addition, my political interests were awakened in reaction to the people who picketed his performances in my youth, and to the blacklist that kept Seeger off network television for many years because of his alleged association with Communists.

My inclusion as a featured speaker at a national waste management conference was more surprising. I was new to the field, having taken a job directing New Jersey's Low-Level Radioactive Waste Disposal Facility Siting Board only 15 months earlier. The convention organizers apparently had decided to embrace new participants into this event which had been held annually since 1980. I began my remarks by referring back to my recent peak experience.

"Two days ago," I said, "I had the honor of introducing Pete Seeger at a concert in New Jersey. One of the songs he sang was 'The Water Is Wide,' a song I have known for years and always thought of as a simple romantic song:

The water is wide, I cannot cross over
And neither have I wings to fly
Build me a boat that can carry two
And both shall row my true love and I

"But, in introducing the song, Seeger spoke of 'the oceans of misunderstandings we all have to overcome.' He went on to say that if the world is going to survive, it will not be as a result of any big actions or changes, but rather because of thousands of smaller ones.

"Low-level radioactive waste management may be 'big' to most at this conference, but our success in solving it will be seen as one, or perhaps several, of the smaller types of actions Pete Seeger spoke of."

My job between September 1994 and August 1998 was to lead one state's efforts to find or create a place at which the low-level radioactive waste produced within its borders over the next 50 years could be disposed of safely and dependably. While I am certain this subject was not on Seeger's mind that December Sunday afternoon, the issue is indeed an island surrounded by oceans of misunderstanding. Finding a way to bridge those oceans could be one of the small actions that teach us better ways to resolve difficult public problems.

The issue crosses most academic disciplines as well as traditional political affiliations. Evaluating a site for a disposal facility requires geology, physics, chemistry, biology, and hydrology. Then politics, history, and sociology must be added to form a process that tries to stand on civics and education, but often is more shaped by journalism and theatre. Even politicians who usually advocate that most important decisions be assigned to their particular level of government tend to prefer that someone else be directed to find a good location for radioactive waste. Some environmental activists who believe each of us should be responsible for our own trash feel differently about waste from our use of nuclear energy. Meanwhile, industries that otherwise seek to avoid taxes and governmental intrusion have often embraced public agencies and new taxes or fees dedicated to the management of the low-level radioactive waste they generate.

This book chronicles how New Jersey addressed this problem that many consider untouchable and unsolvable. It describes how a small agency in a state government tried to rise above the low expectations that had implicitly accompanied its creation. Rather than conceding that the only possible way to build a disposal facility for low-level radioactive waste would be for the state, backed by the courts, to decree where it should be located, the agency — the Low-Level Radioactive Waste Disposal Facility Siting Board — asked towns to consider volunteering a site.

To expect any responses to such an unusual offer, the Siting Board realized that its operations would have to be unusual as well. It would have to communicate clearly about complex scientific and bureaucratic processes, it would have to be creative and flexible, and it would have to prove itself to be completely trustworthy.

For four years, the Board tried to follow this course. Its record may offer insights both about decision-making processes for controversial land uses and about how we discuss and consider other potentially risky substances and behaviors.

The book also provides a picture of life inside a government agency. New Jersey's experience with low-level radioactive waste can serve as a lens through which the gulf between the public's impressions, demands, and expectations for government on the one hand, and government's capabilities, constraints and performance on the other, can be clearly viewed. I hope this story of one agency's attempt to bridge the gulf offers a parable that can contribute to discussions of changes needed to enable government to better serve us all.

Directing New Jersey's Low-Level Radioactive Waste Disposal Facility Siting Board was a wonderful experience. This was largely due to the people I was privileged to work with; the public-spirited, dedicated, and good-humored members of the Siting Board and its Radioactive Waste Advisory Committee during my tenure: Judy Blum, Jim Clancy, Tom Dempsey, Bill Dressel, Don Fauerbach, Francis Faunt, Jan Gottlieb, Mike Lakat, Leslie London, Bonnie Magnuson, Rick McGoey, Bill McGrath, Bob McKeever, Lee Merrill, Dick

Olsson, Lisa Roche, Jim Shissias, Ted Stahl, David Steidley, Joe Stencel, Burt Sueskind, and Paul Wyszkowski; several extremely talented members of the much-maligned staff of the Department of Environmental Protection — Tom Amidon, Jill Lipoti, Karl Muessig, and Gerry Nicholls; Roger Haas from the Attorney General's office; and most of all, my wonderful teammates on the Siting Board staff — Bernie Edelman, Jeanette Eng, Prudy Gaskill, Maryann Kall, Greta Kiernan, Denny Medlin, Fran Snyder, and Ed Truskowski.

I am also grateful to David Kinsey and Rick Sinding, two friends who understood from the beginning and actually encouraged me to take this job; Sue Boyle, Rich Gimello, and Sam Penza who generously offered advice and commiseration; Jan Deshais, Ron Gingerich, and Tom Kerr, longtime national leaders in the field of radioactive waste management, who helped me believe that we could succeed in New Jersey; and Don Graham, Richard Sullivan, and the late Donald Jones and Eddie Moore, who each set ideals of public service I have tried to emulate.

The enthusiasm and support of Mark Magyar and the staff of the now defunct Center for Analysis of Public Issues was most welcome, as was a grant from the National Low-Level Waste Management Program that helped fund preparation of a more formal, less interesting report on New Jersey's siting experience.[1] I also appreciate the willingness of Marie Curtis, Jay Kaufman, David Kinsey, Gregg Larson, and Jack Sabatino to read and offer helpful comments on an early draft.

Most important, I am lucky and proud to share my life with Debbie Spitalnik, and to have as our daughter Molly who, among her many insights, suggested that low-level radioactive waste be disposed of in hot dogs because "people will eat hot dogs regardless of what's in them."

Finally, I dedicate this book to the memory of my mother, Florence Weingart, who believed that government could and should make the world a better place, and that paying taxes was not only an obligation, but also an honor.

PART 1

The Challenge

There is a basic law of political physics, often overlooked:
Nuclear waste tends to remain where you first put it.[2]

- **Richard Riley,** South Carolina Governor, **1982**

CHAPTER 1

Visualizing Success

L ate in the afternoon on March 27th, 1995 I was driving from my office in Trenton to the small town of Roosevelt, thinking how wonderful it was that this was where New Jersey's low-level radioactive waste disposal facility was going to be. I had been on the job only six months as director of the New Jersey Low-Level Radioactive Waste Disposal Facility Siting Board, a state agency that was proposing to find a municipality that would volunteer to provide a permanent home for the low-level radioactive waste generated in New Jersey.

This would be a stunning vindication of government's ability to solve problems and a triumph for a new voluntary siting process that could take the place of the traditional government method of top-down decision-making enforced over the will of local residents through the power of eminent domain. The residents would put up with headlines about "dumps," jokes about not needing lights for their ballfields because they would glow, and fears that government wouldn't keep its promises. Meanwhile, they would learn about the subject, gain satisfactory answers to all their questions, structure a deal that would make this an attractive light-industrial development for their town, and solve New Jersey's low-level radioactive waste disposal problem for the next 50 years.

Low-level radioactive waste disposal was not my field of expertise. I had been serving as an assistant commissioner in the Department of Environmental Protection, where I had worked for 19 years, focusing on coastal zone management, land-use planning, and the administration of a variety of permit programs. I had no scientific training and no knowledge of radiation or programs to manage it.

When the job became a possibility, I spoke with Michael Gallo and Robert Tucker, two prominent environmental scientists I knew who often advocated aggressive actions to combat environmental hazards and threats. At the time, Gallo directed the Environmental, Occupational Safety, and Health Institute at the University of Medicine and Dentistry of New Jersey, while Tucker, the longtime director of science and research for the New Jersey Department of Environmental Protection, had recently become the founding director of Rutgers University's Ecopolicy Institute. With no hesitation, each told me that the technology was readily available to safely handle low-level radioactive waste, and that this was a problem of public perception and education, not of science. Both said they would not be concerned if a disposal facility for this type of material was built near their home.

Comforted by these assessments, I accepted when the job was offered. I was attracted by the challenge of trying to help a public agency solve a problem that most people considered unsolvable. I hoped that success in addressing this issue could enhance the reputation of government in general and perhaps help reinstate some of its rapidly vanishing good name.

I also thought it important that good science be used in public policy and decision-making. Otherwise, all fears would be treated as equal, and resources would be increasingly allocated based on the volume and persistence of public concern rather than on the seriousness of the threat to public health and the environment. In addition, I worried about the increasingly popular attitude that seemed to conclude that everything is getting worse, and that nothing we have done — from anti-poverty programs to changing our diets — has made any difference, so why not simply stop wasting money trying to address public problems?

As I explained my new job to friends, I quickly realized that optimism would be essential, if only to justify my career decision. The dominant reaction was that this was not a doable task, and that I must have really wanted to leave my old job. While the latter was true, I accepted this job believing the disposal facility would be sited and built in New Jersey.

I came by my optimism honestly, for I thought that if I had decided this could be safe, anyone could. My only previous thoughts about the subject of anything radioactive had been to enjoy the records of the highly publicized "No Nukes Concert" held in 1979, and to be very impressed by lectures I had attended by Barry Commoner and Amory Louvins, both of whom argued, as I recall, that the use of nuclear power was dumb and dangerous. I had particularly liked Commoner's comparison of the efficiency of nuclear power with that of aiming a cannon from across the street towards a doorbell you wanted to ring. He said that it would probably do the job, but that it was neither the most efficient nor safest technique available.

Furthermore, this *voluntary* siting idea was intriguing. In the spring of 1994, the Siting Board had formally proposed abandoning its seven-year-old effort to select the "best" location in the state, and then rely on a protracted confrontational legal process with the disposal facility's future neighbors to obtain all necessary approvals. Instead, the Board was seeking public comment on a plan under which locations would be considered only upon the request of local residents or officials. This was a new concept that had been used to find locations for a few controversial projects in other states and Canada, but never in New Jersey.

On February 2, 1995, several months after my arrival, the Siting Board adopted the *Voluntary Siting Plan*. Now, in response, Roosevelt's Borough Council, Environmental Commission, and Planning Board had jointly invited the Siting Board to a public meeting that evening to discuss the possibility of the Borough volunteering to host New Jersey's disposal facility.

A site in Roosevelt would never have been considered under the traditional statewide screening process the Board had abandoned. Environmental consultants had been reviewing maps prepared at a scale of 1:24,000, meaning that an inch on the map equaled 2,000 feet on the ground. It took 178 maps, each 1.5 by 2 feet, to view the entire state. Roosevelt is so small that its 1.96 square miles of land fit in an area of 15 square inches on one of these maps. Within the town is just one 150-acre vacant tract.

The minimum size needed for a disposal facility is 100 acres — 50 acres for the disposal area and another 50 acres for the surrounding 300-foot buffer zone required by the Nuclear Regulatory Commission.

To make a statewide screening process manageable, the Siting Board had

Roosevelt and Millstone
Monmouth County – 1995

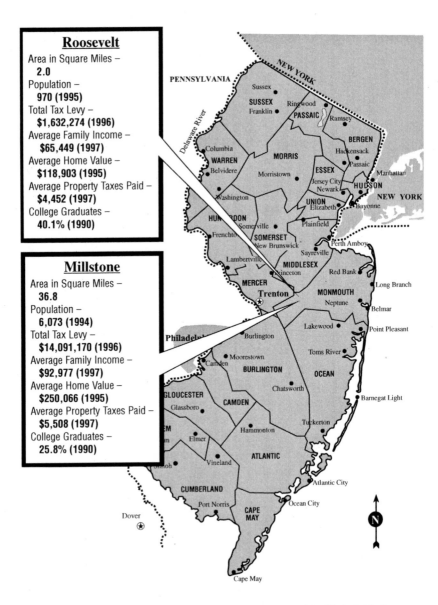

Roosevelt
Area in Square Miles –
2.0
Population –
970 (1995)
Total Tax Levy –
$1,632,274 (1996)
Average Family Income –
$65,449 (1997)
Average Home Value –
$118,903 (1995)
Average Property Taxes Paid –
$4,452 (1997)
College Graduates –
40.1% (1990)

Millstone
Area in Square Miles –
36.8
Population –
6,073 (1994)
Total Tax Levy –
$14,091,170 (1996)
Average Family Income –
$92,977 (1997)
Average Home Value –
$250,066 (1995)
Average Property Taxes Paid –
$5,508 (1997)
College Graduates –
25.8% (1990)

Sources: N.J. Department of Treasury, New Jersey Almanac, NJ Legislative District Databook

set 250 acres as the minimum-size site it would consider. The Board had concluded that examining statewide data would yield a sufficient number of 250-acre sites worthy of further examination. In addition, the larger area could perhaps enable planners to design around environmentally sensitive features that might be found during on-site fieldwork.

Now, with towns and individuals being asked to volunteer specific sites, bigger was still better because it would provide more flexibility, but it was possible that an area between 100 and 250 acres also could be acceptable. If someone brought such a site to the Board's attention, it was well worth examining.

The other reason the Board had planned to limit its map searches to larger areas was to account for drafting inaccuracies. These constraints are not only endemic to paper maps, but also were evident in the computer-based geographic information systems that were then evolving in environmental and planning agencies with much more fanfare than speed. Pockets of wetlands could have been missed by the mapmakers, a new school or shopping center could have been constructed after the aerial photos that formed the basis for the maps had been taken, or the character of hundreds of square feet of land could have been misrepresented by a miniscule drafting error.

The traditional siting methodology would have passed over Roosevelt for lack of open land. The borough would have been even less likely to be considered if the Board had tried to pick possible locations using historical or socio-economic data.

Roosevelt was one of 99 communities created during the 1930s by the federal government under the National Industrial Recovery Act. Originally called "Jersey Homesteads," it was established with a farm, retail stores, and a clothing factory, each run as a cooperative, and was marketed specifically to New York City garment workers and their families.

By the end of the 1930s, Jersey Homesteads had attracted the already famous painter Ben Shahn, and other artists and musicians followed. From its birth in 1933, residents displayed a high level of civic and political activity, and shortly after Franklin Roosevelt's death in 1945, they renamed the town in his honor.

Five decades later, the patron saint is still FDR and the scripture is *The New York Times*. Roosevelt continues to cherish its political roots, with the 970 residents voting at a much higher percentage than is normal in New Jersey, and their votes being disproportionately Democratic. Some residents are still proud that the town supported independent Henry Wallace for president over Harry Truman and Thomas Dewey in 1948, and that it was one of the few towns in

New Jersey in which George McGovern in 1972 and Peter Shapiro in 1985 received more votes than Richard Nixon and Tom Kean in their landslide reelections for president and governor.

In 1983, Roosevelt voters had declared the town a "Nuclear Free Zone." Similar symbolic gestures had been embraced by Berkeley, California; Cambridge, Massachusetts, and a number of smaller communities in the aftermath of the near disaster at the Three Mile Island Nuclear Power Plant in 1979. It was intended to serve notice of the residents' opposition to nuclear weapons and nuclear power.

Roosevelt's declaration had no legal significance, but it made the residents' willingness to consider a facility for low-level radioactive waste all the more surprising and exciting. As I passed through Hightstown and headed down the final miles of two-lane road that lead into Roosevelt, I thought about why this might be occurring.

Roosevelt's residents tend to be well-educated, and include some scientists connected with Princeton, Rutgers and other area colleges and universities. Perhaps most significantly, one of them, Jill Lipoti, heads the state's radiation protection program in the New Jersey Department of Environmental Protection.

Lipoti is well known in the small town and had been an elected member of the Borough Council in years past. In the four weeks since this meeting had been announced, she reported that many of her friends and neighbors had talked with her and seemed impressed that she believed the disposal facility would be safe and might be very good for the borough. Perhaps the key to success was going to be working in a small closely-knit community where people would be able to learn from knowledgeable neighbors they already knew and felt they could trust.

A less important factor, accompanying the town's history, size, voting patterns and occupational mix, was an unusually high interest in music of all kinds and folk music in particular. At least a dozen musicians who lived in Roosevelt had performed on my radio show at one time or another, and I had the impression that the program had a sizeable following in the town. Maybe my involvement in this process was going to add credibility and make a difference. When the disposal facility eventually opened in Roosevelt and analysts tried to quantify the factors that had created this surprising success, folk music, of all things, might be accorded some of the credit.

It was inevitable that the Roosevelt Disposal Facility would be intensely studied, for it would be the nation's first siting of a low-level radioactive waste disposal facility, and one of the few success stories for any type of controversial land use siting. Political scientists and risk communicators would be forced to generalize, with appropriate caveats of course, from the events that had led to this initial town meeting.

They would find that the spark had been struck when reporter Jay Romano wrote a piece that appeared on the front page of the New Jersey section of *The New York Times*. Next to it was a drawing of a person, dressed in a moon suit and standing on the back of a cart, offering mysterious-looking bottles to the skeptical townspeople. Although the headline was, *"Radioactive Waste? NIMBY, Towns Tell State,"* the article accurately described the voluntary siting process and noted that the municipality hosting the facility would receive at least $2,000,000 a year.

Roosevelt's tax assessor, Michael Tickton, discussed the article with other local leaders, all of whom were concerned that their property tax rate had been rising dramatically and showed no signs of stabilizing. Since the town's annual budget was under $3,000,000, the revenue from the disposal facility could enable Roosevelt to choose between entirely eliminating its property tax or making significant improvements to its school and other public facilities and services.

As I parked my car by Borough Hall, I was aware that this was only the first of many times I would be making this drive and parking in this spot as the siting process unfolded over the next two years. I met the other people from my office who had arrived by then, and we walked down the road to Rossi's, Roosevelt's one coffee shop/restaurant. We commented as we were seated that this menu that was going to become very familiar to us.

While I felt I knew Roosevelt pretty well, I had not known that it is one of the several municipalities in New Jersey that are a "hole in a donut." Its 970 people and almost two square miles are entirely surrounded by Millstone, a township of more than 6,000 residents living within 37 square miles.

Some of these geographic anomalies occurred when small clustered developments grew up around railroad stations. The new residents became interested in such non-rural needs as street lights and assuring adequate water for fire protection, and soon formed an identity separate from the other people in their township who were living scattered in the surrounding countryside.[3]

Roosevelt, with no railroad station, had taken a different route. In 1933, federal funds had been used to purchase a 1,200-acre tract for the Jersey Homesteads project. Initially, the project was a part of Millstone Township, but in 1937 Jersey Homesteads was incorporated as a separate borough, and thus became a small island in the Millstone sea.

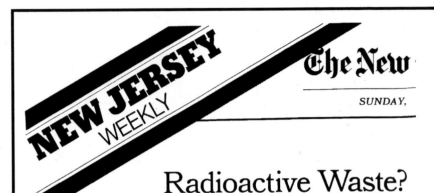

Radioactive Waste?
Nimby, Towns Tell State

By JAY ROMANO

SPARTA

JOHN WEINGART, a state official, spoke before a meeting of the Rotary Club here in Sussex County last week. His mission: to tell some 25 local business leaders about the benefits available to any municipality that will agree to be host to the state's first low-level radioactive waste disposal site.

Among those benefits, he said, would be about $2 million a year in tax revenues and fees, guaranteed open spaces and even a mechanism to compensate neighbors for any

Accept a disposal depot and get $2 million a year.

loss in the value of their property.

About two minutes into the speech, Art Sears, the county's former Tax Administrator, loudly commented, "Sounds good, as long as you stay out of Sussex County." The audience laughed, but as Mr. Sears confirmed after the session, he wasn't kidding.

Mr. Weingart is the executive di-

rector of the state's Low Level Radioactive Waste Disposal Facility Siting Board. He is barnstorming the state, seeking a home for a $100 million, 50-acre repository for thousands of tons of low-level radioactive waste.

Federal law makes each state responsible for the disposal of its own low-level waste, and most experts and local officials acknowledge that such a site might indeed be both essential and safe. But so far, the response has been one or another variation of Nimby — Not in My Backyard.

"Most people are skeptical when

Continued on Page 4

STRIKING A SPARK: It was a lead story in *The New York Times'* Sunday New Jersey section that led Roosevelt's tax assessor to suggest that the town consider becoming the host municipality for the state's low-level radioactive waste

This geography would have been significant in considering the location of any industrial facility because the surrounding town would experience increased truck traffic and any other negative impacts the development brought, without necessarily sharing any of the financial benefits. In this case, it was even more significant because Millstone contained a site that a decade before had come very close to being chosen as the location for a hazardous waste incinerator.

In February 1986, a different commission seeking a site for a hazardous waste incinerator in New Jersey had used a statewide screening process to

York Times

JANUARY 29, 1995

Section **13**

Copyright © 1995 The New York Time

disposal facility. The $2 million a year Roosevelt would have received could have completely eliminated local property taxes or funded significant improvements to its school and other public facilities and services.

identify seven municipalities as "candidate" locations. Millstone was one of them. By September 1988, this Hazardous Waste Siting Commission had concluded that five of the seven did not meet all of the applicable siting criteria. Two months later, it eliminated one more, and on November 29, 1988, the Commission formally proposed the site in Millstone for designation as the home for New Jersey's hazardous waste incinerator. Simultaneously, it granted the town $50,000 to prepare a municipal site suitability study.

Millstone's officials and other residents mobilized to oppose the incinerator. They lobbied and demonstrated using strategies both dramatic and subtle.

The most visible was a gallows, on which the Commission's executive director, Rich Gimello, was hanged in effigy. This was constructed adjacent to the heavily traveled Route 195 that passed through the town.

Later, then-Mayor Seymour Littman proudly recounted how he convinced a merchant to donate two television sets that were offered as raffle prizes on buses chartered to take residents to monitor crucial hearings in Trenton. The legislators and reporters who saw 60 concerned senior citizens in the audience were unaware of the added incentive that might have motivated them to attend.[4]

The hazardous waste incinerator has yet to be built anywhere in New Jersey. When the Commission eventually removed Millstone from consideration, the vehemence of the protests was regarded as a contributing factor. The protests also created a network of people, many of whom still lived in the area, who apparently were prepared to organize again if they perceived any similar threat coming their way.

As the Siting Board staff walked back to the Borough Hall after dinner, we found about two dozen people demonstrating outside against placing a low-level radioactive waste disposal facility in Roosevelt. I introduced myself and had a reasonably pleasant conversation with all but one of them who instead insisted angrily that this meeting should not be taking place. I later learned that he had helped work against the incinerator in Millstone and had returned for tonight's meeting, even though he now lived 100 miles away.

Once inside the Council Chamber where the meeting was to be held, it was clear that the room was too small. When the Board had been invited to this meeting, a date had been suggested that was two weeks away. Since this was to be our first encounter with a community exploring the voluntary siting process, we had wanted more time for our office to prepare, and a date four weeks away had been agreed upon. However, we were not the only ones who used that time for preparations.

As soon as the Borough Council, Environmental Commission, and Planning Board had announced that they were going to sponsor a public meeting on this topic, several residents began to organize against it, holding their own meeting a week in advance. About 75 people attended to hear Diane D'Arrigo, from a Washington-based anti-nuclear group called the Nuclear Information Resource Service (NIRS), talk about the dangers of disposal facilities for low-level radioactive waste.

In retrospect, I am not sure whether it would have been better to hold the meeting on the earlier date the Roosevelt officials first requested, but it certainly was a mistake not to use the largest possible meeting room. More than 350 people were packed into a room with a fire safety capacity of 271, and many others showed up and then went home when they found they would have to stand in the halls. This would have been an advanceman's dream for a political candidate's rally, but for a serious discussion of a complex and controversial issue, it was a nightmare. Before they had heard a word, most people were physically uncomfortable and resentful.

Roosevelt Mayor Lee Allen and the Council members sat behind a table in the front of the room, flanked by Environmental Commission and Planning Board members. With little fanfare, the mayor introduced me. I then stood up, introduced Jeanette Eng, the Siting Board's deputy director, and said that the two of us were going to take about 30 minutes to define low-level radioactive waste, describe how a disposal facility would operate, and explain the voluntary siting process. I added that we would stay as late as necessary to respond to any and all questions.

Immediately, questions and statements came flying at us:

"Why are you here?"

"Why don't you put this in your town?"

"We don't care what you have to say, we don't want it here."

"Let us talk first."

For the next three hours, Jeanette and I took turns trying to speak in complete sentences. I tried to be humorous; I tried to be forceful. As a last resort, I asked for a straw vote of who was in favor of considering the disposal facility and who was opposed, hoping that once the overwhelming opposition was clearly registered, people would settle down and listen. I had once done this successfully with 1,200 people at a public hearing on a pharmaceutical plant proposed near the Jersey Shore in Toms River, but here it only led people to cry out that they had just voted this down and we should go home. I did hear one woman yell to the hecklers, "This isn't talk radio," but she was not persuasive.

The personal credibility I had been so proud of only hours earlier counted for nothing. One man approached me and said, "You do a good radio show and you ought to stick to that."

The low point came when Jill Lipoti bravely stood up and identified herself as someone who had lived in Roosevelt for many years, held a Ph.D. in environmental science, and directed New Jersey's radiation control and safety programs. She said she had two young children and was proud that Roosevelt was considering the possibility of hosting this disposal facility, which she thought could be designed and built to be safe.

As Lipoti finished speaking, a woman on the other side of the room shouted, "You must be a bad mother!" Not one person gasped or otherwise challenged her to indicate that she might have crossed a threshold for acceptable discourse in a civil society.

After the meeting, many people stayed and chatted for another hour. Most apologized for the behavior of their neighbors, and said all or most of the rudest people were from Millstone and not from Roosevelt. Some asked factual questions about the waste, the facility and the process. Some said they were opposed, but appreciated our coming. Others asked how they could learn more. In short, we had the type of conversations we had hoped to have during the meeting.

The next day, the front-page headline in the *Asbury Park Press,* the area's largest daily newspaper, was:

NUCLEAR SHOUT-DOWN; RADIOACTIVE WASTE-SITE PLAN HOT TOPIC

The article provided a comprehensive summary of the information we had tried to convey about the design and safety of the disposal facility, while also giving an accurate flavor of the crowd's sentiments.

Despite the tenor of the public meeting, when Roosevelt's Environmental Commission met later that week, the commissioners voted 6-0 to recommend that the borough seriously explore hosting the disposal facility. That recommendation was sent to the Borough Council, which subsequently ended the process with a 4-2 negative vote. Had one more council member favored the study, Mayor Allen would have had to break a 3-3 tie. Even after the public meeting, he had been quoted as saying, "We are a poor community. I think we should examine anything that might possibly help this community."

An optimist could reasonably say that the siting process in Roosevelt lost by just one vote. But optimism is a word that has rarely proved justified in connection with the nation's quest to build disposal facilities for radioactive waste.

CHAPTER 2

Radiation and Risk

In 1980, one year after the Three Mile Island accident galvanized public fear and effectively put an end to new construction of nuclear power plants, Congress divided responsibility for non-military radioactive waste disposal between the federal and state governments. The federal government would build one national repository for high-level radioactive waste, and the states would take care of the low-level radioactive material. It was a step intended to guarantee permanent safe repositories for the fuel rods, buildings, and radiation-contaminated clothing and gloves that had helped produce a wide variety of goods and services including decades of electric power and would now take decades and, in some cases, centuries to decontaminate.

The terminology associated with this issue is awkward and cumbersome. Neither "low-level radioactive waste" nor "disposal facility" roll off the tongue or lend themselves to a pronounceable acronym. Even calling the New Jersey agency a "Siting Board" seemed to add confusion, with people surprisingly often pronouncing it as if there was a second "t" in "siting."

"Low-level radioactive waste" is the term Congress used in 1980 when it enacted the Low-Level Radioactive Waste Policy Act. Rather than defining the waste in a way that might have helped foster public dialogue and understanding, the act listed what low-level radioactive waste was not: It was not spent nuclear fuel, "high level" waste (including material left over from the making of nuclear bombs), or uranium mill tailings.

What low-level radioactive waste primarily consists of is trash that has become contaminated with radioactive elements. This includes tools, building parts, laboratory equipment, paper and plastic, and gloves and clothing that are contaminated by much lower concentrations of radioactive atoms than are found in the waste defined as high-level. Low-level radioactive waste also includes metals activated by prolonged exposure to neutrons in the cores of nuclear reactors, but it does not include fuel rods.

The waste is quantified by volume and by radioactivity. In New Jersey, 55 percent of the volume and 93 percent of the radioactivity is generated by the state's four nuclear power plants located in Ocean and Salem counties. The rest comes from a mix of about 100 pharmaceutical plants and other industries, colleges and universities, hospitals, and research laboratories.

Everyone agrees that prolonged exposure to some types of radioactive materials can be extremely hazardous. There is, therefore, a consensus that radioactive waste should be handled separately and differently from other garbage. The point of contention is over whether, once separated, low-level radioactive waste can be taken care of in a way that does not threaten public health or the environment.

People who believe there is no safe way to dispose of this material describe the waste by emphasizing that it could include entire demolished buildings from nuclear power plants once they are decommissioned. Those who believe it can be safely disposed of argue that no harm has been caused by the waste that has been generated over the last 50 years and that our knowledge of how to care for it safely increases each year. They tend to describe low-level radioactive waste by first pointing to the booties and other protective clothing worn by workers in research labs and hospitals. Since both descriptions are accurate, and it is difficult to put the two of them together into one easily digestible sentence or news article, large amounts of time are spent arguing over wording and alleged bias in the definitions and examples selected.

The definition is sufficiently contentious that there is no satisfactory way to shorten the phrase. To leave out "low-level" makes the waste sound overly dramatic, and to leave out "radioactive," as the Department of Energy's

National Low-Level Waste Management Program did, might open the speaker or writer to a charge of trying to hide the true nature of the material.

"Disposal facility" is only slightly less problematic. This phrase does come with a readily available alternative, but "dump," while favored by many headline writers, has considerable baggage. It connotes a hole in the ground into which garbage is, literally, dumped. It was a good name for the old town landfills, and accurately reflects the way in which low-level radioactive waste used to be handled. But it does not describe how it is handled now or what would be built in New Jersey or in other states.

Today, when low-level radioactive waste is considered for disposal, it must be in solid form. All institutions that use radioactive materials either have onsite processes or hire brokers to remove liquids before the waste is packaged into drums or boxes and shipped by truck to its final destination. All of the new facilities contemplated in humid climates like New Jersey's include plans to encase the drums in concrete. In most cases, they will rest on a concrete pad built at ground level.

Once the pad is full, an additional thick layer of concrete will be built around the already encased drums. Layers of clay, sand, dirt and grass will then be placed over the concrete, forming a large mound designed to keep the waste isolated from water, the surrounding environment, and people.

Radioactive waste, unlike other hazardous materials, decays naturally without incineration or any kind of treatment. As a result, the mounds of waste would be accompanied only by reasonably modest buildings and parking areas designed to accommodate the trucks bringing in the waste, and the equipment and staff necessary to accept and monitor the waste.

Since this is not a landfill, incinerator, or dump, it has come to be known as a "disposal facility." These two phrases — "disposal facility" and "low-level radioactive waste" — are part of the name of almost every state and regional agency that has been established to address the problem.

Semantics, in this case, are not trivial. Discussing the possibility of anything involving radiation is inevitably controversial and, for many people, emotional. To raise the idea within the phrase "low-level radioactive waste disposal facility" immediately sounds bureaucratic, intimidating, and unfriendly. The acronym LLRW has no vowels, and linking the initials of the Siting Board's full name, which the Board once did on T-shirts, is just laughable — NJLLRWDFSB.

On the other hand, citizen groups formed to oppose any type of development usually find names that can be shortened to appealing phrases that fit easily in newspaper headlines. An opposition group formed in New Jersey called itself CHORD, which stood for "Citizens Helping to Oppose Radioactive Dumps." This was shorter, though not nearly as memorable as an organization rumored to have been formed in northern California, called No Nukes of the North.

In New Jersey, we spent many odd moments trying to devise better names or acronyms. The idea of a "center" seemed preferable to "facility." The "Center for Advanced Radioactive Treatment" (CART) was the most serious contender. The inevitable comments about not putting it before a horse did not doom this suggestion so much as the fact that it would not have been entirely accurate since no waste "treatment" would be taking place. Substituting "entombment" or "placement" yielded CARE or CARP but seemed too forced.

Less serious suggestions included naming the facility PRIEST (for Permanent Repository for Isotopes in an Engineered Safe Terminus) or substituting for "radioactive" the phrase "inertly challenged." Mike Lakat, the Health Commissioner's representative on the Siting Board and a cooking enthusiast, offered a more enticing list of possibilities. These included:

CREPE — Center for Radioactive Element Placement for Eternity;

CUMIN — Center for the Use and Management of Innocuous Nuclides;

GARLIC — Government Agency Responsible for Long-term Isotope Cloistering;

OREGANO — Organization Responsible for the Effective Generation of Alternative Nuclear Options;

SHALLOT — Society for Hospitable Actions towards Low-Level Organic Trash; and

SOUFFLE — Society of Optimal Understanding of our Fission and Fusion Legacy to the Environment

Unfortunately, none of these exercises ever produced an acceptable alternative for low-level radioactive waste disposal facility.

More significantly, 20 years after enactment of the federal Low-Level Radioactive Waste Policy Act, the federal government is not close to accomplishing its task of creating a disposal facility for high-level waste, and not one state, or group of states, has created a new permanent location for low-level radioactive waste. Yet, the use of radioactive materials has become more and more widespread. By the 1990s, more than 12,000 electric utilities, universities, hospitals, laboratories, and other institutions were regularly using radioactive materials, necessitating more than two million shipments of radioactive materials each year.

The increased use has not led to a waste disposal crisis, however, because it has been accompanied by a dramatic reduction in the volume of low-level radioactive waste generated. Repeated threats by the State of South Carolina to close the only disposal facility open to most of the nation led the nuclear plants, industries, hospitals and other users of nuclear devices to adopt techniques to limit the number of articles they exposed to radiation. This business decision was further spurred by disposal fees in South Carolina that soared from 90 cents per cubic foot in 1971 to $79 in 1990 and $336 by 1996. Meanwhile, high-level radioactive waste continues to accumulate at the nuclear plants that generate it.

Each American nuclear power plant was licensed by the Nuclear Regulatory Commission (NRC) to operate for 40 years. While extensions are possible, the nuclear plant closures that started in the 1990s will have a major impact on the need for radioactive waste disposal as more plants across the country begin to close. NRC regulations assume that when a plant is shut down, anything contaminated with radiation will be removed and taken to an approved disposal facility. This includes parts of buildings and major equipment which occupy a large amount of space. The decommissioning of the four nuclear plants in New Jersey, for example, will triple the amount of low-level radioactive waste generated in the state over the next 50 years, from approximately 1 million cubic feet to 3 million.

Federal regulations direct that each closing nuclear power plant be taken apart and removed so that the site can be returned to its previous natural state. It is not clear, however, that utilities and states will follow this directive to pack up as if the plants were a county fair at the end of the season.

To begin with, the United States is not likely to have a permanent repository for high-level radioactive waste in the foreseeable future. If the high-level waste must be retained on site for lack of an alternative, the argument for moving the low-level waste becomes much weaker. Even if existing disposal facilities in South Carolina and Utah are ready to accept the low-level material, utilities and states may well feel that they should avoid the expense of taking the facility apart and transporting the low-level waste if they are obligated to continue caring for the far more dangerous fuel rods and other high-level waste. In most cases, keeping the waste on site will also engender much less controversy than moving it, even if placing it in a permanent repository would present less risk.

The Uses of Radiation

The French physicist Henri Becquerel discovered natural radioactivity in 1896 and, with Marie Curie, won the Nobel Prize for separating and naming radium and polonium in 1903. But it was not until 1942, when the Manhattan Project got underway, that significant amounts of radioactive waste were generated. At first, most was simply buried on federal lands, but after World War II, the preferred method became "diluting and dispersing." This was the term of art for pouring low-level radioactive waste into the nearest waterway. Sometimes this was the ocean, which received large amounts of both high- and low-level radioactive waste before the practice was outlawed in 1967.

While the war had certainly dramatized one use of radioactive materials, the post-war era saw commercial uses become increasingly widespread. Nuclear power plants have been the largest user of radioactive materials and have generated the majority of the country's low-level radioactive waste. Even though the nuclear plants in this country were all built between the end of the war and the 1979 accident in Pennsylvania at Three Mile Island, more than 20 percent of the nation's electricity is still derived from nuclear power.

The figure is much higher in New Jersey, where more than half of the state's energy is supplied by nuclear power. As a consequence, the 55 percent of New Jersey's low-level radioactive waste volume contributed by nuclear plants is larger than the average share in other states.

Medical uses and research-and-development activities at pharmaceutical and academic institutions generate another 35 percent of New Jersey's volume of low-level radioactive waste. Radiation is used to diagnose illness and avoid exploratory surgery, to treat cancer and other diseases, to scan organs to detect disease before symptoms appear, and to test for the presence of antibodies, hormones or drugs in the blood. Nationally, about 100 million procedures using nuclear medicine are performed each year. These procedures are used in the treatment of one-third of all Americans who are hospitalized, as well as many who receive out-patient care. About 10 percent of the procedures involve administering radioactive pharmaceuticals directly to patients, while 90 percent use small amounts of radioactivity to analyze bodily fluids. In addition, radioactive materials are used to develop and test most new medicines and to sterilize medical instruments.[5]

The remaining 10 percent of New Jersey's low-level radioactive waste is generated from a wide range of industrial activities. Radioactive materials are

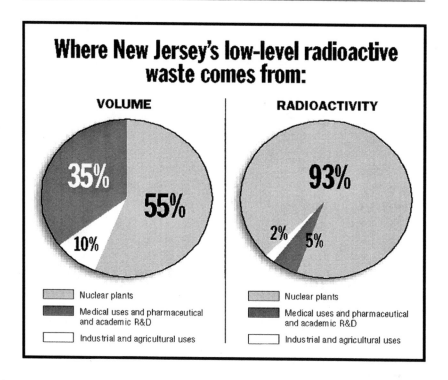

Where New Jersey's low-level radioactive waste comes from:

VOLUME

35%
55%
10%

RADIOACTIVITY

93%
2% 5%

Nuclear plants
Medical uses and pharmaceutical and academic R&D
Industrial and agricultural uses

Nuclear plants
Medical uses and pharmaceutical and academic R&D
Industrial and agricultural uses

used to obtain precise measurements of densities and thicknesses. This is useful, for example, for checking the alignment of steel belts in rubber tires and assuring that the coating adheres properly to non-stick frying pans, as well as for measuring the density of liquids flowing in pipes, the thickness of chemical emulsion on the surface of film, and the amount of air whipped into ice cream. Radioactive materials are also employed to examine welds in bridges and building, detect hidden explosives, date ancient artifacts and increase agricultural productivity.

The national volume of low-level radioactive waste generated by all these enterprises rose steadily from the 1950s through the 1970s. It peaked in 1980 at about 4 million cubic feet and has declined almost every year since. Between 1986 and 1989, for example, the annual average was more than 1.6 million cubic feet, but by the period from 1992 to 1995, it had fallen to just under 700,000 cubic feet. Put another way, since a football field with waste stacked 20 feet high could hold 960,000 cubic feet, the national need for disposal space fell from a little over four footballs fields per year to less than one.

The 50 states do not produce proportionately equal amounts of waste. In most years, for example, Illinois, Tennessee and Pennsylvania have each generated more than 10 percent of the total. New Jersey's contribution has always been well under 5 percent, ranging from a high of 82,305 cubic feet in 1987 to an average of 35,000 cubic feet annually from 1991 to 1994 and 16,000 cubic feet in each of the following three years.

As the New Jersey Siting Board moved to a voluntary siting process in 1995, it projected that the state would generate an average of 20,000 cubic feet of low-level radioactive waste for each of the next 50 years. Thus, the equivalent of one football field, or perhaps two fields with waste only 10 feet high, would be sufficient to contain the state's waste over this period. The Board chose to visualize these quantities in terms of two-car garages. Calculating that one could enclose about 4,000 cubic feet of space, New Jersey needed to locate the equivalent of five such "garages" each year.

If all that is needed to accommodate the nation's low-level radioactive waste over the next 50 years is the combined acreage of 50 football fields, the problem should not be so difficult to solve. A wide range of uses have claimed much larger amounts of land — the chain of 2,200 Starbucks coffee shops opened in recent years, for example — usually without the aid or intervention of any government program.

Radioactive materials, however, are hazardous. Many studies, including those examining survivors of the bombing of Hiroshima and Nagasaki, have shown that exposure to high doses of radiation increases the likelihood of developing cancer later in life, and that very high doses can cause immediate death by limiting the body's ability to manufacture blood cells.

While the best-known measurements for radiation are curies and half lives, the most useful in considering radioactive waste are "millirems." A "rem," short for "roentgen equivalent man," quantifies the amount of energy carried by radiation based on the extent to which it is absorbed by biological tissue. Because one rem is a huge exposure, discussion usually occurs in terms of thousandths of a rem, or millirems.

The average American is exposed to about 360 millirems of radiation per year. About 300 millirems come from natural sources including cosmic rays, radioactive elements in the earth, and the potassium in our bodies. The remainder are the result of specific incidents such as X-rays, airplane travel, and occupational exposure.

The studies that have confirmed the health effects of radiation have shown that increased likelihood of disease has followed exposure to more than 50,000 millirems at one time, with about 450,000 millirems proving quickly fatal. No

Cause and Extent of Average Annual Exposure to Radiation

Construction of a low-level radioactive waste treatment facility would have added a maximum of 25 millirems of additional exposure to radiation for people living near the center — less than the average American receives through medical X-rays each year. Average exposures, as of 1996, were:

200 MILLIREMS — RADON

40 — NATURALLY OCCURRING RADIATION IN HUMAN BODY

39 — X-RAYS FOR MEDICAL PURPOSES

28 — NATURALLY OCCURRING RADIATION IN ROCKS AND SOIL

27 — COSMIC RAYS

14 — RADIATION THERAPY AND OTHER NUCLEAR MEDICINE

10 — COLOR TVS, SMOKE DETECTORS CONSUMER PRODUCTS

1 — NUCLEAR POWER   **TOTAL 359 MILLIREMS**

DANGEROUS RADIATION LEVELS

450,000 millirems	Immediate Death
50,000 millirems	Likelihood of cancer and other diseases
Below 10,000 millirems	No demonstrable ill effects
5,000 millirems	Maximum annual exposure deemed permissible for workers at nuclear power plants under U.S. Nuclear Regulatory Commission standards.
100 millirems	Maximum permissible additional exposure for general public living near a nuclear plant under NRCstandards.

Source: U.S. Nuclear Regulatory Commission

ill effects from exposures of less than 10,000 millirems have ever been demonstrated. Studies by the Pennsylvania Department of Health of people living near Three Mile Island at the time of the 1979 accident, for example, found that residents received no more than 25 millirems in additional exposure and have experienced no increase in disease or birth defects, even among children of women who were pregnant at the time.[6]

The debate about radiation safety is focused largely around the phrasing of the second sentence in the preceding paragraph. No one can say that exposures of less than 10,000 millirems have been proved to be safe, only that they have not been shown to be hazardous.

Federal and state health and environmental officials have responded to this information conservatively by assuming that any level of exposure may be dangerous. For policy and regulatory purposes, government agencies have, in effect, created a graph with a line running from zero, which represents no exposure and no risk, to the various amounts of exposure, all above 50,000 millirem, at which harm has actually been demonstrated. The line is straight and climbs gradually but steadily, indicating that exposure and risk are directly linked at all levels. This is referred to as a "linear no threshold approach" to risk.

The linear no-threshold assumption has led to regulations strictly limiting allowable radiation exposures. Disposal facilities for low-level radioactive waste, for example, must meet a stringent Nuclear Regulatory Commission standard that no more than 25 millirems per year be added to the exposure received by a person spending 24 hours a day near the facility. This would be the equivalent of two additional chest X-rays per year.

Critics of radioactive waste disposal proposals worry that even if a facility is planned to meet this level, once operating it might exceed it, and some feel that even the possibility of adding up to 25 millirems to their annual level of exposure is an unacceptable risk. On the other hand, there are many scientists who believe that low levels of exposure are not harmful and may in fact be beneficial. They question the linear no-threshold model of analysis, often pointing to the many substances we consume regularly such as aspirin and salt that are only harmful in large doses.

Background levels of radiation vary around the world. In the United States, the range is from about 60 to 600 millirems, while in Kerala, India the naturally radioactive soils can expose individuals to 16,000 millirems. Studies have shown no greater incidence of cancer in Kerala and other areas with the higher levels. In fact, Marvin Goldman of the Department of Surgical and Radiological Sciences at the University of California at Davis has noted that "most of the studies show the opposite, giving support to a concept of hormesis, a beneficial effect of a low-level exposure to an agent that is harmful at high levels."[7]

One of the leading proponents of radiological hormesis is Myron Pollycove, a Professor Emeritus of Laboratory Medicine and Radiology at the University of California at San Francisco. Pollycove has concluded that, "Scientific understanding of the positive health effects produced by adaptive responses to low-level radiation would result in a realistic assessment of the environmental risk of radiation. Instead of adhering to nonscientific influences on radiation protection standards and practice that impair health care, research, and other benefits of nuclear technology, and waste billions of dollars annually for protection against theoretical risks, these resources could be used productively for effective health measures and many other benefits to society."[8]

Professors Goldman and Pollycove are far from alone in their views and may even reflect a growing scientific mainstream. But it almost goes without saying that they have neither captured nor influenced the attitudes of the general public. In fact, the scientific community's beliefs about radioactive safety in general are dramatically different from those of most of the general public. The gaps appear deeper and wider than the differences accompanying other public issues. Even people with passionately different views about abortion usually operate from a much larger set of shared factual information.

CHAPTER 3

Multi-State Compacts: Seeking Regional Solutions

I n the 1950s, when several states first applied for authority to regulate the disposal of low-level radioactive waste, the Atomic Energy Commission (AEC) responded that the federal government was better situated than the states for the required long-term management. The AEC changed its position, however, and, between 1962 and 1971, disposal facilities were licensed and opened in six states: Beatty, Nevada in 1962; Maxey Flats, Kentucky and West Valley, New York in 1963; Richland, Washington in 1965; Sheffield, Illinois in 1966; and Barnwell, South Carolina in 1971.

Each of the six facilities was operated by a commercial firm on government land. The Richland site was federal, and the other five were state-owned. Users of radioactive materials from around the country could choose which of these operations was best for them.

During the 1970s, the Illinois facility closed after it had exhausted its 20-acre capacity. Public opposition, spurred by several leaks of radioactive material into adjacent lands, caused the facilities in Kentucky and New York to close as well. As a result, the amount of low-level radioactive waste coming

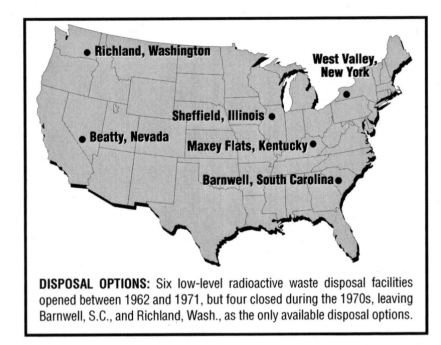

DISPOSAL OPTIONS: Six low-level radioactive waste disposal facilities opened between 1962 and 1971, but four closed during the 1970s, leaving Barnwell, S.C., and Richland, Wash., as the only available disposal options.

into the remaining states increased, and as it grew, so did local political opposition. Soon, Nevada, South Carolina, and Washington were all seeking to end their status as the nation's nuclear dumping grounds.

Public attitudes about radioactive materials and about citizen participation were changing sharply during this decade. The final public hearing for the Barnwell facility held by South Carolina's Department of Health and Environmental Control in 1970, for example, attracted less than 20 people. Officials from the company proposing the facility, Chem-Nuclear Systems, Inc., and the other attendees who all supported the application spent an hour informally waiting to see if anyone wanted to speak in opposition.[9]

Eight years later, with 80 percent of the nation's commercially generated waste being shipped to Barnwell, South Carolina elected Richard Riley governor. During the campaign, Riley pledged that the state would not continue to offer the entire country "the path of least resistance" for disposing of low-level radioactive waste.[10]

Governor Riley pushed for a federal policy to stop the national flow of waste to South Carolina. A few years later, he recalled, "What I said in the very beginning was that I saw a national problem that South Carolina was on the way to becoming the entire answer to. It was my judgment that we could jump up and down and raise hell, but the only way I, as governor, could effectively handle the problem in my state was to try and bring out some response nationally. I tried to do that in a reasonable way. At least, I think it was reasonable — some people seem to have thought otherwise."[11]

In 1980, Congress passed the Low-Level Radioactive Waste Policy Act and made each state responsible for disposal of the low-level radioactive waste generated within its borders. The act created a mechanism intended to encourage groups of states to work together to solve the problem on a regional basis. The states could form "compacts" of two or more states within which one regional facility could be sited.

The law offered two incentives for states to join in compacts. First, the members would be exempted from interstate commerce provisions so that they could exclude waste from states not in their compact. As a result, they would not risk gaining the reputation that Nevada, South Carolina, and Washington were trying to shed. Second, the compacts were authorized to require that all waste generated within their region be sent to the disposal facility they had constructed. This assured them a captive market and, consequently, a reasonably predictable and guaranteed revenue stream.

The idea of compacts, and much of the 1980 law, originated from a Task Force on Low-Level Radioactive Waste Disposal established by the National Governors Association. The group was chaired by Arizona Governor Bruce Babbitt and included Arkansas Governor Bill Clinton — the two men who years later would lead the federal government in thwarting California's plan to comply with the law by building a regional disposal facility.

Babbitt, Clinton, and the other governors argued that Congress did not need to decree the boundaries of the regions, that the states' elected leaders were perfectly capable of accomplishing such a task. While Congress did not specify an optimal number of compacts, and thus of regional facilities, the U.S. Department of Energy had suggested that six would be needed.[12]

In response to the federal law, groups of state representatives in different parts of the country met to consider their options. New Jersey was included in a proposal by the Coalition of Northeastern Governors for a compact with 12 states and the District of Columbia. Representatives gathered in the late 1980s and drafted a Northeast Interstate Low-Level Radioactive Waste Management Compact.

Their discussions bogged down, however, when they tried to agree on a mechanism for selecting the state to host the disposal facility. Massachusetts voters had passed a resolution establishing stringent standards that appeared to disqualify the entire state. Pennsylvania generated the most waste, but it soon dropped out, agreeing instead to be the designated host state for an Appalachian Compact. Formed initially with West Virginia and eventually joined by Delaware and Maryland, membership in this compact obligated Pennsylvania to accept much less waste than would have come its way from the original Coalition of Northeastern Governors aggregation.

Later, other northeastern states also held their own negotiations, with the strangest result being an alliance that Maine and Vermont forged with Texas. Texas wanted to build a facility primarily for itself and was looking to form a compact that would require it to take the smallest possible amount of out-of-state waste.

Soon, the only participants remaining to form a Northeast Compact were New Jersey and Connecticut, states united largely by similarly ambivalent feelings about their proximity to New York and by being the final two states to retain the 55-mile per hour speed limit.[13]

In the end, 43 states formed 10 compacts, ranging from a Northwest Compact with eight members and a Southeast Compact with seven, to bi-state compacts formed not only by Connecticut and New Jersey, but also by Illinois and Kentucky. Seven states and the District of Columbia remained unaffiliated.

As if the vision of regional compacts was not already sufficiently undermined, Connecticut and New Jersey were unable to decide which state should host the disposal facility for the "region." As a result, they agreed that each would build its own facility for its own waste. By virtue of being a member of the Northeast Compact, each would be able to retain waste generated in its state and keep out waste from others.

The Northeast Compact, therefore, became largely a marriage of convenience. Each governor selected a prominent environmental official as his state's compact commissioner: Richard Sullivan, who had served as New Jersey's first environmental protection commissioner, and Kevin McCarthy, who was then directing Connecticut's hazardous waste programs. Throughout the 1990s, the two commissioners shared a staff of one and held an hour-long public session every three months, alternating between motel meeting rooms in New Jersey and Connecticut.

Meanwhile, the three states with disposal facilities were threatening to close their gates entirely or to limit access to waste from just their own state or from their compact region. Worried by the prospect that many states soon would have no place to send their waste, Congress amended the Low-Level

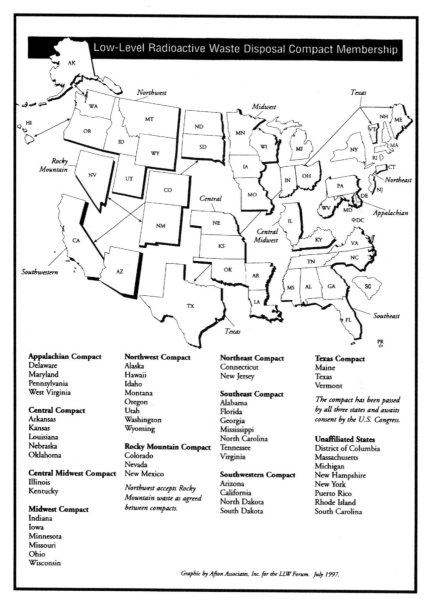

Low-Level Radioactive Waste Disposal Compact Membership

Appalachian Compact
Delaware
Maryland
Pennsylvania
West Virginia

Central Compact
Arkansas
Kansas
Louisiana
Nebraska
Oklahoma

Central Midwest Compact
Illinois
Kentucky

Midwest Compact
Indiana
Iowa
Minnesota
Missouri
Ohio
Wisconsin

Northwest Compact
Alaska
Hawaii
Idaho
Montana
Oregon
Utah
Washington
Wyoming

Rocky Mountain Compact
Colorado
Nevada
New Mexico

*Northwest accepts Rocky
Mountain waste as agreed
between compacts.*

Northeast Compact
Connecticut
New Jersey

Southeast Compact
Alabama
Florida
Georgia
Mississippi
North Carolina
Tennessee
Virginia

Southwestern Compact
Arizona
California
North Dakota
South Dakota

Texas Compact
Maine
Texas
Vermont

*The compact has been passed
by all three states and awaits
consent by the U.S. Congress.*

Unaffiliated States
District of Columbia
Massachusetts
Michigan
New Hampshire
New York
Puerto Rico
Rhode Island
South Carolina

Graphic by Afton Associates, Inc. for the LLW Forum. July 1997.

MULTI-STATE COMPACTS: By 1997, 43 states had formed compacts ranging from two to eight members to plan for low-level radioactive waste disposal. Six states, the District of Columbia and Puerto Rico decided to remain unaffiliated.

Radioactive Waste Policy Act in 1985. The result was that all three facilities would maintain their open access policies, but only until the end of 1992. Notice was served to the other states that if they did not stay on schedule to create new disposal facilities, they would suffer denial of access to the three open disposal facilities.

The 1985 amendments were a response to the lack of action by the states after passage of the federal law in 1980, and they included more specific mandates. New Jersey met the first new milestone in December 1987 when the Legislature enacted the Regional Low-Level Radioactive Waste Disposal Facility Siting Act. This act recognized New Jersey's obligation to establish a mechanism for the timely siting of a disposal facility in the event that it was designated the host state for the regional facility.

Connecticut had enacted a similar law. Once both laws were in place, the two-member Northeast Compact Commission took the formal action that designated both New Jersey and Connecticut as host states for the disposal of all low-level radioactive waste generated within their respective borders.

When the federal deadline arrived at the end of 1992, the Beatty, Nevada facility did close. The Richland, Washington facility began accepting waste only from the seven states that had formed a Northwest Compact with Washington and from three other states that had become the Rocky Mountain Compact. That left the facility in Barnwell, South Carolina as the only waste repository available for the other 39 states and the District of Columbia.

In South Carolina, the future of the disposal facility in Barnwell had become a hot political topic. The state was part of the Southeast Compact, along with Alabama, Florida, Georgia, Mississippi, North Carolina, Tennessee, and Virginia. In 1986, the Southeast Compact Commission designated North Carolina as its next host state and began work for a new facility that would be in operation by January 1, 1993. Once it opened, the South Carolina disposal facility would close.

The South Carolina Legislature set and then extended several deadlines for closing the Barnwell facility. As a condition for continuing to accept waste, the Southeast Compact Commission began requiring that other states demonstrate continuing progress toward realizing their own disposal options. Finally, in 1992, Carroll Campbell, Governor Riley's successor, announced that beginning July 1, 1994, the Barnwell facility would accept waste only from states in the Southeast Compact. It would be closed to waste from the 31 other states that were relying upon it.

"Barnwell" is the universally used term for the Barnwell Low-Level

INTERIM STORAGE: With the long-term future of out-of-state disposal in doubt, every low-level radioactive waste generator, from nuclear power plants to universities and hospitals, was forced to build its own interim storage facility. Staffers at Rutgers University, top, weigh and check in a barrel of low-level radioactive waste material. The old greenhouse on Rutgers' Busch campus, bottom, was one of more than 100 interim storage facilities in New Jersey.

Radioactive Waste Management Facility. It is located in the town of Snelling in Barnwell County, South Carolina. When I came to the Siting Board in September 1994, Barnwell had been closed to New Jersey for 10 weeks.

The years of advance warning from South Carolina, along with impressive, though unheralded, planning in New Jersey, kept Barnwell's inaccessibility from creating a crisis. The Nuclear Regulatory Commission had required waste generators to plan for the projected lack of out-of-state disposal options. Each had had to designate space on site to safely store the waste they expected to generate over the next five years.

In response, each generator of low-level radioactive waste, from the nuclear power plants to the hospitals and universities, had built or modified a building on its property to serve as an interim storage location where its waste would be kept until New Jersey had its own disposal facility, or some other out-of-state disposal option emerged.

While some of the generators clearly had capacity for a longer period, five years seemed like the window we had in which to get the New Jersey facility sited and well on the way to being licensed, built, and in operation.

CHAPTER 4

First Steps: The New Jersey Siting Process Begins

T he federal Low-Level Radioactive Waste Policy Act of 1980 had done little to stimulate action for new disposal facilities. Once Congress amended it in 1985, however, most states decided they could no longer avoid addressing the issue. When the New Jersey Legislature began the deliberations that led to its 1987 Low-Level Radioactive Waste Siting Act, legislators looked to a statute they had proudly enacted in 1981 to address a similar need for a different type of waste.

The Major Hazardous Waste Facilities Siting Act had been drafted with the active involvement of representatives from business, environmental groups, and local government. Its signing by Governor Brendan Byrne was heralded as the dawn of a new cooperative era in environmental decision-making. Hazardous waste, which includes all toxic or dangerous waste except for radioactive material, was a particularly high-profile issue in New Jersey. One of the first industrialized states, New Jersey led the nation in identified toxic waste sites eligible for cleanup under the "Superfund" law. This federal statute was written by James J. Florio, then a New Jersey Congressman and later the state's governor. New Jersey's 20-year effort to site a hazardous waste disposal facility foreshadowed the struggles the Low-Level Radioactive Waste Siting Board would face.

The law created a Hazardous Waste Siting Commission to determine the state's need for hazardous waste disposal and treatment facilities, to decide where the facilities should be located, and to oversee their construction and operation. The nine members of the Commission were to be nominated by the governor and confirmed by the New Jersey State Senate. The Commission included three county or municipal officials, three employees of industrial firms, and three representatives of environmental or public interest organizations.

In general, when legislatures choose to assign tasks to new commissions instead of directing them to ongoing governmental departments, it is because they feel an added level of public respect or credibility will be afforded. The issue may receive more public prominence than would occur if it was to be just one among the many priorities that a major department has to juggle. In addition, the heir apparent for this assignment in New Jersey would have been the Department of Environmental Protection. This agency was gaining increasing unpopularity by enforcing the fairly stringent regulatory programs the Legislature had created in the post-Earth Day enthusiasm of the 1970s, and by becoming a somewhat cumbersome bureaucracy.

The Legislature may also have been mindful of the nation's experience with the Atomic Energy Act, which had assigned one agency, the Atomic Energy Commission, with the dual missions of advocating nuclear power while also regulating it. The perceived inherent conflicts subsequently led Congress to replace the AEC and split its roles between the U.S. Department of Energy and the Nuclear Regulatory Commission. Creating a distinct Hazardous Waste Siting Commission in New Jersey would separate the problem-solving or advocacy responsibility from the safeguards represented by the permit programs in the Department of Environmental Protection.

Setting up a new commission also enables the governor to give unpaid but arguably prestigious roles to people he or she considers worthy, and then gives the members of the State Senate a chance to approve or hold up the governor's choices. In New Jersey, senators rarely actually veto nominees, but delaying action is one of their most effective means for gaining the governor's attention on other issues of importance to them.

To involve even more people in the process and to have even more volunteer positions to fill, the law also created a 13-member Hazardous Waste Advisory Council. As with the Siting Commission, the members of this Council, too, could not simply be people the governor felt would do a good job, but rather had to fit into categories. The Legislature made the membership requirements slightly different for each of the two bodies.

For the Hazardous Waste Siting Commission itself, the governor was to choose people who were in the described categories. For the Advisory Council, he was to include people recommended by organizations representing the listed categories. Thus, the Advisory Council was to include three members *recommended* by environmental or public interest organizations, two by municipal officials, two by county officials, one by community organizations, one by firefighters, one by industries that use on-site hazardous waste facilities, one by industries that use major hazardous waste facilities, one by hazardous waste transporters, and one by major hazardous waste facility operators.

Once constituted, the Hazardous Waste Siting Commission was to determine how many and what type of facilities were needed to handle the hazardous waste generated in New Jersey. It also was charged with developing siting criteria for adoption by the Department of Environmental Protection. Once the criteria were in place, the Commission would hire consultants to objectively identify the best locations in the state for each of the needed facilities.

After these locations were determined, the Commission was to notify the mayor and governing body of each lucky town and provide them with funds to prepare their own Municipal Site Suitability Study. The towns would be given six months to complete this study. Forty-five days later, a hearing would be held before a state administrative law judge.

The Office of Administrative Law is the first stop in the appeals process on most decisions made by New Jersey state government agencies. The judges, who often have prior experience in state government, hold hearings and issue rulings that serve as recommendations back to the agency which then issues its final decision. Occasionally, this process helps resolve contentious or confusing decisions, but more often it simply provides a detailed public record that is then used when one or both parties subsequently bring their disagreement to the Appellate Division of Superior Court.

The administrative law judge assigned to the Hazardous Waste Siting Commission case would hear testimony from the Commission and from the town, and would issue recommendations back to the Commission. At that point, the potential host county and municipality would each be given a voting seat on the Commission. The now 11-member Commission would then affirm, conditionally affirm, or reject the judge's findings.

The Major Hazardous Waste Facilities Siting Act was based on a determination that this waste disposal problem was not an area the private sector was likely to be able to address through market forces. It rested on three other major unstated assumptions:

- First, the Legislature assumed that it is reasonable to establish a new agency and direct it to determine the best single site for a particular type of land use in a state of 4.8 million acres.
- Second, any municipality selected as the possible location was expected to oppose the determination and want to appeal it.
- And, third, the Legislature thought, or hoped, that a process through which a commission would choose the best site and then guarantee the potential host municipality an avenue for appeal and financial resources to help it build its case would eventually result in the state gaining the facilities it needed to handle hazardous waste. The law stated, "The choice of hazardous waste disposal sites is now, all too frequently, made on an indiscriminate and illegal basis... It is necessary to establish a mechanism for the rational siting of hazardous waste facilities."[14]

Six years later, when the Legislature responded to the federal directive to address low-level radioactive waste, it was still possible to view the Hazardous Waste Siting Commission as a success, even though it was moving much more slowly than had been anticipated. After a systematic statewide search by a consulting firm, the Commission had identified 11 New Jersey locations that appeared to meet the siting criteria. Upon further examination, the Commission had eliminated all but two of the sites, which were then undergoing more detailed analysis — one of which turned out to be Millstone, the township that surrounded the Borough of Roosevelt.

The 1981 hazardous waste statute thus became the template for the new low-level radioactive waste law. The Legislature again decided to set up a new agency, this time with 11 members instead of nine, and this time to be called a "board" instead of a "commission," and another 13-member advisory group, this time called a "committee" instead of a "council." Again, the Legislature required that the governor pick certain categories of people for the siting agency, while choosing people recommended by specified groups for the advisory group. This time, the requirements were even more specific.

The Low-Level Radioactive Waste Disposal Facility Siting Board was to include three representatives of industries that generate low-level radioactive waste, one of whom must represent public utilities, one of whom must represent hospitals or other health care facilities, and one of whom must represent the radiopharmaceutical or nuclear medical research industries; three members with training and expertise in disciplines relevant to the management of radioactive waste, at least one of whom must be a physician specializing in nuclear medicine; and three members who would represent recognized environmental organizations or other public interest groups.

Unlike the Hazardous Waste Commission, this board would also have the commissioners of the state Department of Environmental Protection and the state Department of Health as members.

The 13 members of the Radioactive Waste Advisory Committee were to be very similar to the members of the Hazardous Waste Advisory Council except that firefighters were not singled out to provide a representative.

More importantly, the new law adopted the hazardous waste siting process, based on the same unstated assumptions. Again, a new state agency was to pick the best site in the state and then ensure that the selected municipality was given the money and time to mount a good fight against having been chosen. And again, the premise seemed to be that the state would wind up with the site it needed, and the host community would feel that it had lost a fair fight.

Relatively soon after New Jersey's Regional Low-Level Radioactive Waste Disposal Facility Siting Act became law in December 1987, Governor Thomas H. Kean nominated 22 people to fill the nine non-Cabinet slots on the siting board and 13 positions on the advisory committee. The state Senate reviewed their credentials and confirmed them all by August 1988.

Over the next decade, only five of the original appointees left the Board. By 1998, six members had been serving since at least 1990. This unusual continuity was accompanied by remarkable dedication and collegiality. Board members rarely missed a meeting and worked largely by consensus.

The tone was set by Paul Wyszkowski, who was appointed to the Board at its inception and became its chair two years later. A civil engineer who managed environmental affairs at Bell Laboratories, Wyszkowski's responsibilities included ensuring that Bell Labs' waste was disposed of safely and legally. He ran the Board with a gentle and inclusive firmness that kept discussion open but focused. When he perceived proposals from any of his colleagues as micromanaging the staff, he would quash them, often with an anecdote. He told the Board, for example, that he had been in too many meetings in which hours were spent debating whether to buy a $40 shovel, leaving no time to discuss what it would be used for.

Two of the other longtime members came from the nuclear industry. Rick McGoey, also an engineer, directed systems that support nuclear power plant operations for GPU Nuclear. Jim Shissias managed environmental affairs for Public Service Electric & Gas (PSE&G). Together, these utilities operate all four of New Jersey's nuclear power plants.

In addition to having two engineers, the Board had two physicists. David Steidley was chief physicist for the department of radiation oncology at St. Barnabas Medical Center in Livingston. Joe Stencel was a health physicist at Princeton University's Plasma Physics Laboratory when he was named to the Board. He left that job in 1996 to enter a doctoral program in water resources. One could have argued that Stencel should then have been removed from the Board since he had been appointed to fill a seat assigned to a representative of industry, and there were no seats for graduate students. Fortunately, no one raised such an argument.

The other original member was Leland Merrill, who had been the Dean of Agriculture at Rutgers University before he retired. He had also served on the Board of the New Jersey Audubon Society and was often seen by others, and perhaps himself, as the environmental conscience of the Siting Board. Once he became comfortable with the environmental and health implications of a proposal, he often led the Board through the questions he thought other environmental activists would be likely to ask.

When the Board hired me as executive director in 1994, it had several vacancies. One that had been held by Pat Clark from the League of Women Voters (LWV) until she moved to Florida was subsequently filled by another LWV representative, Bonnie Magnuson. In addition to being active with the LWV's Natural Resources Committee, Magnuson was director of religious education for an Episcopal church. She had no familiarity with the issues associated with radioactive waste when she came to the Board, but was interested in helping government to effectively involve citizens and make decisions.

The most specific admonition the Legislature had stated about the Board was the requirement to include "a physician specializing in nuclear medicine." While this is a growing field, few of its practitioners feel able to devote a fixed day each month to a meeting in Trenton. When the first person appointed to that seat, Fred Palace, resigned after serving six years, it took more than a year to find a replacement. Theodore Stahl, who was named to the Board in 1995, was chief of nuclear medicine at St. Peters' Medical Center in New Brunswick.

The ninth member of the Board was also appointed in 1995. Robert McKeever had recently retired as president of Cogeneration Systems, Inc. In that capacity, he had led a successful effort to locate a 200-megawatt power plant in Logan Township in Gloucester County, New Jersey. McKeever's experience siting a large and potentially controversial industrial plant was directly applicable to the Siting Board's mission. Unfortunately, the Legislature's tight drafting of the membership requirements did not leave an obvious slot for him. As a result, Governor Christine Todd Whitman nominated McKeever to one of the seats intended for "people who would represent recognized environmental organizations or other public interest groups." No one objected, probably because no one noticed. The final two seats were assigned to staff representatives designated by the commissioners of the state Environmental Protection and Health departments.

When the Board began, it borrowed staff from the state Department of Environmental Protection. It soon decided to hire a small staff and to rely on consultants for major tasks. In July 1989, Sam Penza was hired as the first executive director, and the next month, the Board contracted with EBASCO, Inc. as site search consultant and Holt, Ross and Yulish as public affairs consultant.

Penza had been serving as an assistant commissioner in the Department of Human Services, responsible for its administrative and personnel programs. He brought to the Board a wealth of knowledge about the inner workings of state government, including the Legislature, but no experience with radioactive materials management. He soon hired Jeanette Eng, a health physicist working in the Department of Environmental Protection, as deputy director, and Denny Medlin, also from Human Services, as office manager. With the addition of a secretary and a community coordinator, they formed a staff of five.

In 1990, the Board adopted criteria to use as a basis for identifying and evaluating potential locations for a disposal facility. The 19 sets of policies ranged from preventing groundwater intrusion to attempting to avoid areas that are meaningful to people because of historic, cultural, religious, ethnic, or racial heritage.

These criteria had been developed by EBASCO and discussed in draft form at public hearings. Little interest had been expressed, however, since no specific sites were being discussed, and the criteria did not otherwise lend themselves to news articles or spirited debate.

Shortly thereafter, the board published an *Area Screening and Candidate*

Site Identification Methodology Report. This document described how the Board planned to have consultants review statewide maps and other data to identify several areas that appeared likely to meet the adopted siting criteria. Under this procedure, the consultants would present the Board with information about possible sites, without identifying where they were located. Board members would have enough information to vote on whether or not the site appeared viable for the disposal facility, but would not know who owned the land, or the town or legislative district in which it was located.

The theory of this "geographically neutral" or "blind" approach was that it would protect the Board from accusations that it picked or excluded sites for "political" or self-interested reasons. This was a process that also had been adopted by the Hazardous Waste Siting Commission and by Connecticut's low-level radioactive waste program. It was intended to increase public confidence in the integrity and fairness of the agency's work.

The Board would then undertake more detailed local studies, called "pre-characterization," of the candidate sites and select one or more of them for extensive historical, environmental, and modeling analysis. This final process, called "site characterization" would determine whether a disposal facility proposed at this location was likely to comply with all of the Nuclear Regulatory Commission's applicable regulations and to receive other required federal and state approvals.

The people living near a candidate site would first learn of the Siting Board's possible interest in their community when the Board selected sites for the pre-characterization process. The Board, anticipating that the local reaction might not be welcoming, met with representatives of environmental groups and local governments to discuss what could be done to make the eventual public discussions as productive as possible. It also contracted with the Center for Negotiation and Conflict Resolution at Rutgers University seeking recommendations on "designing a multi-purpose local assistance program for siting a facility."

By the end of 1990, the Board felt it had all the pieces in place to follow the mandate it had been given by the Legislature. One consulting firm would identify several scientifically acceptable sites for the disposal facility, while another firm worked to increase public understanding of the necessity and feasibility of safely disposing of low-level radioactive waste in New Jersey. The Siting Board and its staff would direct the firms' work and, with the Rutgers Center, prepare to have productive discussions in the selected communities.

CHAPTER 5

Connecticut and Illinois: The Seeds of the Voluntary Siting Process

C onnecticut, which had a siting process very similar to New Jersey's, had the chance to implement it first. It didn't work. In 1991, its siting agency, the Hazardous Waste Management Service, relying on its environmental consultants, identified the three "best" sites in the state. Following its geographically neutral selection procedure, the agency scheduled a press conference after it had heard descriptions of the sites, but before its members were told where they were located.

Just hours before the public announcement, the consultants showed the state officials the maps they would display for the press. That was when they revealed that the three sites, while conveniently located on the same map, were also all in the same legislative district — a district represented by John Larson, the Senate President *Pro Tempore*, a very powerful legislator who was already being mentioned as a likely gubernatorial candidate three years hence.[15]

This coincidence might have helped demonstrate the integrity of the state's process. The Senate president might have said, "Even though I disagree with the outcome, I must applaud the fairness of the process." But such a sentence has probably never before been uttered in any contentious process, and this was not to be the first time.

Instead, within a year, under Larson's leadership, the Connecticut Legislature stopped the siting process, added new criteria designed to eliminate the three sites from further consideration, and directed the Hazardous Waste Management Service to begin again.

The agency's staff regrouped and concluded that only something radically different would have a chance of succeeding. Ronald Gingerich, an anthropologist whose career path had led him to become director of Connecticut's low-level radioactive waste program, and Anita Baxter, a former local elected official working in the program, had learned that several provinces in Canada had employed voluntary approaches to siting with promising results.

In Alberta and Manitoba, communities had volunteered to host hazardous chemical waste disposal facilities, and Ontario was then in the process of working with 14 communities, each of which had established liaison groups to consider volunteering for a disposal facility for low-level radioactive waste.[16] The Connecticut officials decided this could form the basis of a new approach for them.

New Jersey's Siting Board and Advisory Committee watched these developments with interest and started to question whether the statewide screening they had underway would have an outcome any different than Connecticut's. Nothing suggested that it would, particularly not the latest developments at New Jersey's Hazardous Waste Facilities Siting Commission. Millstone Township had now succeeded in getting itself eliminated from consideration and the other municipality the Hazardous Waste Commission had identified as having a potentially ideal site, the city of Linden in Union County, was locked in a fierce, protracted battle with the agency.

During 1992, the New Jersey Siting Board and Advisory Committee explored the idea of voluntary siting and took increasingly definitive steps toward adopting it as their approach. Several of the members had been introduced to the concept through the work of Peter Sandman, who was then a professor at Rutgers University. Sandman, now with the Harvard-MIT Public Disputes Program, has been an observer and participant in a number of land use controversies, and has spoken and written extensively about community involvement and public participation programs.

Siting Board Chair Paul Wyszkowski remembers that, "Mr. Sandman's opinion was that in order to have such a program work successfully, we first needed to remove the perceived hazardous nature of the site by removing the threatening connotation and indicating that it could be sited virtually anywhere, and then offer that there could be substantial financial and other rewards for such a host site. He felt that through such efforts we might even be able to have communities competing with each other to be the victorious community to host such a site."[17]

Siting Board and Advisory Committee members found Sandman's suggestions attractive for pragmatic as well as philosophical reasons. Some worried that a mandatory siting process would never work, while at least a few thought a court might later invalidate a Board decision to impose a site on a municipality if it had not first at least tried a voluntary siting approach.

Near the end of the year, on December 3, Board members adopted a statement of policy that summarized their journey:

> *The Siting Board continues to examine siting efforts and finds that the conventional selection processes are experiencing extreme difficulty. As a result, the Board, through a subcommittee of Board and Advisory Committee members and interested citizens, examined what approach might hold the most promise for a successful siting outcome. On July 9, 1992, the subcommittee recommended that the Board consider a voluntary siting approach. The Board unanimously adopted this alternative as a more workable approach to success in solving the LLRW disposal problem for New Jersey.*
>
> *A central element of this approach is to invite communities to voluntarily explore the possibility of locating the facility in their jurisdiction. The process encourages full consideration of citizens by empowering the community to actively participate in the facility development process. Interested communities must meet all siting criteria, including strict environmental safety standards, as established by the board. A community that enters into preliminary discussions also has the option to discontinue negotiations after full consideration. A host community is entitled to certain benefits delineated in the Siting Act. And, specific needs indicated by that community can be met with additional benefits, in order to provide a host community with appropriate protection, compensation and enhancement for helping to solve a statewide problem.*

During the first half of 1993, the Board continued to develop the concept of offering incentives to encourage a town to volunteer to host a disposal facility. In June, the new approach was outlined in an article in the *Star Ledger* headlined:

STATE LOOKING FOR BRAVE TOWN TO TAKE ON TOXICS

By this time, the Board had become sufficiently committed to changing its approach that at its July meeting, it directed EBASCO to stop the site selection work. At the same time, the Board asked the Advisory Committee to recommend a specific process it could use to implement a voluntary siting approach. The Committee worked collaboratively with the Siting Board, and the two groups made little distinction between their roles. But it was the Board that formally voted to move toward a voluntary process and the Committee that took the lead in developing a specific approach.

In August, the Board sent a letter to the mayors of each of New Jersey's then-567 municipalities announcing that:

> *This Board is investigating the possibility of an alternate approach to siting than the traditional 'decide-announce-defend' method. At the July 9 Board meeting, we endorsed the development of a practical program designed to maximize local government participation by encouraging municipalities to voluntarily explore the feasibility of siting a facility in their jurisdiction.... We will be seeking your input as we develop the details of the volunteer program. We trust you will agree that this is an innovative approach worth exploring. We welcome your participation in the process.*

This letter produced a few written requests for more information, each couched with a sentiment similar to that expressed by the mayor of Hopatcong: *"Please understand that this inquiry is for information purposes only and should not be construed as any indication of the Borough of Hopatcong's willingness to accept such a facility within its boundaries."*

The more common responses were letters making clear the municipality's

Freehold News-Transcript

complete disinterest, even though the Board's letter had been very general and explicitly did not require a reply:

- *Hopefully, you will understand why Washington Township, Morris County, does NOT want a low-level radioactive waste disposal facility to be located within or on the edges of our border.*

- *While your volunteer program is innovative in concept, our position, or rather opposition, to the location of a facility under any circumstances in this Township* (Maurice River) *must be clearly understood.*

- *We have had our share of problems with landfills... Although the township* (Gloucester) *understands that low-level radioactive waste must have a place in which to be disposed, we feel in light of the above mentioned circumstances we would have to resist any effort to place such a facility in our township as the residents would never tolerate any kind of waste to be brought within township boundaries again no matter what assets could be afforded them by being a host community.*

- The mayor and Township Council believe that Florence Township has more than its fair share of waste facilities. Please do not consider Florence Township for a future site.

- After receiving input from the Chairperson of the Mansfield Township Environmental Commission ... the Township Committee voted to vehemently oppose any proposal to site any disposal facility for low-level radioactive waste within our township.

- Accordingly, the township officials do not believe that Cranbury Township would be a feasible site...Furthermore, the officials are not interested in pursuing the facility as an economic development strategy.

- In my opinion and experience, the proposal sounds good but I believe your chances of success will be minimal or nil.

In December, the Board adopted a statement of policy that was less than two pages but embodied its radical shift from traditional siting strategies. The statement represented a conclusion that there was no need to try to find the single, or perhaps several, best locations in the state for a disposal facility. Rather, the advice the Board and Advisory Committee were receiving from EBASCO and from scientists in the Department of Environmental Protection convinced them that there were probably many sites in New Jersey where the facility could be safely located.

The new proposed plan suggested that interested towns or individuals bring forward possible locations. Many would prove unsuitable, but the Board would be able to evaluate each one as it was suggested. Instead of searching a haystack for the best needle, the Board would, in effect, examine each needle brought to it to determine if it could be threaded to meet all the federal and state regulatory criteria.

The Advisory Committee presented a first working draft of a voluntary plan to the Siting Board in June 1993. In October, the Committee held a two-day workshop in Princeton, with presentations from people experienced with alternative siting processes. This group included Barbara Connell from the Manitoba Hazardous Waste Management Corporation; Don Gorber from an environmental consulting firm in Toronto; Tom Kerr, who managed Illinois's low-level radioactive waste disposal facility siting project; and David Leroy, who had served as the U.S. Nuclear Waste Negotiator for a spent-fuel interim storage facility.

After the workshop, the Advisory Committee revised its June draft and, in January 1994, submitted a proposed voluntary siting plan to the Board. The Board accepted the proposal, and five months later, two years after first considering the idea, formally began a public comment period on the proposal.

What is most noteworthy about this chronology is that a government agency on its own decided that the path it had been assigned to follow would not lead to the goal with which it had been charged. The impetus to forge a new path was not an investigative report, media inquiry, challenge from an interest group, or a legislative committee hearing, but rather a group of mostly citizen volunteers, all taking seriously their duties as members of the Siting Board and Advisory Committee, including being open to using new information to question their own earlier decisions.[18]

There are few examples of organizations admitting, "We were wrong." Would such a major change of direction have occurred if the siting responsibilities had been assigned to a large ongoing state agency such as the Department of Environmental Protection? Would a member of the governor's cabinet have concluded, after spending more than $2,000,000, that he or she should essentially start over?

Perhaps. On the one hand, an agency with many functions might have been better able to justify the decision by finding other uses for the maps and data that had already been gathered. In addition, perhaps it could have made the shift with less public deliberation, which might have made it easier.

On the other hand, the fact that all of the Board's meetings were open to the public may have been what made its course correction possible. Representatives of the waste generators were present at every meeting and even participated in one of the workshops on voluntary programs that the Board convened. The siting law had been amended in 1991 to require the entities generating low-level radioactive waste to pay all expenses associated with the siting, design and construction of a disposal facility. As a result, the generators' participation in and support for the new direction was very helpful, and perhaps essential.

While the original siting process was based on unstated assumptions, the voluntary plan grew from extensive analyses of both experience and literature regarding risk communication. This growing field of study wrestled with such questions as why people become terrified of relatively minor risks while voluntarily exposing themselves to demonstrably more dangerous activities. Most of us worry much more when stepping into an airplane than into a car, despite overwhelming evidence that plane travel is safer. Many people continue to smoke cigarettes while losing sleep and joining demonstrations over a landfill or other environmental concern that could not possibly cause them more harm than smoking.

The general conclusions of those studying risk has been that people are more scared when they have less control. An airplane may be safer than a car, but if I am driving, I feel safer. The fact that someone chooses to smoke makes smoking feel safer and less mysterious than the workings of a landfill or factory seemingly imposed upon them.

The members of the Advisory Committee and Siting Board extrapolated from these conclusions to wonder whether a siting process could be designed that would give the residents of an area a controlling role in the decision-making process. What would happen if they said, "We think this will be safe, and we think the financial benefits could make this a good deal for your community, but this isn't going to be in your town unless you decide you want it." Would people react differently than they had in Connecticut, or than they had when New Jersey's Hazardous Waste Siting Commission had introduced itself by sending letters to 11 mayors congratulating them on their town having been selected as a possible location for an incinerator?

As this concept developed, several themes emerged. One was about responsibility or, perhaps more accurately, guilt. Everyone benefits from the use of radioactive materials. More than 60 percent of the electricity in New Jersey is generated by nuclear power plants, and most people know someone whose life has been improved or saved by medical procedures involving radiation. In addition, many of the lesser-known uses of radioactive materials, from smoke detectors to testing stress in bridges, improve our lives every day. If we reap all these benefits, don't we then have an obligation to deal with the waste?

This guilt trip was promoted as a major theme by some of the organizations representing utilities and other generators of low-level radioactive waste, and by some states and regional compacts. A major trade group, the Nuclear Energy Institute, for example, concludes the low-level radioactive waste section of its web page, by noting, "Everyone shares in the benefits of radioactive materials. That's why it's everyone's responsibility to help dispose of the waste – not leave it for our children."

This approach attempted to link the safe disposal of this type of waste with the broader and more popular ethic of environmental protection and care for the planet. The theme, however, did not seem likely to help convince people that their community should provide the home for this disposal facility. If there were any concerns about health and safety, no one, regardless of the depth of his or her civic spirit, would say, "Okay, let's put it here."

The Siting Board decided that emphasizing the benefits the host community would receive was more promising. The Board would publicize all the information, speakers, and organizations of which it was aware, regardless of their views, and recommend that anyone interested read, listen, ask questions, and explore. The Board would also pay all reasonable costs associated with this pursuit of knowledge, including travel to related installations both within and outside the state.

The Board's hope was that a few residents would decide that the facility might be beneficial for their town. They would be sufficiently attracted by the economic benefits to say out loud, "If this thing is safe, it might be a good deal for us, so let's go learn about it." Some might add, "I recognize this is a big 'if,' but ..."

The third theme in creating a voluntary siting process was that residents of a town had to be able to explore the possibility of hosting the disposal facility without incurring any obligation. The plan adopted by Connecticut went so far as to say that if a town volunteered and dropped out after passing a prescribed and fairly early stage in the process, the disposal facility would never be placed within its borders. This promise was designed to address the fear that stepping forward, however briefly, might alert the state government to a good site it could return to if no other towns volunteered. New Jersey's Siting Board felt its process could be open and non-threatening without including such a provision.

On the other hand, New Jersey's Board gave interested communities more time before requiring a commitment to host than did Connecticut or any of the other states that considered similar processes. At various times during the 1990s, Illinois, Massachusetts, Michigan, Ohio, Pennsylvania, and Vermont also floated voluntary siting plans.

While the other states would not begin full-scale site characterization until a community had signed a binding facility development agreement, municipalities in New Jersey would be able to review the results of this feasibility study before having to make a commitment. Thus, if towns in the two states expressed interest the same day, and their explorations followed the same schedule, the town in New Jersey would be considering a binding agreement some 18 to 24 months later than the town in Connecticut. New Jersey's Board also would have invested several million dollars more than Connecticut's during the period in which the town could still withdraw.

Consistent with that approach, flexibility was a recurring theme in New Jersey's voluntary siting plan. The Board did not want to prescribe the steps a town would have to follow to explore any interest it might have in hosting the disposal facility. This was based, in part, on a belief that potential host communities would feel more in control if they could design the local community siting process themselves. The Board also recognized that New Jersey's many municipalities are far from homogeneous in their geography, population, and governance, and that the same process would not be attractive to all of them.

In February 1995, the Board adopted the *Voluntary Plan for Siting A Low-Level Radioactive Waste Disposal Facility*. This plan described the siting process as three broad phases called "Exchange of Information," "Community Exploration and Consideration," and "Decision-Making." The Board's commitment to flexibility was apparent when the most contentious issue debated was how far the process could go in a particular town without any written expression of interest from a municipal official or agency.

The majority view was that something written — Board members didn't want to limit flexibility by suggesting whether it had to be a letter or resolution or who would have to sign it — would have to be submitted for the town to move from the "Exchange of Information" to the "Community Exploration and Consideration" stage. This communication would in no way bind the town to host the disposal facility, but it would indicate some level of local interest to warrant the Board investing money in a comprehensive characterization of the site in the town that was under consideration. The minority view on the Advisory Committee was that even this requirement was too onerous, and that there could be situations in which the Board would want to proceed even if local officials didn't want to put anything in writing.

The final reason that the Board did not precisely define how the siting process would proceed was an unstated sense of humility. No matter what site-selection and decision-making processes were used, a disposal facility for low-level radioactive waste would be extremely controversial. The Board knew it was putting forward a site selection process that had never been tried in New Jersey, and had only rarely been employed elsewhere.

The Canadian experiences with voluntary siting were instructive and encouraging, but the differing national government structures prevented the

Siting Board from attempting to directly replicate them in New Jersey. As a result, much of the inspiration for New Jersey's strategy came from the only known successful use of the process in this country. It had occurred in Martinsville, Illinois.

Illinois was the host state for the Central Midwest Compact it had formed with Kentucky. The Illinois Low-Level Radioactive Waste Management Act of 1983 had specified that the disposal facility had to be located in a volunteer community.[19]

In 1987, the director of the Illinois Department of Nuclear Safety wrote to the state's 102 county boards inviting them to learn more about the low-level radioactive waste program. Twenty-three counties expressed interest and were included in a study to map potentially acceptable sites.

Early in 1988, 17 counties withdrew, and the department eliminated four others. At the same time, the Martinsville City Council in Clark County and the Wayne County Board each passed resolutions asking to host the disposal facility if a mutually agreeable site could be found. That November, opponents succeeded in placing non-binding referendums about the "dump" on the ballot in both counties. Sixty-eight percent of the voters in Wayne County and 58 percent in Clark County voted against the facility. Within the City of Martinsville, however, 58 percent voted in favor.

When the referendum was again on both ballots a year later, the negative vote in Wayne County was 70 percent, leading the county board to drop out. Again, however, the resolution was supported in Martinsville by 56 percent of the voters, although the no vote in Clark County climbed to 74 percent.

Following the vote, the Illinois Department of Nuclear Safety and local officials in Martinsville worked together to choose and analyze a potential site for the disposal facility. Eventually, they reached agreement and on June 3, 1992, Martinsville Mayor Truman E. Dean and the six city aldermen signed a 35-page "Community Agreement" document describing how the facility would operate, the city's role in its oversight, and the benefits the city would receive. The agreement began, "The City of Martinsville approves the location within the City of a facility for the disposal of low-level radioactive waste and the city agrees to provide customary city services to the facility on the same basis as such services are provided to other businesses within the city."[20]

As the state department prepared to hold the final public hearing required before designation of a preferred site, it convened a series of informal public sessions with the authors of their studies analyzing sites within Martinsville. Among those attending the forum on hydrology were representatives of the

Illinois State Geological Survey and State Water Survey who questioned some of the terminology in the report. The next day, the agency representatives met together and agreed to revisions that satisfied their concerns. The major change was to substitute the word "aquifer" for "water-bearing zone" in 22 locations, and "water-bearing zone" for "aquifer" in 19 other sentences.

Even though the objections of the Geological Survey and Water Survey were quickly addressed, news accounts of the hydrology forum enabled the opposition that had been intensifying elsewhere in the state to focus public discussion on the technical merits of the siting process. Soon, the Legislature created a new review agency, the Illinois Siting Commission, with the responsibility to hold hearings on any site recommended by the Department of Nuclear Safety.

The new Illinois Siting Commission eventually held 71 days of hearings. On October 9, 1992, it ruled that the Martinsville site was neither safe nor suitable for the location of a low-level radioactive waste disposal facility. The City of Martinsville appealed, asking that the Commission consider additional information and reverse its decision, but in December, the Commission denied the appeal.

Although a disposal facility was never built in Martinsville, the lesson taken by New Jersey's Siting Board was that an American municipality, after learning about the disposal facility, had decided it would be a beneficial addition to their city. Not only did the residents twice vote to host it, but local officials negotiated and agreed to the terms of a final agreement. This saga did nothing to address the nation's need for disposal options for low-level radioactive waste, but it did demonstrate that a voluntary siting process could work.

Unfortunately, however, Martinsville offered no buildings or disposal operations that could be toured or reproduced in photographs to help allay the fears of local residents in other states considering hosting a disposal facility. The one success story that proponents of voluntary siting programs could cite had no happy ending. The good news was that New Jersey could supply the first one that would.

CHAPTER 6

'One Tough Job: Peddling Radioactive Waste'

V ery few people know very much about the origins and management of low-level radioactive waste. From its first meeting, long before a voluntary plan was on the horizon, the Siting Board had recognized that any program to site a disposal facility could succeed only if many more people became aware of the problem and issues associated with it.

The outreach efforts began to grow in 1992 as the Board became more intrigued with the potential for voluntary siting. The public relations firm the Board had first engaged in 1989 helped to prepare written materials and to find opportunities for the Board members and staff to meet with a variety of individuals and groups. The Board also arranged a contract with the state Office of Dispute Settlement to engage the services of a trained mediator with a strong interest in processes for public participation.

The Advisory Committee made clear its belief that input from "constituencies with a stake in the outcome of the process would be critical."[21] Among the constituencies the Advisory Committee identified were organized labor, emergency response providers, environmental groups, land-use planners, and generators of low-level radioactive waste. It convened separate meetings with representatives of each group, which provided immediate feedback and suggestions, and also created a small constituency that would be interested in at least observing the course of the voluntary siting program.

Once the *Voluntary Plan for Siting a Low-Level Radioactive Waste Disposal Facility* was adopted, outreach was ratcheted up. The Board decided that public involvement would be more fully integrated into its work if it was completely directed by the staff, instead of being farmed out to consultants. The Board members also felt that the trust they needed to establish with the residents of communities would come more easily if communication was not filtered through a consulting firm or perceived to be "public relations."

As a result, the Board amicably terminated its contract with the public relations firm and hired two new staff members, both of whom had extensive experience communicating with the public. Bernard Edelman, an accomplished author, had most recently been communications director for the New Jersey Department of Health and had previously worked in the Office of New York City Mayor Ed Koch. Greta Kiernan was a former state Assembly representative who also had worked for a number of New Jersey legislators, as well as for the Port Authority of New York and New Jersey, the Board of Public Utilities, and the Department of Environmental Protection.

The Board now had a staff of six, each of whom not only believed in the importance of involving the public in public policy, but also enjoyed doing so. In addition, Fran Snyder, the mediator from the Office of Dispute Settlement, felt similarly and continued to work for the Board on a close to fulltime basis. Only the deputy director, Jeanette Eng, had a scientific background, but all had come to believe that working to resolve New Jersey's low-level radioactive waste disposal problem was an important and honorable calling.

Low-level radioactive waste management is one of the many issues that have very low public visibility except when a particular location is under scrutiny as a possible solution. That is rarely the best moment for calm consideration of a complex issue.

The Board initially worked to introduce the topic to as many people as possible before any areas were identified as potential sites for a disposal facility. The introduction sought to emphasize that the facility would be safe — much safer than most people would expect. The Board emphasized that the voluntary siting process was different, and that it was responsive to some of the criticisms that have been voiced about past programs to find locations for controversial developments.

The Board wanted to increase general awareness of the issue and also to stimulate people to think about whether the disposal facility might be worthy of consideration for their town, or perhaps another town with which they were familiar. When specific sites would eventually be proposed, the goal was to have some people in that town and neighboring towns who would have been to a meeting or read some literature that might lead them to support the proposal, or at least help them counter some of the misinformation that would inevitably be aired.

The Board's commitment to finding a solution was not the only approach available. There were suggestions that a proper role for the agency was only to lay out the information without conclusions: to say, for example, "Some people think this disposal facility would be safe and some don't. Here is lots of literature, and a list of speakers and organizations. We'll help you get any information you need to make up your own mind." The Board's function would be educational and objective in the sense that it would give no opinion about the health and safety impacts of the disposal facility.

The League of Women Voters had, to some extent, adopted this perspective while taking on radioactive waste disposal as a major issue of concern to them. On a national level, the LWV produced booklets and videos on the subject, and also sponsored public workshops. In New Jersey, first Governor Kean and later Governor Whitman had selected an LWV official as one of the three Board members representing "recognized environmental organizations or other public interest groups."

As an organization, the League's interest and position on the issue was focused much more on the decision-making process than on a particular outcome. In a concluding paragraph of a 150-page *Nuclear Waste Primer* published in 1993, the LWV wrote:

> *The cleanup of contaminated sites and the siting, construction and operation of facilities for the disposal or storage of radioactive waste may have significant effects on communities, public and worker health, the environment, and the local and national economies. Similarly, the delay or failure to provide safe and permanent disposal sites may adversely affect the communities in which waste is now held in temporary storage – in facilities that were never intended for such long-term use. Thus, it is important*

> *not only that citizens have roles in deciding how things will be done, but also that they contribute to the process of making things happen, addressing significant problems rather than delaying action indefinitely. Deciding when action is called for is sometimes the most difficult decision. Those who have faith in the democratic process believe that the public will help ensure that the right decisions are made about how and when to act.*[22]

The interest and activism of the League of Women Voters added credibility and energy to the Siting Board's efforts. But while the Board emphasized the process for decision-making, it also explicitly advocated a particular outcome.

The *Voluntary Siting Plan* proposed in January 1994, for example, listed five primary objectives. The first four were to "help communities explore the possibility ..., broadly disseminate and discuss facts about radioactivity..., promote constructive dialogue, and address concerns of affected citizens." The fifth objective was to "establish a disposal facility for low-level radioactive waste generated in New Jersey that protects public health, safety and the environment."

The Board was comfortable with this position because each member, including the LWV representative, agreed that with proper siting, design, and operation, a disposal facility would be safe. Board members concluded that their role was to get the facility built and not just to preside impartially over an interesting public policy debate.

The Board was stuck, however, with the reality that some people with impressive academic and professional credentials believed that any disposal facility for low-level radioactive waste would inevitably become a hazard to public health and the environment. This position was certainly the minority viewpoint among knowledgeable professionals, but how could the Board say so without sounding self-serving and unconvincing?

Many recent surveys show that government is not trusted. The Pew Research Center, for example, found the level of trust in state government at 9 percent, city or local government at 14 percent and the federal government at 6 percent. In this particular analysis, the highest rated institutions, at 78 percent and 46 percent, respectively, were the fire department and the police department.[23]

Other polls that focused on job categories, as opposed to institutions, reported that the word of doctors and teachers was highly valued. The Board decided to ask individual New Jersey doctors, scientists, and educators to sign a statement attesting to the safety of a disposal facility.

In the spring of 1995, Board staff contacted five prominent individuals who were likely to be supportive. When all five agreed to sign the statement, the Board began to distribute it more widely, gaining additional signatures and reprinting it periodically as more names were added. Most of those who signed were people approached directly by a Board or staff member, but some signed after picking up a form at a meeting at which the topic was discussed.

Surprisingly unsuccessful was an expensive full-page, tear-out advertisement the Board purchased in the October 1995 issue of *New Jersey Medicine, The Journal of the Medical Society of New Jersey*. The ad listed the 60 names that were on the statement at the time. Although the magazine has a circulation of 19,000, only one physician returned a signed card.

Eventually, the statement garnered 275 names, although contrary to Board expectations, adding new signatures did not become appreciably easier as the list grew. Nevertheless, it was extremely useful. Board members and staff could refer to it in talks and hand it out at public events. It was also particularly helpful when talking with reporters. Faxing it to them after — or sometimes during — telephone conversations would encourage them not to simply succumb to a notion of objectivity that would answer a quote from the Siting Board with one from a person who disagreed. They could add that many scientists believe the facility would be safe, and they sometimes mentioned and even contacted signatories who lived or worked in their circulation area.

The Board also employed most of the standard public participation techniques. Staff actively sought opportunities to speak before virtually any civic organization, municipal and county agency, and environmental group anywhere in the state. If the organizer didn't want to devote an entire meeting to low-level radioactive waste, we would ask to speak for 15 or 20 minutes.

As a result, some of the people who received an often unexpected introduction to low-level radioactive waste included members of the New Jersey Association of Counties and the New Jersey County Planners Association; the New Jersey Farm Bureau, Agricultural Development Board, Public Health Council, and Health Officers Association; various Chambers of Commerce and Development Councils; the New Jersey Business and Industry Association; the Treasurers and Tax Collectors Association; and the New Jersey chapter of the American Society for Public Administration. Board members and staff also spoke to several League of Women Voters chapter

New Jersey Doctors, Scientists & Educators Agree: 'A disposal facility for low-level radioactive waste would be a safe neighbor.'

Although the word 'radioactive' raises immediate and understandable concerns and fears, we are convinced that the low-level radioactive waste generated in New Jersey can be packaged, transported and disposed of in a safe manner. We believe that a disposal facility for this material would cause no health or safety hazards if care is taken in the selection of a site and in the design, operation, monitoring and eventual closure of this facility.

Furthermore, we know that low-level radioactive waste generated by New Jersey's nuclear power plants, hospitals, research labs, pharmaceutical companies, and colleges and universities is an unavoidable by-product of the energy production, medical procedures, research and manufacturing we have come to depend on.

Accordingly, we would suggest that New Jersey communities and municipalities, including those in which we and our families live, consider whether they have a site that might be suitable for a disposal facility, and evaluate the benefits they would receive if they choose to volunteer and are accepted as the host community for New Jersey's Low-Level Radioactive Waste Disposal Facility.

We recommend that there continue to be an open, public, voluntary process in which residents of potential host communities, their neighbors and other interested individuals can gain sufficient information to evaluate the potential impacts, and benefits, of the disposal facility, specific sites that might be suitable, and the health and safety measures that will be integral to the facility.

We further recommend increased research and management efforts to continue conservation efforts to reduce the volume and radioactivity of the waste generated in New Jersey.

STATEMENT OF SUPPORT: Recognizing that public trust of government officials was low, the Siting Board collected and distributed a list of 275 physicians, scientists and educators endorsing the safety of a disposal facility.

meetings and Exchange Clubs, as well as to some 50 Rotary and Ruritan Clubs; to college and high school groups; and to meetings of employees arranged by several of the institutions generating low-level radioactive waste. In addition, the Board displayed an exhibit each year at the annual conventions of the New Jersey Conference of Mayors and state League of Municipalities, and at a statewide Environmental Exposition. The display was also showcased at meetings of the state Medical Society, the Society for Environmental and Economic Development, and the Association of New Jersey Environmental Commissions. For several years, the exhibit was also part of a "Super Science Weekend" held at the State Museum in Trenton and, for one week when the legislature was in session, it stood in a well-traveled corridor in the State Capitol. The display had panels explaining the goods and services that generate low-level radioactive waste, the siting criteria for the disposal facility, and the benefits that would accrue to the host community. On the floor adjacent to the panels were two full-size half-drums with plexiglas windows filled with the types of trash that constitute the bulk of low-level radioactive waste, and a table with a three-dimensional model of a disposal facility. The display could be self-explanatory, but at most of these events one or two Board staff would be available to answer questions and collect business cards from people who seemed interested.

The League of Municipalities convention is one of the largest in New Jersey. Scheduled each year in the Atlantic City Convention Center for a week between Election Day and Thanksgiving, it attracts thousands of local officials and literally hundreds of exhibits. Most exhibitors are private vendors who give away some type of trinket to help them and their service be remembered. Pens, mugs, refrigerator magnets, and paperclip holders are popular, but there are also more unusual items such as potholders and stress-relieving synthetic rubber balls. Several offer plastic litter bags imprinted with their name and phone number, and many attendees wander the aisles of exhibits picking up the bags and then filling them as if it were Halloween. Most parents who have attended the convention in years past have children who expect to see some good loot when they get home.

The Siting Board also wanted to be remembered and in its early years ordered several different giveaway items, each printed with the Board's long name, and some with the address and phone number as well. The T-shirts, caps and pens were fairly standard, but other items required some explanation and were intended to lead to more conversation around the exhibit. The most successful was a map of New Jersey on a small, six-piece cardboard puzzle. The puzzle was surprisingly difficult, and people would leave the booth and bring their friends back to watch them be stumped.

The Board's other items were a cloth frisbee that was mistaken by some for a yarmulke, and a small mysterious-looking teardrop of hard black plastic that was very effective both for removing ice from windshields and for scraping pots. These toys were sufficiently interesting that each year they were picked up by many people, most of whom undoubtedly had no interest in the words emblazoned on them.

Several weeks before the 1995 convention, the Board staff, finding that its supply of trinkets was shrinking, decided to create a giveaway item that people would not only want to pick up and give to their kids, but that would also tell them something that might pique their interest in the voluntary siting process. The new product, produced cheaply and quickly, was a $2 million bill representing the annual benefits a host community would receive. The staff considered putting Marie Curie's picture in the center, but eventually selected Albert Einstein because he was more recognizable and had much more of a link with New Jersey.

The production of this bill was one of the moments, albeit a minor one, when the Siting Board clearly benefited from being an independent agency. In a large department, the number of approvals required to print something so unusual probably would have doomed the initiative through delay, the need to incorporate the ideas of a hierarchy of directors and assistant commissioners, or outright rejection.

The bills did serve their purpose, although a few supporters of the siting process worried that they were in poor taste. They seemed to be particularly appreciated by high school and college classes. Also, less than a month after their debut at the League of Municipalities convention, they became the centerpiece for a large article in *The New York Times*.

The Board assumed that anyone who became at all interested in the siting process by picking up a $2 million bill, viewing the display or listening to a talk would want written materials they could read at home and perhaps show their friends. It had printed 10,000 copies of the *Voluntary Siting Plan* and distributed it widely, but soon found that that document didn't specifically address many of the questions people were asking. As a result, the Board prepared a booklet of questions and answers called *An Introduction to the Issues: Siting a Low-Level Radioactive Waste Disposal Facility in New Jersey*. Fortunately, the unavoidably lengthy title was less noticeable than an unusually eye-catching and user-friendly format Bernie Edelman borrowed

from a publication on AIDS that he had produced for the Department of Health.

The inside cover of the brochure quoted Charles Peters, the editor of *The Washington Monthly* magazine, saying, "We face so many complicated and difficult issues that we need to attain new levels of citizen responsibility for learning about public problems and participating in their solution."[24] A large rendition of this quote was also added to the top of the display.

In subsequent years, the Board also prepared a series of five colorful pamphlets. This reader-friendly format was able to incorporate a large amount of information. The first one posited, *If Your Town Has Contaminated Land, Here's How You Might Turn an Eyesore into an Asset.* Another, called *We Live in a Radioactive World*, listed many of the ways in which radioactive materials are used in energy, diagnosis and treatment of disease, medical research, industry, agriculture, and science.

Transporting Radioactive Materials in New Jersey was written in response

to often-expressed public concerns. This booklet pointed out the extent to which radioactive materials are already shipped around the country every day, and noted that they are carried to and from almost 500 institutions in New Jersey. These trips, conducted mostly by truck, occur without incident.

What Have We Learned about Disposing of Low-Level Radioactive Waste described the leakage problems that had occurred at the first generation of disposal facilities, and tried to explain why New Jersey's facility would be different.

The fifth booklet asked, *Would a Disposal Facility for Low-Level Radioactive Waste Affect Daily Life in My Community?* Inside, one column was headed, *It Would*, and one said *It Wouldn't*. The former listed benefits that would accompany the facility, while the latter listed possible problems and tried to explain why they wouldn't occur.

The Board also contracted with the Environmental Sciences Training Center at Rutgers University to provide additional public information. The center, headed by Alan Appleby, a Rutgers professor with extensive expertise in radioactivity, prepared 18 *Fact Sheets* on topics ranging from *What Is Radioactive Material?* to *What Lessons Have Been Learned From Early Low-Level Radioactive Waste Disposal Facilities?* Each sheet was three or four pages, and the set fit into a simple folder.

The *Fact Sheets* were based on similar material prepared by Ohio State University Extension with a grant from the Midwest Compact Commission. While the Midwest Commission and the New Jersey Board had paid for these reports, they were the work of two academic centers and not of two radioactive waste siting agencies. The Board hoped that this distinction would give them a credibility that its own publications inevitably lacked for some readers.

The second task requested of the Rutgers Training Center was to assemble a roster of experts in various disciplines who could speak to interested groups around the state about radiation in general, and the public health and safety questions that would likely be raised by a proposal to site a low-level radioactive waste disposal facility in any community. Eventually, the center printed a two-sided card advertising a speakers' bureau with expertise in physical and chemical properties, health effects of radiation, public health, geology and hydrology, environmental impacts of waste management, medical radioactive waste management, radioactive waste transport, and hazardous waste treatment.

The Board also financed a *Sourcebook* compiled by Fran Snyder of the Office of Dispute Settlement. This document provided a brief description of every group known to have an interest in the siting of a low-level radioactive waste disposal facility in New Jersey. This included federal and state agencies, academic institutions, industry groups, and civic and environmental organizations. Prominent in the last category were pages about each of the groups actively opposing the Siting Board's work. The *Sourcebook* used verbatim descriptions provided by each organization and provided their addresses and phone numbers. The Board hoped that this information would be helpful to people trying to learn more about the issue, and also that it would help demonstrate our good faith in advocating that interested people examine all points of view before forming their own conclusions.

The two Board creations that were most useful were the *Question-and-Answer* booklet and an 18-minute videotape called *Hosting New Jersey's Low-Level Radioactive Waste Disposal Facility: Could it Be Right for Your Community?* In addition to reiterating information that was available in writing, the video showed the operating disposal facility in Barnwell, and included a computer simulation dramatizing what New Jersey's disposal facility would probably look like and how it would work. The video incorporated a short clip from an old monster movie to touch on some of the less scientific inputs that have helped form public opinions about radiation. The Board arranged for *New Jersey Network*, the state-owned public television station, to produce the video. This was quicker than choosing a private firm because the Board could use the talent available there without seeking other bids.

The Board wanted to have a well-respected person introduce and close the tape, and had read that retired NBC television anchorman John Chancellor now lived in Princeton and had recently undergone radiation treatments for cancer. When we wrote to ask if he would do it, he replied that his doctors had advised him to limit his activity. In closing his warm letter, he added, "I can think of no public problem more daunting than yours."[25]

The next person the Board approached said yes. Richard Sullivan was the first commissioner of the New Jersey Department of Environmental Protection and has been New Jersey's commissioner to the Northeast Compact Commission since it was formed. While less well-known than Chancellor, of course, he too is widely admired.

The Board eventually distributed about 5,000 copies of the videotape. Although it had cost $32,000 to produce, each additional copy cost less than $2. While many people watched them at home, some local officials showed them at planning meetings, finding them less likely to provoke controversy than inviting a speaker from the Siting Board. Often, an individual trying to raise the issue locally would ask for multiple copies to distribute to members of the town council or planning board and other area residents.

Although one cable television station used about a third of the video in a half-hour program devoted to the issue, the Board had little success gaining TV coverage for low-level radioactive waste disposal issues, and fared only slightly better on radio. Print media, however, did give attention to the issues.

During 1995, I wrote slightly different articles for four publications, each aimed at groups that could be helpful, and might be essential, to the Board's success. The Association of New Jersey Environmental Commission's *Environmental Quarterly* suggested a format in which I would interview myself. Among the questions I asked and then tried to answer was, "Why should environmental commissioners care about this issue?" I concluded by asking, "So, will I glow in the dark?" and responded that the readers would have to decide for themselves, and that the voluntary process was designed to help them do so.

For the *New Jersey Planners Journal*, I emphasized the type of sites that would be suitable for the facility and suggested various ways in which land-use planners could use their skills to contribute to the process. In the *New Jersey Conference of Mayors Quarterly Magazine*, I tried to write in a style that might be used to attract a private developer of a less controversial development. And finally, for New Jersey's largest and most prominent newspaper, *The Star-Ledger*, I submitted a more general column to a "Speak Out" section it had recently inaugurated on its op-ed page. The paper accepted it without change, except for giving it the headline:

"LOW-LEVEL WASTE AND YOU: PERFECT TOGETHER"

Mistakenly, they ran it twice, 10 days apart. Three people told me they had liked both of my articles.

The Board staff also wrote letters to newspaper editors, inspired in part by the folk wisdom that they are among the most well-read items in many papers. Often, the letters seized upon the opportunity to correct a recent article as an opening to introduce readers to the Siting Board and invite their participation.

We also wrote to the local newspapers each time a town considered and then rejected the possibility of volunteering, to both thank the people who had become involved and to make known the Board's availability to work with people from other towns. Most of these letters were printed, often with positive headlines, such as:

INPUT WELCOMED ON LOW-LEVEL WASTE PLAN
- The Sunbeam, June 22, 1997

GET WORKED UP OVER NUCLEAR DISPOSAL PROBLEM
- Gloucester County Times, June 22, 1997

TOWNS CAN MAKE RESPONSIBLE DECISIONS
- Bridgeton Evening News, September 12, 1996

The Board also distributed to the press all the mailings it sent to local officials. Reporters received copies of each document the Board produced plus advance notice for any meeting at which a Board representative was to speak. Some became interested in the topic as a result, while others worked for papers that had decided to track the issue and write about it from time to time. Still others responded to specific events. Altogether, the effect was that the siting process was featured in a large number of informative newspaper articles, although the headlines and sub-headlines varied:

STATE TRYING TO SWEET TALK SALEM COUNTY INTO TAKING WASTES; OFFICIALS ARE CONFIDENT THAT WITH PUBLIC EDUCATION A COMMUNITY WILL AGREE TO HOST RADIOACTIVE MATERIAL
- The Philadelphia Inquirer, June 8, 1995

STATE PITCHES WASTE DISPOSAL TO SPARTA; TOWNSHIP UNINTERESTED IN LOW-LEVEL RADIOACTIVE FACILITY
- New Jersey Herald, January 25, 1995

A substantial number of articles focused on me as an individual with an odd job:

THINK YOU HAVE A TOUGH JOB? MEET JOHN WEINGART . . .
- The Star-Ledger lead, October 4, 1994

DIRECTOR FINDS WASTE SITE A HARD SELL
- Asbury Park Press headline, April 16, 1995

A RELOCATION SPECIALIST FOR RADIOACTIVE WASTE
- Hunterdon Democrat headline, June 8, 1995

HIS LIFE'S LIKE AM AND FM;
DISC JOCKEY JUGGLES RADIO SHOW, STATE POST
Delaware Valley News, October 12, 1995

TRYING TO SELL RADIOACTIVE WASTE; FINDING TAKER IN
NEW JERSEY IS JOB OF A NUCLEAR-AGE WILLY LOMAN
- The New York Times, December 9, 1995

JOHN WEINGART SAYS HE IS AN OPTIMIST. HE HAS TO BE.
- Daily Record lead, March 4, 1996

WOOING TOWN TO TAKE NUCLEAR WASTE BIG JOB
- Asbury Park Press headline, March 7, 1996

ONE TOUGH JOB: PEDDLING RADIOACTIVE WASTE
"Imagine being a sales representative assigned to sell something that no one wanted.... As a guy trying to get New Jerseyans to welcome nuclear waste, albeit low-level nuclear waste, Weingart might have an easier task trying to sell a Brooks Brothers suit to Dennis Rodman"
-- Tom Perry, Courier-News headline and lead,
January 23, 1997

While these articles were complimentary and sometimes funny, they seemed to do nothing for the siting process. If anything, some of them may have reinforced the notion that any municipal official would be crazy to consider volunteering.

The most unpredictable of the Board's media forays were its requests to meet with the newspaper's editorial boards. We hoped that they might write a story shortly thereafter, but mostly wanted them to have a better understanding of the issues and know who they could contact in Trenton if the possibility of volunteering was raised by a town in their circulation area.

Sometimes these requests were turned down, but most were accepted. Sometimes the "editorial board" was one person, and other times it was a room full of reporters and editors. Sometimes the conversations resulted in several articles over the next few days, and sometimes no story ever appeared.

On a Thursday in March 1995, shortly after the *Voluntary Siting Plan* had been adopted, the editorial board of *The Times of Trenton*, one of the state's most important papers, agreed to see us. The Siting Board was represented by Joe Stencel, the Board member who had worked at the Princeton Plasma Physics Laboratory when he was appointed to the Board and was now pursuing a doctorate in water resources; Dick Olsson, chair of the Advisory

SURPRISE ATTACK: A friendly meeting with *The Times of Trenton's* editorial board was followed by a scathing front-page article two days later by a reporter who had not been in the room for the meeting.

Committee and the head of the Geology Department at Rutgers University, and several staff members.

The meeting with *The Times,* as they like to be called, lasted for more than an hour. The half-dozen reporters and editors seemed interested in the voluntary process and appreciative of the difficult mission faced by the Siting Board. We were, therefore, amazed to open the paper on Saturday morning and find a banner headline and front page article completely at odds with the meeting's substance and tone.

The article, credited to a staff writer who had not been in the room two days earlier, began:

> *How's this for an enticing prospect: a state agency arrives in your town, clears a 50-acre tract of land, erects a row of ugly concrete warehouses, and then — here comes the beauty part — begins loading tons of low-level radioactive waste into them. As the lucky owner of the state's Low-Level Radioactive Waste Disposal Facility, you will also receive; (1) About 100 visits a year from trucks bearing yet more of this highly dangerous material ...*

I wrote a letter of complaint to the man who had chaired the editorial board meeting, but received no reply. But almost a year later, and only a few days after my op-ed column in the *Star-Ledger* had appeared for the second time, *The Times of Trenton* ran a full column editorial titled:

AN OFFER WORTH HEARING; FACTS, NOT NIMBY, SHOULD GUIDE DECISIONS

An editor wrote:

> *Any public official who attempts to find a location for any kind of disposal facility — so-called hazardous waste, non-hazardous waste, whatever — quickly runs into a stone wall called NIMBY.... So great is citizen mistrust of government, and so effectively are the opponents of disposal facilities able to exploit the mistrust — often with the assistance of the media, which love conflict — that*

officials can search in vain for years and even decades for polit-
ically acceptable places to get rid of the unavoidable byproducts of
a technical and industrial society.

The editorial clearly and accurately described the need and requirements for a disposal facility in New Jersey, and the benefits the host community would receive. It concluded by quoting the final four sentences from my *Star-Ledger* column:

People may well laugh at you, as they do at me, if you are the
first to suggest that your town consider volunteering. But if they
take some time to compare the risks and benefits with those of
other types of development, the laughter may fade. Some will
decide a disposal facility doesn't make any sense for their area.
But others may find it to be a safe opportunity well worth exploring.

We mailed copies of the editorial to every mayor in the state, distributed it at every meeting, blew it up into a poster that became part of the Board's display, and reprinted it in the Board's next annual report to the governor. I wrote to thank the editor, but just as when I had complained the year before, I received no reply.

CHAPTER 7

Culture

T he Siting Board's outreach efforts aimed to provide accurate information about radiation, radioactive waste, and the governmental processes involved in managing the waste. The Board members focused on trying to respond directly to questions and concerns they heard at meetings and in individual conversations. While these comments were often quite specific, it soon became apparent that underlying them was extreme skepticism that seemed immune to factual rebuttal.

Radioactive materials are used in virtually every hospital in the country and routinely trucked on major highways and local roads, past schools and churches and housing developments. One out of every three hospital patients is treated with radiation or with devices that use radioactive materials. Yet any discussion of siting any sort of radioactive disposal facility sparks an instant public furor.

The two decades since Congress assigned responsibilities for radioactive waste disposal have been years in which public levels of distrust have mushroomed, with government suffering the heaviest casualties. At the same time, environmental consciousness has permeated the culture. While the latter has led to beneficial changes in individual and corporate behavior, it has also led many to believe and expect the worst imaginable scenario for just about any environmental or public health issue. As a result, the person with bad or scary news to report is usually given more attention and credibility than someone announcing that a problem either will not occur or has been solved.

The New Jersey Department of Environmental Protection sponsored a study in the late 1980s which found that more than 75 percent of the people surveyed would believe the agency if it told them there was an environmental problem in their community. Less than 25 percent, however, said they would believe the same agency if it then reported that the problem had been addressed and was no longer of concern.

This skepticism is demonstrated, and perhaps exacerbated, by the prejudice in popular culture against the possibility of radiological safety. The extent to which this prejudice has affected the siting process has not been studied, but its one-sidedness is sufficiently dramatic to warrant a short cultural detour before exploring how New Jersey municipalities responded to the invitation to host a disposal facility for low-level radioactive waste.

In 1958, Pete Seeger put out a record called *Gazette* accompanied by a booklet containing his comments about each song. To introduce the song "Talking Atom," Seeger wrote:

> *August 6, 1945 ushered in a new era in the history of mankind — the age of atomic energy. On that day, an American super-fortress flew over the city of Hiroshima, Japan, its bomb doors slid open and a single bomb went hurtling downward. A few moments later, the world's first atomic bomb exploded, destroying 60% of the city.*
>
> *Less than a week later, a second atomic bomb was dropped on the city of Nagasaki. Within a few days, Japan surrendered and the Second World War was over.*
>
> *But the mighty weapon that opened the atomic age and brought the holocaust of 1941-45 to a close offered many more problems than solutions. Scientists, ministers, intellectuals, ordinary people throughout the world, fearing that at last man did have the means to destroy the world, began to express their concern over the possibility of an atomic war. And then scientists*

became aware of a new danger — radioactivity in the atmosphere, which threatened the normalcy and the very existence of future generations.

Many people, like Vern Partlow, who wrote "Talking Atom," felt that at last the world had come to a crossroads where Hamlet's age-old soliloquy applied to humanity — To be or not to be? [26]

These words neatly summarize the dramatic ambivalence with which much of the public greeted the atomic age: Gratitude for the end of the war and admiration for the ingenuity that made it possible combined with fear that the nuclear genie will someday get out of control and again cause horrible, though perhaps this time accidental, damage.

Finding places to safely dispose of the nation's low-level radioactive waste is a problem only because the fear has grown over the last 50 years, while the admiration has disappeared. This is not to judge whether or not the fear is merited, but to acknowledge its importance either way. Without it, enough disposal facilities to meet the nation's long-term needs would have been constructed years ago.

The 1979 Three Mile Island accident and the 1986 Chernobyl explosion greatly increased public concern about anything involving nuclear power. The fear has risen out of the interplay between those events and the bombings at the end of World War Two, the change in public attitudes towards science and expertise in general, and the influence of popular culture.

In 1958, the Danish physicist Niels Bohr wrote, "The goal of science is the gradual removal of prejudices."[27] For this lofty goal to be achieved, the vast majority of us who are not scientifically literate need to have some respect for those who are. We need to recognize that consensus among scientists will eventually be reached on many questions, but that there will always be individuals, some well-trained and articulate, who will disagree. We need to be open to accepting such a consensus even if it conflicts with our heartfelt, but not well-informed, opinions.

Almost 40 years after Bohr wrote, David Schwarz, chairman of the Center for A Science of Hope, delivered a paper at the New York Academy of Sciences called *Science: From Hero to Villain in One Generation. What Next?*[28] The title alone captures the problem. For many people, a scientific consensus is credible only if it confirms what they already believe.

Many analyses agree that this transformation stems in large part from the Vietnam War and Watergate. Experts were discredited while people who doubted the experts were proven correct. These wrenching sets of events, however, resulted in a serious lessening of the respect afforded to experts in all areas.

At the same time that expertise was being discredited, popular culture was bursting forth. The media explosion that began in the 1960s increased the impact of music, movies and television in particular, all of which devoted some attention to societal issues.

Concern about nuclear power comprised a small part of this tide, reaching prominence in the "No Nukes Concert" of 1979 and the popular 1983 movie *Silkwood*. The concert featured many wonderful and well-known musicians including Jackson Browne, Ry Cooder, Bruce Springsteen, Bonnie Raitt and James Taylor. It was immortalized with a film and best-selling double album.

The film *Silkwood*, starring Meryl Streep, was based on the true story of Karen Silkwood, who worked at the Kerr-McGee nuclear facility in Oklahoma. Silkwood contended that her job had exposed her to large doses of radiation. She was on her way to meet with a reporter from *The New York Times* when she was killed in a mysterious car accident.

Almost 20 years later, when New Jersey was asking communities to consider volunteering to accept the state's low-level radioactive waste, the most pervasive relevant cultural icon was *The Simpsons*. Even supporters of the siting process would mention the popular animated TV show and laugh about Homer, the bumbling idiot in charge of security and safety at a nuclear power plant.

As the voluntary process continued into 1996 and 1997, more and more people were also watching *The X Files*. Viewers now had ominous images, often associated in some murky way with radiation, to add to their slapstick views of Homer Simpson.

These cultural images did not create public attitudes about radiation. They did, however, help shift public opinion from widespread skepticism about nuclear safety to cynicism toward any suggestion that it was possible. During the voluntary siting process, we heard many more references to *The Simpsons* and *The X-Files* than to health benefits community residents and their families had received from nuclear medicine.

There have been many other less prominent cultural references to radioactive materials. Most often, they are in the context of death, explosions, fear, and malevolence. Sometimes, they are a basis for humor, but in popular culture radiation and radioactive materials are never the hero.

Each example is trivial, but cumulatively they serve to reinforce public doubt and dread. Typical was a made-for-television movie called *The Atomic*

Dog aired early in 1998 just as the siting process was ending. The network synopsis was:

> *A highly intelligent stray dog infected with radiation is on the loose. And he's stalking the family that adopted his puppies. Isabella Hofmann and Cindy Pickett star in this USA Pictures Original.*

Later that year, when the movie *Armageddon* opened, one of its commercial tie-ins was a new Nestles candy bar called *Nuclear Chocolate*. The wrapper featured an exploding planet circled by the phrase, "The Chocolate Chain Reaction."

The Duck's Breath Mystery Theatre comedy troupe writes a column called "Ask Dr. Science." It appears in occasional newspapers and was included for several years on *All Things Considered* on *National Public Radio*. The short question-and-answer features always ends by reassuring listeners or readers that Dr. Science has a "masters degree in science."

In one column, a mock reader writes to ask, "Whatever happened to Fizzies?" Dr. Science replies:

> *Is it mere coincidence that Fizzies hit the market at the same time our manufacture of nuclear warheads was at an all-time high? Or that radioactive waste disposal suddenly became less of a problem when the bitter fuzzy pills became the latest consumer craze? No, there are no coincidences when the Military Industrial Complex and the Marketplace meet. I had the good sense to stock up on Fizzies, and have warehoused several hundred thousand of them in my lab storage areas. Whenever I need a cheap source of radioactive Iridium, I simply pop open a raspberry flavored fizzie, and drop it in a beaker of hydrochloric acid.*[29]

The cultural references that mean most to me, although even *The Atomic Dog* may have had a larger audience, are those found in folk music. The song *Talking Atom* on Pete Seeger's 1958 album is probably the first to indicate concern about radiation. Heard by small groups at concerts and summer camps, and by the relatively few people who bought his records at the time, it begins:

> *I'm gonna preach you a sermon 'bout Old Man Atom*
> *Now, I don't mean the Adam in the Bible datum,*
> *No, I don't mean the Adam that Mother Eve mated,*

I mean the thing that science liberated.
You know, Einstein said he's scared;
And if he's scared – boy, I'm scared [30]

There were other songs in the 1950s that also reached small, though often different, audiences. Fifteen were later collected in a record that accompanied the documentary film *Atomic Café*. Coming from the fields of blues, country, gospel and rock and roll, their titles included *Atom and Evil, Jesus Hits Like an Atom Bomb, Atomic Sermon, Atom Bomb Baby, Atomic Cocktail, Atomic Love,* and *Atomic Telephone.* [31]

Most of these songs were not at all political, but were simply using atomic energy as a readily understood metaphor for something large and powerful. In 1952, however, Tom Lehrer wrote one of the few songs to mention a federal agency by name, in this case the Atomic Energy Commission or AEC:

Along the trail you'll find me lopin'
Where the spaces are wide open
In the land of the old A.E.C.
Where the scenery's attractive
And the air is radioactive
Oh, the wild west is where I want to be. [32]

Shortly after I arrived at the Siting Board, I read of a forthcoming "Fourth Annual NJ Conference on Environmental Music." I wrote to the organizer suggesting that someone try to write a song about New Jersey's voluntary siting process for a low-level radioactive waste disposal facility. In my letter, I expressed admiration for the ways in which country music had evolved to reflect much more complex feelings and relationships than unambiguous love, misery, betrayal or loneliness.

I summarized the siting process, and then wrote:

Where does folk music enter this picture? It is easy to envision the groups in opposition to a particular site holding rallies and benefits at which talented musicians would perform, perhaps with songs they had written specifically about this controversy. It is also easy to envision the songs suggesting that our society is wasteful and therefore full of waste, that we do not take good care of our garbage and our habitat, and that nuclear power was a mistake and we should quickly move away from it.

But is it possible that people could write songs in support of this kind of facility? This song would not be of the 'head over heels in love' genre, but more like, 'It's not love, but it's not bad.'

To my surprise, my letter helped shape an evening's activities at the conference. I was told that the musicians in attendance spent several hours trying to write songs and laughing about the results, and probably about the voluntary siting enterprise as well.

Two songs emerged. One, written jointly by a group of singers, described the problem in terms of nuclear medicine:

WHAT DO WE DO WITH IT [33]

The doctors prescribed diagnostic tests
Cause Johnny got sick and they didn't know why
Down at the lab they searched and scanned
With X-rays, tracers, radioactive dye

CHORUS: What do we do with it?
Low level radioactive waste
Who's gonna deal with it?
We need a solution post haste!

This battery of tests did the trick
Johnny got better. His problem went away
The machines are quiet and the lab is dark
But the radioactive waste is here to stay

Many many Johnny's have been helped this way
In labs around the country, it happens everyday
But we haven't figured out just what to do
With the waste that's left when we're through
Low level radioactive stew
Don't want it in my backyard! How 'bout you?

The other song came from just one writer, Tom Callinan, a singer and environmental educator in Connecticut. Callinan successfully reduced a complex government program to three minutes of words and music:

THE NEW JERSEY WASTE-LAND [34]
Words & Music by Tom Callinan

One down-side to modern technology,
Is low-level-waste from nuclear energy.
If we want to be on-line and act responsibly,
Then we've got to put that waste somewhere.

Today, when there's waste in a hundred locations,
Tomorrow, there could just be one.
All it needs is a volunteering site in New Jersey,
Where consolidating could be done.

CHORUS: They only want to waste one location
In the beautiful Garden State.
They want to pack it up, cart it off, and stow it away,
Until it de-activates.

"Experts" claim this waste could be managed,
Until its danger disappears.
Some will take nearly no time at all,
While some will take hundreds of years.

They're accepting applications for a long-term commitment,
Don't miss this opportunity —
To store all that low-level radioactive waste
In your community!

Since uncertainty lurks, they've prepared some perks,
To grease the cogs in opposition wheels.
With jobs and money promised, and open space preserved,
Who could resist such a deal?!

Sure, there'll be conflict, crisis, and controversy
From those pesky NIMBY know-it-alls,
Who'll proclaim that a prime location for a wasteland,
Might be Trenton's governmental halls!

Callinan's song provided some amusement and perspective for those of us in the siting process. We used a tape of it to begin a number of informal meetings in New Jersey and played it at one of the periodic gatherings of representatives from the states trying a voluntary siting approach. Thor Strong, the radioactive waste program director from Michigan and also a guitarist and singer, added it to his repertoire and subsequently performed it at several events in his home state and around the country.

New Jersey's voluntary siting process for low-level radioactive waste also inspired part of a column by Calvin Trillin in *Time* magazine. Trillin wrote about the unfortunate men who used to approach cars stopped in traffic in Manhattan, unsolicitedly wash their windshields, and then ask for money. New police policies instituted by New York Mayor Rudolph Giuliani led these so-called "squeegee guys" to leave the streets and seemingly disappear.

Trillin wondered where they had gone and offered several possibilities. One was "the theory that New York City has given a grant to some down-on-its-luck former mill town to take in squeegee guys rather than become the site of a nuclear-waste dump."[35]

PART 2

1995
The First Year of the Voluntary Process

*We face so many complicated and difficult issues . . . that we
need to attain new levels of citizen responsibility for learning
about public problems and participating in their solution.*

- **Charles Peters,** *The Washington Monthly,* **January 1995**

CHAPTER 8

Lessons from Roosevelt

F rom the start of the voluntary siting process, I thought optimism was strategically important. If those of us involved in the process of finding a place to build a disposal facility acted as if we expected it to happen, we might help bolster the local officials who thought it safe but worried about the public reaction and its attendant political consequences. As I began work at the Siting Board, I made a point of saying, "When a town volunteers . . ." instead of "If," and only somewhat facetiously added a task to the office's work plan, assigning a staff member to plan the ribbon-cutting for the New Jersey Low-Level Radioactive Waste Disposal Facility.

After the public uproar in Roosevelt, however, it was clear that shopping for the appropriate ribbon was not a pressing priority.

The first lesson the Board learned from the Roosevelt experience was that large meetings are not the optimal forum to introduce two-way communication on issues that raise fears about health and safety. The majority of people, who tend to have little knowledge of the specific proposal, get no opportunity to ask for basic information because the discussion is dominated by those who have already formed opinions.

I had read and heard this lesson expressed by others, but I had to learn it for myself. I understood that poorly run large meetings could be disastrous, but I thought that I would do it well.

A second lesson was that local officials can have a similar hubris. While they undoubtedly know their towns well, they don't appreciate the extent to which issues involving radiation are unlike anything else they may have had to address. When Board staff sensed before the Roosevelt meeting that a large crowd would be attending, we should have insisted on moving to a bigger space. The packed room with people turned away at the door helped create a feeling of crisis which was the opposite of the lengthy deliberative process we had hoped to start.

The third lesson was a pleasant surprise. Newspaper coverage of Roosevelt's consideration of the disposal facility was generally accurate and informative. The level of controversy was clearly what caught the attention of editors and reporters, but the articles made clear that the Siting Board had a viewpoint and a proposal that was worthy of at least some consideration.

An omission in one article, however, did have significant ramifications. The *New York Times* piece that had first attracted the interest of Roosevelt officials had not listed nuclear power plants as a source of the waste. At least one major proponent of inviting the Siting Board to a public meeting had assured his neighbors that only hospital waste was involved and felt betrayed, not by his beloved *Times* but by the Siting Board, when he learned of the newspaper's omission. This experience led Board staff to redouble our efforts to ensure that any conversations or writing about the voluntary siting process made clear that the largest source of waste, by volume as well as by radioactivity, would be nuclear power plants.

Another article, appearing after the public meeting, raised an interesting question of perception. In a thoughtful commentary titled *"Intolerance"* in the *Roosevelt Borough Bulletin,* local resident Brad Garton criticized the tactics of the people who had been disruptive and said that Board staff had had their lives threatened. When I later asked him when this had occurred, he said it was before the meeting when the man standing in front of the Borough Hall had yelled "Watch your back!" as I walked away after talking with him.

I remembered the comment, but had not heard it as a threat. I chose to continue to view it as part of the give-and-take of a vibrant public participation process. In subsequent months, when I described Roosevelt's consideration of the issue, I ignored the "watch your back" comment. I didn't feel my life had been threatened and, while saying it had would have added drama to my presentations, it also would have added to the image that constructive debate on public issues, or at least this public issue, was not possible.

The public discussion in Roosevelt, which ended less than two months after it began, led several residents to call the Siting Board with reports of a "surprisingly large" number of people in town who felt the disposal facility had not received adequate consideration. They said they were going to bring it up again when the time was right. That time, apparently, never arrived.

Several other residents with a different point of view subsequently formed a group that was to be the only lasting opposition group formed in New Jersey during the voluntary siting process. This was the group with the clever acronym, CHORD, for Citizens Helping to Oppose Radioactive Dumps. It had several active members who showed up in many of the towns where the possibility of volunteering to host the disposal facility was raised over the next three years.

The final lesson offered by Roosevelt was that radioactive waste management is a very volatile, emotional, and difficult issue. This was obvious to most people in advance, but Roosevelt was a sobering reminder that optimism is only one of the ingredients necessary to solve this problem.

CHAPTER 9

Elsinboro

While the Board participated in the drama in Roosevelt, we were also having much quieter discussions with people from a small town in southern New Jersey. Shortly before the Siting Board formally adopted the *Voluntary Siting Plan*, Jack Elk, the mayor of Elsinboro Township in Salem County, told his Township Committee that he had been reading the information the Board sent periodically to all municipalities.

When Mayor Elk called to ask us to meet with his township engineer to discuss possible locations, he said that he had already raised the possibility of volunteering at a Township Committee meeting and had been quoted for two weeks in the local newspaper saying it might be a good idea. He added, "The roof hasn't fallen in yet."

We reviewed maps of the township with its engineer, Carl Gaskill, and with staff from the Department of Environmental Protection and the Board's environmental consultant. It became apparent that the few vacant sites of more than 100 acres were in areas considered by the Federal Emergency

Management Agency (FEMA) likely to be flooded during a 100- or 500-year storm — that is, the largest magnitude storm projected to occur during that time period.

The Board's *Voluntary Site Evaluation Methodology* included a criterion on surface water that read: "The Disposal Site shall generally be well drained and free of areas of flooding or frequent ponding. The Disposal Units shall not be placed in the 100-year floodplain, high hazard coastal areas, or wetlands."

While some of the potential sites in Elsinboro were outside the 100-year floodplain, they were all within the 500-year floodplain. The Board staff and consultants agreed this was sufficient cause to exclude the town from further consideration. The presence of the floodplain made it extremely unlikely that the Board would be able to definitively demonstrate to the Nuclear Regulatory Commission that a disposal facility on any of these sites would isolate the waste from all surrounding surface and ground water.

This was not a clear-cut call. On behalf of the town, the engineer argued that since the 500-year floodplain was not mentioned in the Board's siting criteria, the town's sites should not be excluded. For a more conventional type of development, he might have been able to convince a regulatory agency or a judge that a project meeting the letter of the criteria was entitled to a permit. But this was different. To the Siting Board, it was as if a storm had become visible just as the first step was being taken on what would be a long and difficult path. Turning back seemed the prudent thing to do.

While the Board's rejection letter was worded carefully to say that the Board would be open to revisiting this conclusion if data showed the maps to be in error, we also immediately added the short story of Elsinboro to the introductory talks we gave about the siting process. We were disappointed to have to turn away a potentially interested town, but also welcomed a tangible example of the fact that a voluntary siting process did not entail any sacrifice of environmental rigor.

The mayors of Elsinboro and Roosevelt apparently emerged unscathed from their encounters with the Siting Board, but it became clear that other politicians might well want to know whether there were potentially suitable sites in their towns before they engaged in any public discussion. In Roosevelt, we had felt we were demonstrating the integrity of the voluntary process when Jeanette Eng and I stood in front of hundreds of people and said we had no idea if the borough had an acceptable site. Many in the crowd, however, seemed to think that should have been determined, at least to some extent, prior to any public discussion and the resulting community unrest. Moreover, as

Connecticut had found with its use of the geographically neutral selection methodology, the finer points of process design seemed to impress very few people.

The call from Elsinboro was one of a number the Board was receiving from individuals asking about the possibility of locating the disposal facility in a specific town. Most often, the contact was a phone call, but sometimes it was at a meeting or in a letter. Usually, it was from an elected or appointed local official, but sometimes it came from a landowner or interested resident.

Part of the call was usually to invite a Board representative to a local meeting, but after Roosevelt and Elsinboro, we would suggest instead that the callers first describe any specific sites they had in mind so the staff could do a preliminary check on their acceptability. About half the people contacting the Board were thinking of the disposal facility as the answer for what to do with one or two specific locations, while the others just thought it might be a locally beneficial land use if the Board could find a good site in their town.

A major difference between voluntary siting and the traditional approach is that, in a voluntary plan, talking to local residents precedes looking at the land. Nevertheless, it now seemed sensible to have some site investigation before public discussion. No one would want to have the Roosevelt experience repeated if all locations in the town were going to be quickly eliminated on environmental grounds. On the other hand, when local people first talk with Board staff, some of their neighbors will believe that improper talks were secretly taking place if they learn that the Board has already looked at land in the area.

While we were sensitive to that complaint, I now wonder whether most of those who raise concerns about government operating in secrecy at such a preliminary stage are really looking primarily for yet another argument to augment their instant opposition to studying the disposal facility. Isn't it reasonable to expect that conscientious local officials are quietly going to explore many ideas and proposals, and involve their constituents only when one starts to look potentially promising?

The Board decided to draw the line between looking at mapped information and inspecting sites. Staff and consultants would review maps before holding any public meetings but would not conduct inspections.

The Board had an ongoing contract with a large international firm, Foster Wheeler Environmental Consultants, which could have conducted these reviews. We had found, however, that it was the staff from the state Department of Environmental Protection (DEP) who had been most useful in enabling us to give Elsinboro a quick response. Based on that experience, the

Elsinboro
Salem County – 1995

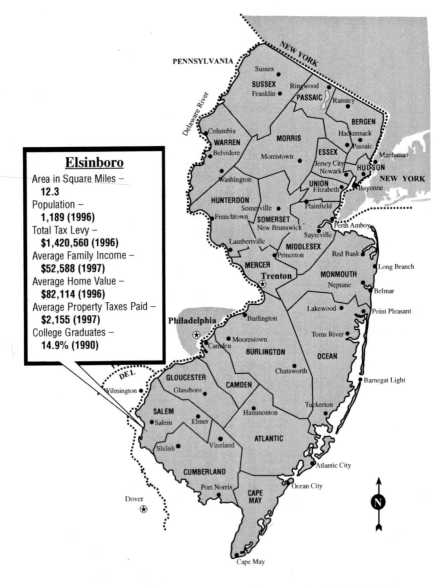

Elsinboro

Area in Square Miles –
12.3
Population –
1,189 (1996)
Total Tax Levy –
$1,420,560 (1996)
Average Family Income –
$52,588 (1997)
Average Home Value –
$82,114 (1996)
Average Property Taxes Paid –
$2,155 (1997)
College Graduates –
14.9% (1990)

Sources: N.J. Department of Treasury, New Jersey Almanac, NJ Legislative District Databook

Board developed a protocol that relied upon the DEP and in particular the New Jersey Geological Survey.

The lead staff member the Geological Survey assigned to work with the Board, Karl Muessig, was knowledgeable about both the Board's work and the state's environmental characteristics. He was also comfortable making quick, informed judgments. He worked directly for State Geologist Haig Kasabach, who was supportive of the Board's work and had even signed the Board's support statement. Since the Board had already included money in its annual budget for staff support from the Geological Survey and the DEP's Radiation Protection Program, it was easy to establish a process for paying the DEP for the time and resources they would devote to future similar preliminary municipal reviews.

After Elsinboro, we would systematically relay each suggestion received for a specific municipality to the Geological Survey. Within two or three days, Muessig would determine whether there were sites in the town that seemed to have potential, or whether any specific sites that had been suggested were in or out of the ballpark. His analyses would be based on four criteria.

First, New Jersey's Siting Act prohibited the Board from locating the facility in the Pinelands National Reserve. This 1.1 million-acre area that dominates the southern part of the state is being preserved by both federal and state law as a major water supply and spectacular ecological wonder. The federal act defines the Pinelands Reserve as being 173,000 acres larger than a 927,000-acre Pinelands Area defined in the state Pinelands protection law. Since most of the land use regulation associated with preservation of the Pinelands occurs under the state law, many people consider that area to be "The Pinelands." As a result, the Board heard from a number of people, including local officials, who were quite certain — and incorrect — that the prohibition against siting the disposal facility in the Pinelands National Reserve would not apply to their town.

This criterion proved to be a major impediment. For one thing, people in other parts of the state often asked two questions: If the disposal facility is so safe, why *can't* it go in the Pinelands? And, why isn't the health of my area considered as important as that of the Pinelands?

For another, by excluding the area defined by federal law, rather than using the state boundaries, several areas were taken out of consideration that might well have provided a safe home for the disposal facility without in any way harming the Pinelands. These include Fort Dix and McGuire Air Force Base, both of which have large contaminated areas that perhaps could have been cleaned up as part of an agreement for hosting low-level radioactive waste. The National Reserve also includes land adjacent to the Oyster Creek Nuclear Power Plant in Lacey Township, where many area residents are knowledgeable about nuclear materials, concerned about the economic and tax impact of the projected closure of the power plant, and aware that there were no conceivable risks from this disposal facility that could measurably add to the risks already present from the plant.

Photo courtesy of GPU Nuclear

LOGICAL SITE EXCLUDED: Land near the Oyster Creek Nuclear Power Plant in Ocean County was barred from consideration for a low-level radioactive waste disposal facility when the area's state senator made sure that the entire Pinelands National Reserve was ruled off-limits in the 1987 enabling legislation.

But this criterion was fixed in the state law. It had been added to the siting act not by a champion of Pinelands protection, but by state Senator Leonard Connors, a legislator generally opposed to state environmental and land use statutes and programs. When the Legislature was considering the Siting Act in 1987, Connors felt that Richard Dewling, the Commissioner of Environmental Protection at the time, had lied to him about the possibility that radioactive contaminated dirt from northern New Jersey would be stored in Jackson Township. Jackson, which is in his legislative district, is outside of the state Pinelands Area, but within the Pinelands National Reserve. Connors' response was to add to the pending legislation a phrase that excluded the Pinelands National Reserve from consideration for a disposal facility for low-level radioactive waste.

If Senator Connors and Commissioner Dewling had had their disagreement six months later, or maybe even six months earlier, the Siting Board probably would have been able to consider sites in 173,000 more acres, since at the time no one else was suggesting that the larger Pinelands definition be used. But soon after the larger area was excluded by the law, it was referenced in the Pinelands Comprehensive Plan which was then approved by the state Pinelands Commission and the federal Department of the Interior. Amending the Siting Act to alter this restriction was considered by most observers to be politically impossible.

Politics was less of a factor in deriving the other three criteria the Board used for preliminary site suitability determinations. Sites that were less than 100 acres were excluded. A half-dollar coin covered just over 100 acres on the 1:24,000 scale maps. The Board received some inquiries that staff could promptly dispatch when maps of the town in question displayed no available sites as big as the coin.

The third criterion was wetlands. While the disposal facility buffer area could include wetlands, the Board wanted to avoid selecting a site that would require filling any wetlands. Wetlands permits are sometimes issued if there is no alternative to filling a small amount, but this was one subject of controversy the Board wanted to completely avoid for both environmental and practical reasons.

To assess the wetlands on site, Muessig looked at wetlands maps prepared by the state Department of Environmental Protection, and at the National Wetlands Inventory maintained by the U.S. Fish and Wildlife Service. While going out on a site could expose errors and resolve conflicts in the maps and would be necessary to determine exact boundaries, just reading the maps gave a good sense of whether or not wetlands would likely be a constraint for most sites.

Finally, Muessig looked at the floodplain maps compiled by the Army Corps of Engineers and printed by the Federal Emergency Management Agency. Like the wetlands maps, these were not perfect but they did provide a good indication of where flooding would and would not be a major issue. Since any site receiving serious study would become extremely controversial, the Board wanted to err on the side of eliminating locations that might appear to be, in the language of the day, "environmentally challenged," even if someone could eventually have demonstrated that the maps were wrong.

This quick analysis was very effective, inexpensive, and uncontentious. The person contacting the Board almost always accepted the response that his or her town appeared to have no suitable land. For other towns with more promising sites, some of the people inquiring went on to discuss the idea with friends and neighbors, while others filed the information away for a later date.

During the three-year course of the voluntary siting process, this protocol was used to preliminarily evaluate sites in about 50 municipalities. Board staff honored the request when callers asked that their inquiries be kept confidential. At each month's meeting of the Siting Board, I would report on the number of towns from which we had inquiries without mentioning their names. Sometimes, I would note whether they were from the northern, central, or southern parts of the state.

Some opponents of building disposal facilities for low-level radioactive waste, as well as others who reflexively always favor more openness in government, would argue that the participants and contents of each of these discussions should have been publicly available in the Board's minutes and perhaps on a web site updated each week. With the support of the Board, I did not do so.

Philosophically, I believe that people, including representatives of local governments, should be able to get information from a state agency in the course of researching a possibility without then finding their name in the newspaper. Had these individuals convinced some of their neighbors to consider the disposal facility, all decisions that could in any way commit the town would have been discussed and decided in public through lengthy and wide-ranging deliberations that would have been known and open to all residents of the town and surrounding region.

These initial inquiries did not warrant that level of public attention. For the local people calling or writing the Siting Board, this was only one of many probable dead-ends that diligent public officials and concerned citizens must pursue in the quest to find good ideas for their community.

Strategically, the Board assumed there were some people in every town who would panic if they heard that anyone was expressing interest, no matter how conditional or preliminary, in considering a specific location for a disposal facility for low-level radioactive waste. The resulting organizing, petitions, and press coverage could not only quickly end any further exploration in that town, but also discourage people from other parts of the state who might otherwise be tempted to look into the idea.

Policies governing public access to information have evolved in many areas from close to strict secrecy to almost strict openness. When in doubt, many now say, let the public know everything immediately.

In New Jersey, the Board, its staff, and its Advisory Committee members found no inconsistency in holding some information back while claiming credit for conducting an open public process. We also felt it was essential if the Board was to fulfill its mission. This issue was one that we revisited often.

New Jersey's policy on this point differed from that adopted by Pennsylvania. Officials there believed that the names of municipal representatives and other individuals who contacted state agencies were part of the public record. As a result, the Pennsylvania Department of Environmental Protection delegated much of its outreach to the nonprofit State Association of Township Supervisors. Local officials and residents interested in suggesting a town or specific site were encouraged to call the association, rather than the DEP, and the association then promised to keep the contact confidential.

CHAPTER 10

Selling Snake Oil at Rotary Clubs

E lsinboro's interest had been sparked by the Board's mailings to local officials, and Roosevelt's derived from a newspaper article. The Board's rejection of Elsinboro and Roosevelt's rejection of the Board ended brief flurries of media attention, but they certainly didn't solve New Jersey's low-level radioactive waste disposal problem.

Informing the public about any aspect of government operations is daunting. When surveys consistently show that few people can name their Congressional representative, the path to creating broad public awareness of a novel approach to waste management is far from obvious.

An article called "Bowling Alone"[36] had appeared in January 1995 just as the Siting Board was formally adopting the *Voluntary Plan*. Written by Robert Putnam of the Harvard Business School, the article received considerable attention for its thesis that Americans are involved in fewer and fewer activities in which they are likely to talk with others about issues affecting their community, state, or country. They might belong to several organizations, but their involvement often extends only to writing an annual check to show their support.

The article's title was a reference to what Putnam called "the most whimsical yet discomfiting bit of evidence of social disengagement in contemporary America." Between 1980 and 1993, the number of people bowling increased by 10 percent, while participation in organized leagues decreased by 40 percent. Bowlers now are much more likely to go to the lanes with family members or close friends than to belong to a league in which they might form friendships with people who have different experiences and points of view.

One set of social institutions that remains quite vibrant, although it too has been affected by this trend, is service organizations. Each is a network of many small community associations formed to foster friendship and to raise money and support for needy causes. As part of its mass mailing after the adoption of *New Jersey's Voluntary Siting Plan*, the Board sent copies to each local Chamber of Commerce, Kiwanis Club and Rotary Club in the state, with a cover letter offering a speaker.

While the three organizations are similar gatherings of community and business leaders, Rotary Clubs have one important distinction that made them much more open to the material they received from the Siting Board: They meet weekly, almost always with a guest speaker. As a result, they tend to be less selective about the speakers and topics they are willing to entertain. Their meetings may include a talk by an award-winning local high school student or perhaps a gardening columnist one week, followed by a state bureaucrat in charge of low-level radioactive waste the next.

The clubs operate with precise rules. Meetings begin promptly, the guest speaker is introduced an hour later, presentations last 20 minutes and after another 10 minutes for questions and answers, a bell is rung to end the meeting.

I had already spoken at several Rotary Clubs when the chapter in Montgomery Township responded to the Board's offer. Montgomery is a rural but rapidly developing suburb of Princeton, and its club meets for breakfast. As family values have changed, many Rotary Clubs have switched from dinner to lunch meetings, and a few to early morning, since so many members now prefer to be home in the evening.

I had learned at an earlier meeting that Rotarians have a credo, called the Four-Way Test. They take it very seriously, but also tease each other for not remembering it. It seemed a good enough summary of the philosophy behind the voluntary siting process that I included it in my talk in Montgomery.

Adopted by Rotary International[37] in 1943, the "test" asks:

Of the things we think, say or do:
  1. Is it the Truth?
  2. Is it Fair to all concerned?
  3. Will it build goodwill and better friendships?
  4. Will it be beneficial to all concerned?

In my presentations, I said that the first two points in the credo should be goals of all governmental endeavors and that I thought our unusual voluntary process offered an opportunity to respond positively to the other two questions as well. The siting process we had in mind would bring people in a community together to address difficult and important issues. It could lead to goodwill and better friendships, and we believed the decisions reached by the community and Siting Board together would be beneficial to all concerned.

In addition to reciting the Four-Way Test, I described where low-level radioactive waste comes from, how it can be managed in a disposal facility, the economic benefits that will come to the host community, and the Board's plan to help New Jersey towns consider volunteering to be the host. The Montgomery club was larger than most, and the 60 members listened attentively and raised some interesting questions.

One asked if any towns had expressed interest, and I told them of Roosevelt and the meeting that was scheduled for the following week. Others asked about the possibility of putting the facility on already contaminated land, on a military base, or perhaps in another state. One member thanked me for reciting the Four-Way Test and so many others then applauded that I incorporated it into my subsequent talks with other Rotary Clubs. When the bell rang, they applauded again, presented me with a Rotary paperweight, and quickly went off to work, having had a good discussion and a good breakfast.

The next day, we were pleased to see that *The Times of Trenton* summarized my talk in a small but informative story. This was just the way the voluntary process was supposed to work. No one had suggested locating the disposal facility in Montgomery Township, but the Rotary Club meeting had enabled 60 people to learn about the issue, and the newspaper article had brought it to more people's attention.

One Montgomery resident reading that article, however, was not versed

Montgomery Township
Somerset County – 1995

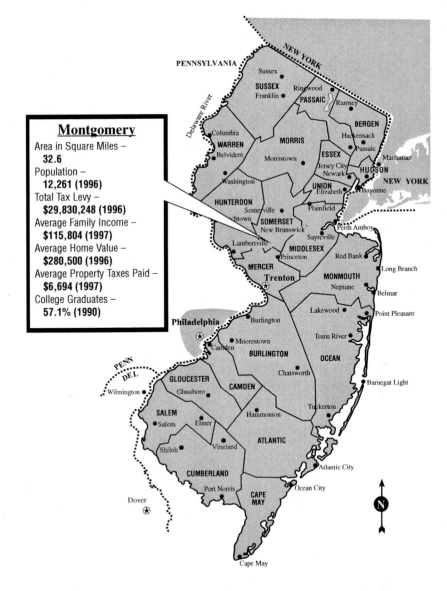

Montgomery

Area in Square Miles –
32.6
Population –
12,261 (1996)
Total Tax Levy –
$29,830,248 (1996)
Average Family Income –
$115,804 (1997)
Average Home Value –
$280,500 (1996)
Average Property Taxes Paid –
$6,694 (1997)
College Graduates –
57.1% (1990)

Sources: N.J. Department of Treasury, New Jersey Almanac, NJ Legislative District Databook

in the ways of Rotarians, and decided that a group in his town would only be discussing a disposal facility if that was where someone was planning to put it. He took it upon himself to quickly print up a flyer headed, "Nuclear Waste Dump In Montgomery?" It urged readers to call the Township Committee "to express opposition to this plan. Don't wait until the ink is on the contract." The page was unsigned except for a reference at the bottom to "NJ-Speakout!" with a post office box address.

Over the weekend, copies of the flyer were plastered on car windshields at the local supermarkets and movie theatre, making the idea of a radioactive waste disposal facility in Montgomery a major subject of conversation. Monday was the night of the Roosevelt meeting, and the flyer's author surfaced there and made a short statement voicing his "solidarity" with the people fighting the disposal facility.

Later that week, the flyer was successful in leading the Montgomery Township Committee to pass a resolution expressing no interest in something for which they had never had interest. More surprisingly, it led the *Princeton Packet*, a major, respected local weekly newspaper, to write a long editorial headlined:

HOT TRASH; LOW-LEVEL RADIOACTIVE WASTE DUMP DOES NOT BELONG IN NEW JERSEY

The editorial began by noting that I was once an assistant commissioner in the Department of Environmental Protection, but that "a background peddling snake oil might have been better preparation" for my present job. It then accurately, though snidely, described the voluntary siting process, and even named local scientists who had signed the statement the Board was circulating, endorsing the safety of disposal facilities for low-level radioactive waste.

The editorial concluded:

> *Their 'endorsement' of a state radioactive waste site is prefaced with the following: '...we would suggest that New Jersey municipalities, including those in which we and our families live consider whether they might have a site that might be suitable.'*

Frankly, we don't think any community in New Jersey is suitable ... Scientists may or may not be right about the risks associated with low-level waste. But we would feel more comfortable knowing that if they were wrong, their mistakes would be played out in the middle of a desert somewhere rather than in someone's backyard in New Jersey.

CHAPTER 11

Nibbles

R oosevelt, Elsinboro, and Montgomery were not the first New Jersey towns in which the possibility of locating the disposal facility was raised. Three years earlier, in 1992, the East Amwell Township Environmental Commission had asked the Board to make a presentation about low-level radioactive waste and the siting process. Officials in this mostly rural town in Hunterdon County had become aware of the Siting Board's new idea for a voluntary approach and were open to learning more about financial benefits that might enable them to save 500 acres of open space and reduce local taxes.

Sam Penza, the Board's director at the time, and Jeanette Eng attended a Township Environmental Commission meeting, described the waste and how it would be handled, and explained the concept of the voluntary plan the Board was then devising. After the meeting, the *Hunterdon Democrat*, *The Times of Trenton* and the *Star-Ledger* all ran articles quoting both Mayor Barbara Wolfe and Environmental Commission Chair Pat Hill expressing interest in studying and evaluating the program, and considering whether it would be appropriate to volunteer East Amwell for the disposal facility.

Although no negative voices were noted at the meeting or in the articles, several local residents formed a group called ACE (Active Citizens for the Environment) to oppose the facility. They gathered signatures on a petition, wrote letters to the newspaper and packed township government meetings to protest against the idea of the disposal facility in East Amwell. Several weeks later, the local Environmental Commission voted 5 to 1 to end the study. Mayor Wolfe then said, "Our township will not continue to investigate this issue any further. The Environmental Commission opened the subject and now they've closed it. It's closed, as far as I'm concerned."

Two years later, in the fall of 1994, the Environmental Commission from Mansfield Township in Warren County invited the Board to send a speaker to its next meeting. Before the meeting could occur, petitions, outraged letters to the editor, and call-in shows on a local radio station led the Township Committee to instruct their attorney to write to the board:

> *A motion at the October meeting of the Planning Board to make arrangements for someone from your organization to speak to the board regarding your project resulted in a telephone call from Board Secretary Pool to you on October 19, 1994. In response, you were kind enough to forward copies of the booklet describing the Voluntary Siting Plan.*
>
> *After review of that material and after considering input from the Mansfield Township Committee and residents of the municipality, the Planning Board, at its November 21, 1994 meeting, rescinded its prior motion and directed me to write to you to advise that the board is not interested in receiving further information on this project.*

In February 1995, after the *Voluntary Siting Plan* was adopted, the Siting Board mailed copies to all of New Jersey's mayors, municipal clerks, planning boards, and environmental commissions. While events in Roosevelt and non-events in Montgomery were played out loudly in local newspapers, people from other towns were contacting the Siting Board more quietly.

East Amwell and Roosevelt are in the center of New Jersey, Mansfield is in the north, and Elsinboro is in the south. The new contacts also came from all over the state. Some were in response to receiving the plan in the mail; others were from reading news articles about it.

East Amwell
Hunterdon County – 1992

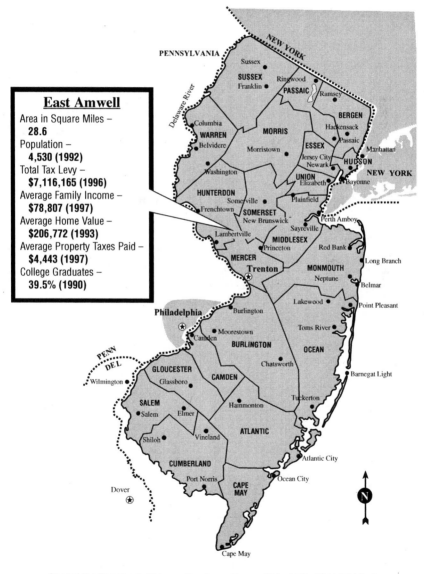

East Amwell

Area in Square Miles –
28.6
Population –
4,530 (1992)
Total Tax Levy –
$7,116,165 (1996)
Average Family Income –
$78,807 (1997)
Average Home Value –
$206,772 (1993)
Average Property Taxes Paid –
$4,443 (1997)
College Graduates –
39.5% (1990)

Sources: N.J. Department of Treasury, New Jersey Almanac, NJ Legislative District Databook

From northern New Jersey, a lawyer sent a map of a 294-acre site in Hopatcong in Sussex County. He had already done enough research to tell us in his cover letter, "It is not in the flood zone and there are minimal wetlands." Another piece of land in Sussex County was suggested by the owner, who called the Board and then sent maps of his 168-acre site in Hampton Township.

The deputy mayor of a town in the next county, Warren, called to ask about the feasibility of a local site bordering a highway. He told us that only he, the mayor, and the town clerk had discussed it, and asked that we send him ten copies of all the literature we had available.

From the central part of the state, the Board received a letter from the Franklin Township clerk asking for a presentation before a regular meeting of the Township Committee. Board staff also heard from several towns in southern New Jersey. A council member from Newfield in Gloucester County attended Siting Board meetings for several months and each time stayed afterwards to discuss the possibility of using a 176-acre tract adjacent to a Superfund site in his town.

Another councilman, this one from Little Egg Harbor in Atlantic County, had taken a course in radioactivity and told us he believed the disposal facility would be safe and a good ratable for his town. A man in Clayton Township in Gloucester County sent a detailed handwritten letter suggesting that we come to a school board meeting and explain that the disposal facility would enable his town to build a new school.

These and many other contacts were greeted warmly and taken seriously. For one thing, each helped the Board feel vindicated for having switched to the voluntary siting process. In addition, since the goal was to build one facility for the state, we felt we were always just one town away from success, and each of these inquiries could be it.

It was quickly apparent that none of them were. Most came from individuals trying to decide whether the idea was too crazy to mention to others and, if not, how they should do it. To each one, based on the experience in Roosevelt, we stressed that before they considered holding a public meeting, they talk with their friends, neighbors, and local officials and try to form an informal group that would support further pursuit of this idea. Most of them then asked for multiple copies of the *Voluntary Siting Plan* to selectively distribute.

Whatever local conversations then ensued, communications between the initial contact and the Siting Board generally dissolved after one or two letters and phone calls. Even as they did, though, new conversations would begin with representatives of other towns. The voluntary process seemed quite vibrant.

The most promising call in these first months after adoption of the *Voluntary Siting Plan* came one afternoon from the school superintendent in Fairfield Township in Cumberland County. That day, I had spoken to a lunch meeting of the Trenton Rotary Club. Trenton does not have a contiguous area large enough to accommodate the disposal facility, but its Rotary needed a speaker. It turned out that the architect for the Fairfield Township Board of Education is based in Trenton, 90 miles from Fairfield, and is a Rotarian. He was sufficiently interested in what he heard to immediately call the Fairfield superintendent.

While the subsequent conversation did not lead very far, at least in the short term, this contact alone further validated the Board's strategy of reaching out to the entire state and talking to anyone who would listen. An alternate approach would have been to concentrate our efforts on the parts of the state that had the most flat open space and, therefore, the most possible locations for a disposal facility. Speaking to a group in Trenton might then have been viewed as a waste of time.

During this time, the Board also had to turn down inquiries from local officials of half a dozen towns within the Pinelands because of the Siting Act's prohibition against locating the facility in that National Reserve. The Board was also sent occasional unsolicited rejections. Its letter offering a speaker to all Chambers of Commerce, for example, led the Vineland Chamber in Cumberland County to write back, saying that its "Business and Industry Committee and Board of Directors have discussed the possibility of this site being in our city and concluded that there were many reasons it would not be advisable."

The most interesting call came from a man who introduced himself as the attorney for a particular township that he named. He said that the Township Committee had met the previous evening and instructed him to inform the Siting Board that this township had no interest in volunteering. He paused, making me wonder why he had bothered to call since the Board's literature explicitly stated that we would only respond to inquiries and would not initiate explorations of siting the disposal facility in any town. Then he went on to say that he also had been instructed to tell us that, "Should the Board give up on the voluntary process and decide to victimize a town, (this town) would not fight too hard against being so victimized."

After I agreed to treat the information in confidence, he told me the specific location the town had in mind and sent maps for our review. A week later, I was able to tell him that the site looked potentially acceptable. That was, however, our last contact with the town. Their apparent comfort with risks scientific but not political did not lend itself to any imaginable next step. Even if the Board ended the voluntary process, how could it justify selecting, or "victimizing," this particular town when no local representative would be willing to admit having requested that it be considered?

CHAPTER 12

Hamburg Helper, or 'Dump Dangles Big Carrot'

T he strangest saga to unfold during 1995 occurred in the small towns of Hamburg and Hardyston, located in Sussex County in northern New Jersey. A man who owned 120 acres that straddled both townships thought the disposal facility might be a good use for his land. Following a suggestion from the Siting Board office, he arranged an informal meeting with Hamburg Mayor Richard Clark, Councilman Hugh Snyder, and Fran Snyder (no relation), the Board's consultant from the state's Office of Dispute Settlement.

The meeting in Hamburg took place in early June. The Board performed no advance analysis of the suitability of the man's land, and was not optimistic since the site was only slightly larger than the 100 acres needed. If any environmentally sensitive features were found, there would be insufficient space to design around them and the site would have to be eliminated. In addition, more than half of the land was located in Hardyston, so the two towns would have to agree to volunteer to gain serious consideration for the site.

Hamburg and Hardyston
Sussex County – 1995

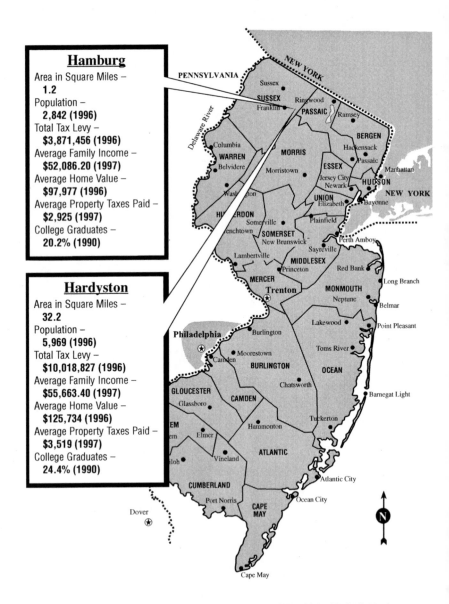

Hamburg

Area in Square Miles –
1.2
Population –
2,842 (1996)
Total Tax Levy –
$3,871,456 (1996)
Average Family Income –
$52,086.20 (1997)
Average Home Value –
$97,977 (1996)
Average Property Taxes Paid –
$2,925 (1997)
College Graduates –
20.2% (1990)

Hardyston

Area in Square Miles –
32.2
Population –
5,969 (1996)
Total Tax Levy –
$10,018,827 (1996)
Average Family Income –
$55,663.40 (1997)
Average Home Value –
$125,734 (1996)
Average Property Taxes Paid –
$3,519 (1997)
College Graduates –
24.4% (1990)

Sources: N.J. Department of Treasury, New Jersey Almanac, NJ Legislative District Databook

After the meeting, Mayor Clark and Councilman Snyder talked to a reporter, leading to an article in *The New Jersey Herald* headlined:

HAMBURG EYED FOR RADIOACTIVE WASTE DUMP; BUT MAYOR SAYS BOROUGH AQUIFERS WOULD PROBABLY HEAD-OFF SITING

The next day, the headline was:

RESPONSE IS LOW-KEY TO RADIOACTIVE DUMP

This article quoted Councilman Snyder as saying that while he had "expected to be tarred and feathered," he had been "stunned to find everyone he met yesterday thought the proposal had some merit."

Nevertheless, we thought the meeting was both the beginning and the end of Hamburg's consideration of volunteering. For two weeks we heard nothing further, and then we learned that a new local group called Citizens Against Radioactive Waste had held its first meeting, attracting 11 people. Three weeks after that, one of the newspapers reported that the Borough Council was "mixed on dump," with three members opposed and three undecided.

An established local group called OUCH (Organized United Citizens of Hamburg), led by the vice-chair of the Planning Board, then announced that it and the county League of Women Voters were inviting the Siting Board to an informational forum in the Hamburg municipal building to be held in the afternoon and evening on August 9. Mayor Clark announced that flyers would be circulated to publicize the meeting and that he had invited the county freeholders to attend.

As the Hamburg meeting was being announced, however, the Borough Council in Hardyston voted to oppose considering the disposal facility. I was then quoted accurately in the paper describing the voluntary process and saying that as a result of the Hardyston vote, this was "no longer a viable site."

Nevertheless, the possibility of siting in Hamburg and Hardyston continued to be the subject of prominent daily newspaper articles. One quoted the landowner urging people to approach the meeting with an open mind. Another reported that the county freeholders had passed a resolution opposing location of the facility anywhere in Sussex County.

These events also captured the interest of a quasi-news program on WOR-TV from New York City. Its reporter created an inflammatory five-minute story by showing scenes of Gingerbread Castle, an amusement park for young children. After a few seconds of watching kids happily playing, viewers were told that only miles from this idyllic scene the government was considering placing a radioactive waste dump.

That was the only television coverage, but other newspapers were now following the story. Articles appeared in the *Star-Ledger*, and also in a wonderfully named, very local community monthly called *The Hamburg Helper*. The best day-to-day coverage, however, came from *The New Jersey Herald*, the regional daily paper. In addition to reporting on the events in each town, its enterprising reporter, Jim Lockwood, called people in South Carolina to ask what it was like to live with a disposal facility. He then wrote a story headlined:

NEIGHBORS GIVE GLOWING DUMP REVIEW

The next day, he researched the financial implications of hosting the facility in New Jersey and reported that it could enable the town to "wipe out the average $700 municipal tax" in the Borough. The headline for this article was:

DUMP DANGLES BIG CARROT

The same day, however, Lockwood also reported Hamburg Mayor Clark was now opposed. The mayor said, "It's one thing to listen when they're dangling $2 million in front of you, but another when you consider everything such as the 500 years it will take for some of the waste to decay to acceptable levels of exposure."

He went on to say that he had been "naive" to think that informal discussions could be held without it appearing that the borough was formally inviting the Siting Board to town. Four days later, on August 7, the Borough Council met and passed a resolution stating that Council members believe "... the Hamburg site to be inappropriate for the siting of a low-level radioactive waste disposal facility and harmful to the interests of the residents of the Borough of Hamburg and the surrounding communities." The resolution went on not only to state their opposition to siting the facility "on or near" their borders, but to also "designate the Borough of Hamburg a Nuclear Free Zone." While this action was neither surprising nor particularly disappointing to the Siting Board, an accompanying resolution was startling: The Council, noting the public session scheduled for August 9, voted to bar "the use of any municipally-owned facility for such a meeting."

Since the meeting arranged by the League of Women Voters and OUCH was scheduled to begin less than 48 hours later, the two groups felt they had no choice but to postpone the meeting until a new location could be selected.

The Siting Board had been looking forward to the public meeting in Hamburg as another forum for helping to make the facts about radioactive waste and the voluntary siting process better known and understood. We thought a calm, useful discussion might ensue, particularly since we would confirm that, because of the opposition by local officials in both towns, the disposal facility was not going to be in Hamburg or Hardyston. We also hoped that the co-sponsorship by the local organizations might attract more people and perhaps inspire residents of another municipality in the area.

As it happened, on August 9, I was in no mood for a public meeting. Jerry Garcia, lead singer for the *Grateful Dead*, died that morning, and I was one of many whose life had been greatly enhanced by his music. In the days that followed, I told a number of people that the session had been canceled out of respect for Jerry. That reason seemed far less bizarre than my subsequent elaboration explaining that a local government had actually prohibited the League of Women Voters from using a public building for an already scheduled public discussion.

The Council's action reinforced the extent to which the normal rules of civic behavior do not always apply to consideration of radioactive waste. It brought to mind the incident more than 50 years earlier when the Daughters of the American Revolution cancelled a concert by Marian Anderson that had been scheduled for Constitution Hall in Washington, D.C., because they reserved the facility for use by white performers. In that case in 1939, Eleanor Roosevelt intervened and the concert took place in front of the Lincoln Memorial. In this case, the postponed meeting was never rescheduled.

CHAPTER 13

Welcome to Allowayste

W hile the uproar in Hamburg was building, the Board was engaged in much more promising conversations with representatives of Alloway Township in Salem County. This was the first town since Roosevelt to seem to have real potential. Alloway, which is located about 10 miles from the three Salem and Hope Creek nuclear power plants in Lower Alloways Creek Township, is a rural community with 3,100 people spread over about 33 square miles.

In April, I received a call from Carl Gaskill. We had met him as Elsinboro's engineer several months earlier, but he and his firm, were also the consulting engineers for a number of other towns including Alloway.

He was calling to say that Alloway was interested in learning more. He told us that the mayor, Les Sutton, worked at the nuclear plant and that many people in the town were quite knowledgeable about radioactive issues. Adding that there are a number of sites located near the county landfill that he thought met the Board's criteria, he asked for a time that he and some of the local officials could come talk with us.

We met the following week in our office in Trenton. We explained the process and they expressed cautious interest. The next day, Mayor Sutton talked to a reporter, resulting in a front page story in the major Salem County paper, *Today's Sunbeam*, headlined:

ALLOWAY CONSIDERS LOW-LEVEL RADIOACTIVE WASTE SITE

The mayor was quoted as saying, "Everything is preliminary right now," but that "on a purely monetary record, the rewards are outstanding and we'd like to have some income." He added, "Just as a layman, I don't see any cons, because we live so close to the nuclear plant and work there, we understand these things."

In an article the following day, Mayor Sutton said, "It's probably pretty unpopular right now. I feel it is our duty to research all possibilities to help out with taxes in the township." The only people quoted in either article were the mayor and the Siting Board staff.

The following Sunday, the *Sunbeam* conducted a "People Poll," asking six people the following question:

The township may be considered by the state as the possible site for a low-level radioactive waste facility in return for monetary and other benefits. Are you in favor of having such a facility here?

All six said no, with comments including: "We already have the county landfill;" "I, for one, don't want the radiation;" "I want my son to play here and I wouldn't want something like that around;" and "I'm pretty big into the environment."

We continued to talk with Mayor Sutton and Carl Gaskill by phone and jointly agreed on a list of groups to contact to begin wide public discussion about the possibility of Alloway volunteering to host a low-level radioactive waste disposal facility. The list included the area's Congressional representative and three state legislators; county officials both in Salem and neighboring Cumberland counties, including health, emergency management, planning, and economic development directors; mayors of the five surrounding towns; municipal and county environmental commissions; Alloway's Board of Education; environmental groups, chambers of commerce and other civic groups; regional planning organizations; Rotary and other service clubs; the League of Women Voters; colleges and universities; generators of radioactive waste in the area, and editorial boards of local newspapers.

The various government officials would be contacted so they would have accurate information about the preliminary nature of Alloway's exploration and a point of contact should their constituents raise questions to them. We included the school board because its members might be willing to take a second look at the disposal facility as a means of providing money to maintain and improve education in the township.

Most of the other organizations would be asked if they had opportunities for Siting Board representatives to meet with their members and talk. Perhaps professors at Salem County Community College and other schools would consider arranging courses, forums, or debates on the disposal facility proposal.

We hoped the organizations that generated radioactive waste and their employees who lived in the area would be willing to share their experience and expertise through individual conversations and perhaps by writing to newspapers, participating on panels, and offering tours of their facilities. Mayor Sutton was only one of many employees of the Hope Creek and Salem I and II nuclear generating stations who lived in and around Alloway.

In addition, Salem and Cumberland Counties were home to E.I. DuPont, Wheaton Glass Industries, and the Elmer Community Hospital. Each entity had disposed of low-level radioactive waste at Barnwell at least once since 1990 and should, therefore, be interested in Alloway's exploration.

Our approaches to the waste generators, both directly and through the small statewide trade association they had formed, were not productive. One official at the nuclear plants offered to arrange for a Board representative to speak before the Salem County Chamber of Commerce, but then had to tell us that the Chamber's chairman opposed any consideration of the disposal facility and would not agree to a program on the issue.

Mayor Sutton was more successful and arranged for me to speak at the August meeting of the Alloway Ruritan Club. Ruritans, we quickly learned, are an unaffiliated rural offshoot of Rotary. Their goals are "to create better understanding between rural and urban people, to aid in charitable work and disaster relief efforts, and to promote industrial and agricultural growth and economic development."[38]

The meeting, which followed a country supper, attracted about 40 members of the club and another 15 people invited by the mayor, who also was a member. I spoke for 20 minutes and then answered questions for more than an

Alloway Township
Salem County – 1995

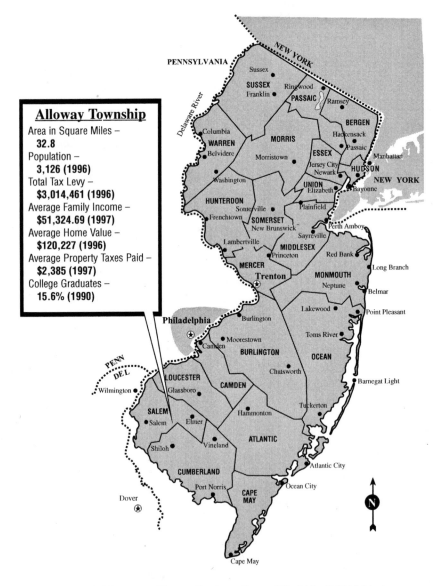

Alloway Township

Area in Square Miles –
32.8
Population –
3,126 (1996)
Total Tax Levy –
$3,014,461 (1996)
Average Family Income –
$51,324.69 (1997)
Average Home Value –
$120,227 (1996)
Average Property Taxes Paid –
$2,385 (1997)
College Graduates –
15.6% (1990)

Sources: N.J. Department of Treasury, New Jersey Almanac, NJ Legislative District Databook

hour. People asked if the state government was targeting southern New Jersey as a dumping ground, if earthquakes could be a problem, and if the promised benefits could be negotiated to be greater than $2 million a year. One person said that promises made when the county's solid waste landfill was proposed had not been kept, and asked how they could be sure this would be different. Another asked if local workers could get hiring preference for jobs at the disposal facility.

Most of those in attendance seemed very curious and to have not yet formed a definite opinion about the facility. In fact, in the two-and-a half months since the Township Committee's meeting in Trenton had been front-page news and the newspaper's "People Poll" had appeared, there had been little press coverage, and we could count the negative comments we had heard.

Several weeks before the Ruritan meeting, one Alloway resident had placed a sign on his property reading, "Welcome To Allowayste." After reviewing the material supplied by the Siting Board, he had concluded that a parcel of land near his house would be the perfect spot for the facility. When a reporter talked to him about his sign, he said, "I know very little, only that I don't want a facility across the street from me."

Because so few concerns were being raised, we were able to give immediate individual attention to each. The man with the lawn sign, for example, had come to the Ruritan meeting, and Greta Kiernan of the Board's staff talked with him afterwards, sent him a letter providing more detailed information, had a lengthy phone conversation several days later, and then wrote him again to address his follow-up questions and concerns.

After the Ruritan meeting, our optimism about Alloway grew. Even the news article about the lawn sign had begun, "Some people may think it's not such a bad idea to consider having a low-level radioactive waste disposal facility located here, but at least one township resident disagrees..."

The next step was to set up some type of structure to continue and expand the public discussion. Earlier, Mayor Sutton had drafted a letter to township residents and we suggested this would be a good time to send it. The letter explained that the Township Committee members "would like to continue to explore the possibility of hosting the facility here in Alloway and thereby obtain the benefits that hosting the facility would bring." The draft went on to ask for volunteers to serve on a committee "to investigate the pros and cons that this facility would bring to our community."

We proposed that shortly after the mayor's letter was received, we would mail residents copies of the Board's brand new publication, *Questions and*

Answers on Siting a Low-Level Radioactive Waste Disposal Facility. Our thinking was that the mayor's letter would raise public awareness and interest so that residents might then welcome the more detailed information they could find in the Siting Board's booklet.

But the mayor never sent his letter. On September 14, after a month of very little communication between the Board and the township, Mayor Sutton called to say that the Township Committee was going to vote that evening to stop pursuing the facility. He said he was disappointed, but that "people are on my back." He added that he might have a basis to reopen consideration in the future if other residents complain about the Committee's action.

The next day, *The Sunbeam* reported that the vote was "greeted with a round of applause from a standing-room only crowd ... of 60 residents." Unbeknownst to us, the man with the lawn sign had formed a group called Alloway Township Against Contamination which had met the week before and now formed much of the Township Committee's audience. Mayor Sutton was quoted as saying, "The more info I've had lately, I'm not sure I want my name on something that's going to be around for a long time. There's not enough assurances to make it worth the risk for me."

Three years later, Sutton said he felt the Township Committee had erred in not forming a local study committee at the very beginning, instead of just allowing time for opposition to build. While he wasn't aware of any previously established groups getting involved, he did feel that many of the critics came from surrounding towns and counties. Among Alloway residents, the most common concern he had heard was fear of the stigma of becoming known as the "waste capital of New Jersey."[39]

Eighteen months after the Township Committee's vote, a Salem County attorney called to say that he represented some people who wanted to begin the process of consideration in Alloway again. He sent the Siting Board a five-page draft resolution that would have the township establish a study group to explore the issue and report back within a year. While his cover note said they "are prepared to move quickly," no further movement ever occurred.

CHAPTER 14

More Nibbles

W hile the Board focused on Alloway, first with growing excitement and then with serious disappointment, we continued to converse with people from throughout the state. Although it was the end of the year before another promising town stepped forward, the conversations in the interim helped maintain the enthusiasm of the Board members and staff and of the representatives of waste generators monitoring the Board's activities.

In August, an industrial developer in Rockaway Township in northern New Jersey wrote because he had followed the debate in neighboring Hamburg and Hardyston. His company owned more than 200 acres zoned for industrial use and located in a "relatively remote area adjacent to the U.S. Picatinny Arsenal Army facility." He enclosed a map of the site.

We replied that his request for a meeting seemed premature and suggested it be delayed until after he discussed the idea with his local officials. A month later, we received a copy of a letter from the Mayor of Rockaway to the developer. After stating that "issues of public policy and welfare are a specific concern of my administration," he wrote that, "It is my belief that this use would not be in the best interests of the residents of Rockaway Township, regardless of any tax benefits which might result from such a facility."

The same month, a member of a local environmental commission in a central New Jersey town not far from Trenton asked for multiple copies of each Board publication. Also in August, we received a phone call from a man representing the mayor and clerk of another southern New Jersey town. After discussing the voluntary process and the 170-acre site they had in mind, he asked that the call be confidential and said they would arrange a meeting shortly after Election Day. While the meeting never occurred, the conversation served as a reminder that we were least likely to receive serious inquiries during the months leading to up to November elections.

A man who had attended several Siting Board meetings without ever saying a word or indicating why he was there finally wrote to suggest a site owned by his former employer. Company and local officials subsequently indicated interest in considering volunteering, and the 585-acre site in South Jersey seemed potentially suitable for the disposal facility, but after a few phone conversations, they never pursued it further.

More promising, and ultimately more frustrating, than the contacts in these municipalities was the consideration by Pennsville in Salem County. Several township officials and residents had attended one of the Board's public meetings in 1994 when a draft of the voluntary plan was being discussed. They thought the facility could be good for the town and arranged for me to address their Rotary Club. All the questions and conversation were focused on how, rather than whether, Pennsville would volunteer. After the meeting, however, when the New Jersey Geological Survey reviewed maps of the township, all vacant land appeared to be in areas that could be flooded in a 500-year storm.

The criteria we had used earlier to reject Elsinboro necessitated that we also reject Pennsville. But the town would not take no for an answer. Its engineer, Carl Gaskill, now back with his third town, believed the flood maps were inaccurate and asked that the Board reconsider. More than a year later, he wrote on behalf of the Township Committee to "formally request preliminary site assessments for the potential siting of the proposed New Jersey Low-Level Radioactive Waste Disposal Facility on the following parcels..."

By that time, the Township Committee had discussed the disposal facility at a number of public meetings. Their interest had been noted in several short local news articles, and apparently generated no objections. Nevertheless, after

Pennsville
Salem County – 1995

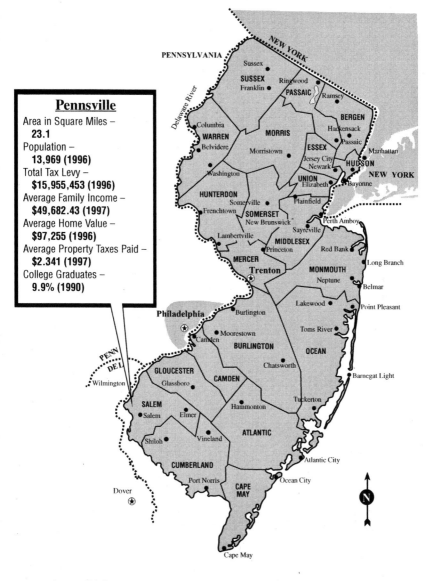

Pennsville
Area in Square Miles –
23.1
Population –
13,969 (1996)
Total Tax Levy –
$15,955,453 (1996)
Average Family Income –
$49,682.43 (1997)
Average Home Value –
$97,255 (1996)
Average Property Taxes Paid –
$2.341 (1997)
College Graduates –
9.9% (1990)

Sources: N.J. Department of Treasury, New Jersey Almanac, NJ Legislative District Databook

further Geological Survey review, we had to respond, "Although the federal Emergency Management Agency (FEMA) maps for Pennsville show several areas in excess of 50 acres that are outside of the 100-year flood plain, a maximum probable precipitation or hurricane surge could cause flooding over these areas as indicated on the FEMA New Jersey Hurricane Evaluation Study. As a result, we believe that it would be very complex to study these sites and be able to demonstrate that one of them met all applicable requirements."

CHAPTER 15

Planning for the General Election

As 1995 ended, the Siting Board could look with satisfaction on the progress of the voluntary siting process. It had received significant positive press coverage, and the Board had been able to discuss the issue with a fairly wide variety of groups and individuals.

Moreover, the statewide outreach seemed to be working. Leaders in four towns — Roosevelt, Elsinboro, Alloway, and Pennsville — had been sufficiently open to the possibility that the disposal facility could be safe and provide their residents with substantial benefits to have publicly suggested they consider hosting it. In addition, residents, including some officials, from more than 20 other municipalities had approached the Board staff about perhaps suggesting that it be explored in their town.

While the Board focused on finding a willing host municipality, we also needed to be prepared to move ahead when the search succeeded. The site would need to be "characterized" or exhaustively analyzed, the facility designed, and adequate financing and legislative authority assured.

I had once worked for a candidate for the U.S. Senate from Connecticut who won an upset victory in the primary. He went on to lose the general election, at least in part because the primary had been such an exclusive focus for the campaign that there had been little attention devoted to preparing for the general election. We did not want to make the same mistake in this campaign.

This work was quieter than the search for interested municipalities, and fell into two major categories. The first was to develop a methodology to determine whether suggested sites were suitable for the disposal facility. The second was to examine New Jersey's Siting Act to determine if changes would be needed to fully implement the voluntary process.

Site Investigation

Analyzing a site for a large light industrial development is a major undertaking. It was more complex in this case because extensive monitoring and modeling would be required to convince first the Board, and eventually the Nuclear Regulatory Commission, that this site was sufficiently dry and stable to safely accommodate radioactive waste for centuries.

This was the primary task for which the Foster Wheeler Environmental Corporation had been hired. In fact, drafting of the request for proposals began when the Board was still expecting to screen the entire state to identify the sites it would consider. The state bidding process took so long that by the time the contract was awarded in the spring of 1994, the Board was well on its way to switching to an entirely different approach to site selection. Moreover, the bidder the Board had selected, EBASCO, had undergone two corporate mergers since it submitted its proposal, first changing its name to Enserch and then being bought and absorbed by Foster Wheeler.

This ultimately created some difficulty. The Board's bid evaluation committee had based its recommendation in part on the experience offered by the EBASCO/Enserch team. When Foster Wheeler took over, key individuals left, one of whom formed his own firm. At the Board's request, he was made a subcontractor, but the relationship was awkward and never provided the committed partner the Board had anticipated.

Even though sophisticated technical expertise is required to analyze a site, the Board wanted to describe the process, and later conduct it, in a way that interested community residents could understand and follow. Foster Wheeler, therefore, was directed to draft a set of documents that would explain how they would proceed when presented with sites to evaluate. These reports were not only for the public, but also to show the Board how the consultants planned to accommodate the required work with the voluntary siting plan.

In June, four months after adopting the *Voluntary Siting Plan*, the Board approved the 26-page *Voluntary Site Evaluation Methodology for Siting a Low-Level Radioactive Waste Disposal Facility in New Jersey*. This updated and replaced the Board's earlier methodology.

"In 1990," the report explained, "the Board adopted Siting Criteria and a statewide screening approach to site selection, whereby the Board would examine the entire State of New Jersey for potential sites. In order to accomplish this goal, a methodology was adopted that was primarily exclusionary in nature; statewide information and maps were to be used to screen out large areas of the state to focus on those land areas which appeared to have the most potentially favorable characteristics."

The criteria were divided between hydrological and geological categories which included surface water, ground water, and geology and geologic stability; and environmental and public health concerns, which were made up of demography and land use, natural and cultural resources, and transportation. For each of these criteria, various circumstances were identified as "exclusion factors" which "prohibit site selection within specific identifiable land areas," or "suitability factors" which "may affect the overall suitability or favorability of specific land areas as sites for the disposal facility."

In 1990, the goal had been to describe a feasible strategy for the consultants to examine maps of New Jersey's 4,800,000 acres, and to identify a small number of sites which would then be examined in detail to determine if they met all federal and state siting requirements. They would use the exclusion factors to eliminate large areas where the potential for finding an acceptable site seemed small.

Now, with people suggesting specific sites, some of the factors could be moved from "exclusion" to "suitability." For example, the earlier document had excluded areas with "land upon which limited access highways are presently located, major active shopping centers and office complexes, and active federal and state military installations." The *Voluntary Methodology*, however, explained that "current land use can be addressed jointly by a landowner or community with the Board. As a result, current land use on potential sites will be addressed as a suitability factor." In other words, a town volunteering might be open to locating the facility adjacent to an existing shopping center or they might know the shopping center was near bankruptcy and could perhaps be replaced by the disposal facility.

The Board also moved to define the site characterization program with much more specificity. The two resulting reports explained the various types of technical investigations that would take place during preliminary site investigation and site characterization. As with the *Voluntary Methodology* booklet, these documents were aimed at municipal officials and the general public, but the process of preparing them helped the Board and its consultants gain a more secure common understanding of the steps that would be taken when specific sites were presented for evaluation.

The *Preliminary Site Investigation Program for Evaluating a Potential Site for New Jersey's Low-Level Radioactive Waste Disposal Facility* described the steps that would follow the quick informal review that the New Jersey Geological Survey was performing. This would be the first time that Foster Wheeler would become actively involved in the review of a particular site.

The environmental consultants would compile a map of surface geology and conduct a limited number of borings to ascertain the layers of soil and rock at various depths on the site. Its staff also would install groundwater monitoring wells and would walk over the area looking for wetlands, habitats of protected or endangered species, and "identification of any evidence or features at the site that suggest unusual or unique climate or air quality conditions." Their review also would include a preliminary investigation of general background radiation and preliminary searches for potential archaeological and historical resources.

This 29-page report explained that none of the work on-site would occur without an access agreement with the landowners. Foster Wheeler estimated that the review of available written information and the preliminary fieldwork would take a total of eight months. A companion publication explained what would happen next if the preliminary site investigation uncovered no reasons to exclude the site, and if the community was willing to continue considering hosting the facility. In 51 pages, The *Site Characterization Program for Evaluating a Potential Site for New Jersey's Low-Level Radioactive Waste Disposal Facility* described the technical site investigations that Foster Wheeler would undertake. In addition to further exploration of the background levels of natural and man-made radiation in the area, most of the work would fall within the categories of geoscience or environmental science.

The geoscience investigations would focus on geology, geological hazards, geochemistry, groundwater hydrology, subsurface work, and surface water quality and hydrology. The environmental science work would add terrestrial and aquatic ecology, meteorology and air quality, natural resources, noise, cultural resources, land use and aesthetics, socioeconomics and demography, and transportation.

The site characterization would take approximately two years. This would include monitoring the site for one year to gather baseline information about its characteristics during each season. Altogether, therefore, completing the preliminary site investigation and site characterization would take more than two-and-a-half years. At that point, the Board would have the information about a site needed to prepare a license application to the Nuclear Regulatory Commission.

The length of time required for all this work was a great concern to the Board. This clock would only start at about the point we had been approaching in Alloway. Since the *Voluntary Siting Plan* did not anticipate a signed agreement with a municipality until site characterization was completed, it would have been almost three more years before we and Alloway could have formally walked to the altar together. That would have been a very long time to maintain municipal support for a controversial idea that was not yet providing any economic benefits to the town.

As an additional cause for concern, Foster Wheeler had resisted stating the amount of time required for the studies and had added their estimates only at the Board's insistence. Their reluctance, combined with the general laws of life that cause most things to take longer than anticipated, further raised fears that the duration of the site work could erode support in even the most progressive municipality.

Legislation

The other issue that began to occupy the Board during 1995 was the adequacy of New Jersey's Siting Act for carrying out a voluntary siting program. In the 1987 law, the Legislature provided a fairly detailed description of a process based on statewide screening. While the phrase "decide, announce, defend" sometimes used as shorthand for that approach was not in the law, neither was the word "voluntary" nor any hint that this approach may have been something it had in mind.

During the three years the Board had been considering and planning the switch to a voluntary approach, the question of amending the act had sometimes been raised. However, it was never pursued or seriously debated.

One reason was that while the voluntary concept was still being developed the specific desired legislative changes were not yet clear. Also, the Board was aware that the two times the Legislature had followed its recommendations to make minor changes, the process had taken several years. In addition, there was concern that initiating any proposal to amend a law could lead to the enactment of additional unwanted changes. In this case, for example, legislators could have added language to exclude parts of their districts from consideration.

With the formal adoption of the *Voluntary Siting Plan* in February 1995, followed closely by the addition to the staff of former state legislator Greta Kiernan, we began to closely compare the premises and promises of the new plan with the intent and authority of the law. The language of the two was dissimilar, but it was possible to argue that the Board could reach an agreement with a town in a way that might be allowed under the act. This argument rested on one sentence, Subsection (g) of Section 17, that read, "The board may offer financial or other incentives to the host municipality, as may be made available to it by the operator or the state."

Winning this argument would depend upon the final agreement going unchallenged, which seemed inconceivable, or having the case assigned to judges who would recognize the importance and difficulty of the Board's mission and rule sympathetically. Even then, the law contained a number of provisions that would make the voluntary process more cumbersome and time-consuming than necessary without providing any additional public benefit.

The Siting Board had formed a legislative subcommittee several years earlier when it successfully sought the amendment that enabled it to raise operating funds by assessing fees from the generators of low-level radioactive waste in New Jersey. This subcommittee, which included Board and Advisory Committee members, was chaired by Jim Shissias, the Board member who worked for Public Service Electric & Gas, the largest waste generator, and was familiar with the legislative process.

The subcommittee now considered the problems the law could raise in implementing the voluntary process. In August, it recommended a set of amendments that the full Board voted to endorse in concept. The theme of the changes was to give the Board more flexibility to respond to the needs and desires of towns that might be willing to consider volunteering, and to the one eventually selected. Thus, the suggested amendments would:

- *Permit the Board to offer financial or other incentives to potential host municipalities, or a designated host municipality, from fees and other sources. This would allow a town to receive benefits earlier in the siting process;*
- *Add "public outreach and education, benefits, incentives and compensation for persons impacted by the facility" to the list of categories for which the generators may be assessed. This would both clarify and expand the types of benefits the Board could negotiate with municipalities;*
- *Remove the automatic hearing before an Administrative Law Division judge if a site is volunteered by a municipality. The act assumed that the host municipality would be an adversary and object to having been designated by the Siting Board. If a municipality wanted to host the facility, however, this hearing, intended to help protect it, would add at least a year and probably more to the overall process;*
- *Add the signing of an agreement by the Board and a volunteer municipality as the trigger point for allowing the town and the host county to each appoint an additional member to the Board. Under the act, this would not occur until the Board was considering the recommendations of the administrative law judge; and*
- *Give the Siting Board the power to assess civil penalties from waste generators who do not comply with fee assessments or other requirements. Now, the Board's only recourse was to seek action by a court.*

Among the other changes the Board recommended was to add a requirement that interest on the money collected from the waste generators be credited to the Low-Level Radioactive Waste Disposal Facility Fund, rather than to the state's general treasury.

The Board's conceptual approval of the amendments enabled discussions to begin with staff of the Governor's Counsel's Office to seek support for requesting legislative consideration. The Board could have directly approached individual legislators, but concluded that because the governor's position would be sought before a bill for such an arcane matter was seriously considered, it would be better to include her staff from the beginning.

In this case, including the governor's staff early also produced some comments that improved the wording of several of the proposals. The staff was generally supportive and then circulated the revised proposal to the other state agencies that might have an interest — the Attorney General's office, and the Departments of Environmental Protection, Health, and Treasury.

The one idea they did not want included was the last suggestion — allowing the Board to keep the interest from the fund. This was not surprising, since the Treasury Department historically has taken a similar stance on other funds collected from a specific group and, presumably, dedicated in its entirety to a specific purpose.

It was, however, very disappointing to those representatives of waste generators who served on the Board or monitored its work. The Governor's Office was not swayed at all by the argument that siphoning off the interest funds would serve to raise the generators' distrust of government.

As 1995 ended, the legislative proposal was under review by the other state agencies and the Board anticipated having it considered by the Legislature in the spring. Since New Jersey legislators are elected in odd years, 1996 could be expected to be a good year for seeking legislative action. We hoped the amendments would be considered quickly as housekeeping changes to an existing law, rather than as a significant policy proposal in need of extensive debate. After all, we were already implementing the voluntary siting process and no one, outside of the Board and its immediate family, was suggesting that the current law was less than adequate.

CHAPTER **16**

Barnwell Returns

While the Siting Board focused on New Jersey, far more significant activities had been taking place in South Carolina. The state's new governor, David Beasley, was one of the many Republicans throughout the country elected in November 1994 on a pledge to reduce taxes without significantly lessening services. Lobbyists for Chem-Nuclear Systems were suggesting that reopening the company's low-level radioactive waste disposal facility to the rest of the country could help him accomplish that goal.

The company's lobbying was supplemented by appeals from local officials in Barnwell County and the town of Snelling, where the facility is located. They argued that the facility had a safety record as good or better than any other industrial operation and provided jobs and significant economic benefit to their area. Snelling Mayor Tim Moore, who was first elected before the facility was approved and built in 1970, was among those who also pointed out that throughout Barnwell County the disposal facility was universally perceived as a good and welcome neighbor.

Photo courtesy of Chem-Nuclear Systems

LOCATION OF LAST RESORT: For years, the continued availability of disposal at Chem-Nuclear's facility in Barnwell, South Carolina was all that kept the nation from a low-level radioactive waste disposal crisis.

In New Jersey, the Siting Board did not take seriously the rumors we heard about these conversations. South Carolina politicians of both parties had been fighting for almost two decades to be able to stop accepting the nation's waste. Their position had been one of the major factors leading to enactment of the federal Low-Level Radioactive Waste Policy Act in 1980 and, as the statute's deadlines for states to find other disposal options were missed, South Carolina had worked hard to not let them slip too far. With their wish granted and the facility finally closed to states outside their eight-state Southeast Compact in July 1994, it seemed inconceivable that only months later the state would choose to relinquish that victory.

But the policy was reversed and on July 1, 1995, exactly one year after it had stopped, the Barnwell facility again began accepting low-level radioactive waste from around the country. This time, however, the rules were different.

The policy reversal was accomplished in the budget approved by the South Carolina Legislature and signed into law by Governor Beasley in June. South Carolina's major daily newspaper, *The State*, reported that the budget bill not only cut property taxes by $195 million, but "also puts several controversial programs into state law. It keeps the Confederate flag flying atop the State House; reopens the Barnwell low-level radioactive waste landfill to the nation; and provides money to keep The Citadel all male."[40]

Harriet Keyserling, a former South Carolina legislator who had worked for years to close Barnwell, wrote later that the Legislature:

...reneged on a commitment to close the site on the day set by the compact legislation of all the states and Congress. They were easily influenced by Chem-Nuclear to not only keep it open for the Southern members but to open up the site to the whole country AND to withdraw from the compact. In withdrawing from the compact, we lost our right to restrict wastes from everywhere and anywhere. Ten years of carefully molded public policy went out the window.

This action was taken in a disturbing way — by amendment to the appropriations bill, tacked on by the Senate at the last minute. There were no public hearings, always required for such a drastic change of direction of a state policy, no third readings, always required for passage of a bill, and the House never voted on it as a stand-alone issue. The public had said, year after year in polls, that it wanted Barnwell closed, but the leadership did not allow a vote. If House members wanted to vote against this action they would have to vote against the whole appropriations bill, which is politically difficult; most legislators have something in the budget they must vote for. They were handcuffed. Beasley gave cover to the many uncomfortable legislators by promising that the additional revenues from Chem-Nuclear would go toward education. We of course felt this was one-sided economics, that there would be untold costs of being the nuclear waste capital of the country.[41]

Barnwell was actually reopened to waste from 49 states. South Carolina's neighbor, North Carolina, was excluded out of some combination of pique that it had not yet built the disposal facility it had promised, and ill will between the states stemming from differences on several contentious issues. Promised completion dates for the new facility North Carolina had been designated to build for the Southeast Compact were long past, and the opening was thought to be at least three years off. South Carolina's exclusion of North Carolina was probably not legally enforceable, but it wasn't challenged.

The South Carolina Legislature made three significant changes from the way Barnwell had operated previously. First, it required that all waste coming into the disposal facility be packaged with concrete overpacks, providing an extra layer of protection around the metal containers.

Second, the law added a surcharge of $235 per cubic foot, more than tripling the cost of disposal from the $95 that Chem-Nuclear had been charging. The goal was to raise $137 million a year for government programs. Five percent of this amount was directed to Barnwell County and the rest was earmarked to support higher education in the state. As the only disposal facility available to most of the country, no one seemed to worry that the added cost would keep customers away.

The third change was more subtle. No longer would states and compacts have to demonstrate that they were making progress towards siting and building their own disposal facilities. Now, Barnwell would be available to anyone willing to pay the new fees.

The 1985 amendments to the federal Low-Level Radioactive Waste Policy Act had specified a set of milestones intended to assure that each state, either individually or as part of a regional compact, had a new disposal facility in operation by January 1, 1993. As the governors of Nevada, Washington, and South Carolina worked to close or restrict the disposal facilities in their states, they had required that compliance with the deadlines be demonstrated as a condition for retaining access for waste disposal. This had added an oddly moralistic tone to deliberations about garbage and money.

New Jersey was able to meet the first deadlines for 1986 and 1988 by joining a regional compact and passing enabling. siting legislation. To meet the 1990 requirement, Governor Tom Kean had signed a letter certifying that "the State of New Jersey will be capable of providing for, and will provide for, the storage, disposal, or management of any low-level radioactive waste generated within the State and requiring disposal after December 31, 1992."

But, a year later, officials from Nevada, South Carolina, and Washington wrote to Governor Jim Florio, Kean's successor, to express concern: "This communication is intended to notice New Jersey that the sited States[42] have serious concerns that efforts to date are inadequate to enable it to develop disposal capacity or otherwise manage its waste... As a result of these concerns, Nevada, South Carolina, and Washington require persuasive evidence by December 7, 1990 that New Jersey's efforts are sufficient to guarantee its wastes will not constitute an involuntary burden on other states. Without such persuasive evidence the sited states ... have no alternative other than to invoke the associated sanction of denial of access to the sited states' disposal facilities to all waste generators within the state."[43]

The subsequent response by New Jersey's Commissioner of Environmental Protection, Judith Yaskin, left the representatives of the three states "less than satisfied with New Jersey's overall progress toward development of a disposal site." They reiterated to Governor Florio, "Obviously, failure to make significant progress in the immediate future jeopardizes New Jersey's generator access to our disposal facilities."[44]

A year later, Roger Stanley of Washington State's Department of Ecology had informed Governor Florio that the disposal facility in Richland, Washington was adding a penalty surcharge of $120 per cubic foot for low-level radioactive waste originating in New Jersey. This surcharge was imposed because "Washington has... determined that New Jersey is out of compliance with the Act. Specifically, New Jersey was to have filed an application for a license from the Nuclear Regulatory Commission by the beginning of 1992, and instead was just considering starting a voluntary siting process." In his letter, Stanley added, "Please be reminded that beginning January 1, 1993, the Washington site will become a regional facility and will no longer accept waste from outside the Northwest Interstate Compact Commission."[45]

Also planned for the first day of 1993 had been the complete closing of the facility in Barnwell. Late in June of 1992, however, as South Carolina legislators were wrestling with a budget for the fiscal year that would start in four days, they decided to keep the facility open for two more years. Supporters of the extension, projecting that it would yield $37 million per year, were able to gain enough votes for passage only when Governor Carroll Campbell agreed to ban waste from outside of the Southeast Compact after July 1, 1994. *The State* newspaper added that, "In a move that probably helped landfill supporters, votes were not recorded."[46]

As staff of the South Carolina Governor's Office officially notified other states and compacts of the new dates, they added, "The importation of out-of-region waste will not be approved for states or compact regions which are not making adequate progress toward providing for disposal of their own low-level waste."[47]

By 1993, the Southeast Compact seemed to be seriously considering denying access to Barnwell for waste generators from New Jersey and Connecticut. In October, Rick McGoey, the member of the New Jersey Siting Board who worked for GPU Nuclear, joined Ron Gingerich and others from Connecticut's siting program for a trip to meet with the Southeast Compact Commission in Atlanta, Georgia. They traveled as supplicants, hoping to convince the commissioners that, appearances to the contrary, both states were on the way to siting and building their own disposal facilities.

The trip to Atlanta led to more requests for documentation of progress. Even as the date for ending out-of-region access to Barnwell approached, South Carolina continued to force the out-of-region governors to grovel by mail. As he was leaving office in December 1993, Governor Florio sent another letter to the chairman of the Southeast Compact Commission: "I am writing to reconfirm that the State of New Jersey is fully committed to and has been working to fulfill its responsibilities under the Low-Level Radioactive Waste Policy Act."

Four months later, Christine Todd Whitman became the third successive New Jersey governor required to reassure South Carolina and the Southeast Compact Commission. "As you know," she wrote, "the Siting Board has been working very diligently to develop a siting process and implement a siting program consistent with the unique demographic characteristics of New Jersey. This is to confirm that New Jersey remains committed to this effort..."

The progress reports demanded by South Carolina, and earlier by Washington and Nevada as well, were a result of the conflicting goals of the sited states. They wanted the substantial taxes that waste generators from other states were prepared to pay, yet they also sought to insulate themselves from critics who wanted their state to get out of the waste collection business as quickly as possible. While the 1985 federal amendments gave them authority to exclude waste from other states, the sited states apparently feared that Congress might be pressured to again reevaluate the dates it had established unless new facilities were on the horizon.

On July 1, 1994, Barnwell was closed to generators from all but the eight states that comprised the Southeast Compact. When the facility reopened one year later, all preconditions for acceptance of waste were gone. Thus, when the first budget signed by New York's new governor, George Pataki, included no funds for dealing with low-level radioactive waste and thereby abolished its siting agency, New York waste generators remained as welcome to use the Barnwell facility as anyone else, as long as they paid the new fees.

In retrospect, perhaps, one would think the reopening of Barnwell would have caused New Jersey's Siting Board to reexamine the need for its siting process. But in New Jersey, as in all the other states and compacts struggling to build new disposal facilities, few thought South Carolina's change of heart represented enough of a commitment to warrant any significant change in their activities.

The Barnwell situation seemed very precarious. For one thing, the state's action was being challenged in a lawsuit contending that it was not legal to accomplish such a major policy change through an amendment to the budget. "This was bobtailing," writes Keyserling, the former South Carolina legislator, "a power-driven practice which ignored the voice of the people."[48]

The lawsuit was eventually unsuccessful, but the court decision was not rendered until many months after the facility had reopened.

A related and more powerful concern was that the state budget amendment required the new policy to be reapproved each year. Because the change was not accomplished through separate legislation that would have engendered public hearings and debate, the positions of most legislators on this issue could only be guessed. The available estimates were that the Legislature was about evenly divided, with support for reopening coming from legislators in districts close to the facility and opposition coming from those farthest away. Finally, Governor Beasley, Barnwell's staunchest ally, had said Barnwell would be kept open for no more than seven years.

In New Jersey, the reopening of access to Barnwell meant that interim storage areas were no longer needed at each site where low-level radioactive waste was generated. No longer could the Board display a map with 100 dots representing places in 17 of New Jersey's 21 counties where the waste was currently being stored. And no longer could we explain that the Board planned to replace those 100 radioactive waste sites with one statewide disposal facility built with much more advanced engineering, site design, and safety and monitoring procedures.

While New Jersey continued its siting process, the board game we had been trying to master had suddenly become three-dimensional. No longer was the goal simply to strike an agreement with a town willing to host a disposal facility for low-level radioactive waste. Now, we also had to keep an eye on whether a facility was truly needed. No longer could we tell civic groups, newspaper editors, and others that a five-year clock was ticking with a potentially serious crisis at the end. Now, we had to say that we were working to resolve a problem, not a crisis, but that a crisis would erupt when Barnwell closed, as it undoubtedly would.

Now, New Jersey needed to continue its siting process to prepare for the next time Barnwell closed. Without knowing when that would be, we were running a political campaign with election day not yet scheduled. And since the Siting Board had no idea which New Jersey municipalities would eventually volunteer to consider becoming the facility's host, our campaign had to continue without identified candidates.

PART 3

1996
The Second Year of the
Voluntary Process

I can think of no public policy problem more

daunting than yours.

- John Chancellor

CHAPTER 17

The Saturday Paper

As 1996 began, the most promising location for New Jersey's disposal facility had become Springfield Township in Burlington County, another largely rural municipality less than an hour south of Trenton. I had been the invited speaker at the Planning Board's December meeting and felt that my remarks were well-received. Moreover, we knew that township officials had been quietly exploring the siting process for some time. The township administrator had heard Jeanette Eng speak at one of the Siting Board's public forums almost two years earlier, and the township attorney had advised us to expect the invitation a full eight months before it actually arrived.

Springfield looked like a great location. Fewer than 3,300 people live on 30 square miles that are largely flat and dry. Adjoining the town on its eastern boundary is Fort Dix, which seemed likely to be the most agreeable potential neighbor imaginable, far more so than a municipality with adjacent residential or recreational areas.

The Planning Board's discussion of the disposal facility was a calm and

Springfield Township
Burlington County – 1996

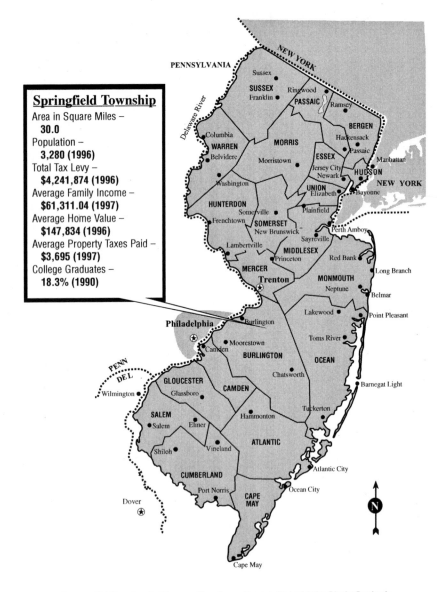

Springfield Township
Area in Square Miles –
30.0
Population –
3,280 (1996)
Total Tax Levy –
$4,241,874 (1996)
Average Family Income –
$61,311.04 (1997)
Average Home Value –
$147,834 (1996)
Average Property Taxes Paid –
$3,695 (1997)
College Graduates –
18.3% (1990)

Sources: N.J. Department of Treasury, New Jersey Almanac, NJ Legislative District Databook

150

thoughtful exchange of questions and information. Far more excitement had been raised earlier in the evening when it reviewed proposals for cellular telephone towers. Despite being presented with a variety of possible designs intended to disguise the 200-foot towers as metallic trees, several Planning Board members and the people in the audience seemed to think they would be unattractive.

Several days before the meeting, Jon Nordheimer, a *New York Times* reporter, had called me as part of research he was doing for a story about how New York, New Jersey, and Connecticut were addressing the low-level radioactive waste disposal problem. After I mentioned the upcoming Springfield meeting as an example of how the voluntary siting process was intended to work, he decided to attend.

The following Saturday, his lengthy article appeared on the front page of the *Times'* "Metro" section.[49] Its focus was almost entirely on the Springfield meeting, which he described as "cordial and attentive." The story appeared under a prominent headline proclaiming:

TRYING TO SELL RADIOACTIVE WASTE; FINDING A TAKER IN NEW JERSEY IS JOB OF A NUCLEAR-AGE WILLY LOMAN

Nordheimer had arrived in Springfield a few hours before the meeting started, and reported that he found "few in town outside the planning board members were aware a state official would be there advocating the storage of radioactive waste." He quoted a clerk in the general store as saying, "They would have one heck of a fight on their hands if they tried to bring something like that here," and the owner of an exotic fish distributing company as saying, "I don't see why they just don't put this facility over in Fort Dix, where there's 30,000 acres and already there's Nike missile sites."

The article was accompanied by photographs of the two quoted residents as well as two pictures of me making my "pitch to Springfield," plus a copy of both sides of the $2 million bill we were distributing. The caption under the bill photos said, "Not-so-funny money is part of the campaign by New Jersey's search group. While the offer is genuine, there are no takers."

Two weeks after the article appeared, we had heard nothing further from anyone in Springfield. We wrote to the Planning Board, asking whether representatives might be interested in visiting the low-level radioactive waste disposal facility in Barnwell, South Carolina. We told them that they could join a trip already being planned for February by several members of the Siting Board and its Advisory Committee, and the Board would pay their expenses. To our delight and surprise, they said yes.

The Springfield contingent was to include all nine members of the Planning Board and the county engineer. Rick McGoey from the Siting Board, Jeanette Eng from the staff, and Fran Snyder from the state Office of Dispute Settlement would accompany them.

Arranging for the state government to pay for 13 people to fly to South Carolina for a two-day trip was its own challenge, even with about six weeks between the start of itinerary planning and the day of departure. The Siting Board had more than enough money, and the *Voluntary Siting Plan* and annual budgets each explicitly stated that the Board would pay for municipal representatives to go to Barnwell. Nevertheless, officials in the Department of Treasury still had to ask many questions about the entire siting process and check with staff in the Governor's Office before they would sign off on the expenditures.

In the end, the only reason the travel arrangements were approved in time was that the Siting Board's administrative officer, Denny Medlin, had Radar O'Reilly-like knowledge of the state paperwork systems and great working relationships with the people who labored in them. One of them told her that approval would never have been granted if the disposal facility had been in Florida or some other destination considered more desirable for winter travel.

On February 22, the travelers flew from Newark through Atlanta to Augusta, Georgia. There they rented cars and drove to the Winton Inn in Snelling, South Carolina, the only place to stay and the favored place to eat when visiting the Barnwell facility. That afternoon and the next day, they toured the facility, drove through the surrounding area, chatted with people they met, and spent time talking among themselves about how they could best bring the possibility of hosting a disposal facility to public discussion in Springfield.

With one possible exception, all of the Planning Board members seemed impressed with the safety of the facility and excited about the potential it offered for Springfield. They may have grown a little more cautious when one of their members confessed that he hadn't told his wife where he was going because he thought she would worry if she knew he was visiting a place with radiation. Apparently, whatever she might have imagined he was doing on an overnight trip would have been better than the truth.

Some members wondered if they were violating provisions of New Jersey's Open Public Meetings Act by having not only a quorum, but the full Planning Board together, without any prior public notice. The chairman responded that they had taken "fact-finding trips" in the past to look at the work of developers seeking approvals in Springfield, and that — other than being a longer trip — this was no different.

When one member said the disposal facility had not yet attracted any press attention, another corrected him. "No," she said, "there was the big story in that Saturday paper." Apparently, *The New York Times* was not the paper of record in Springfield. In fact, the *Times* article had generated no local conversation or follow-up stories. No one outside the Planning Board, including the reporters who covered the town for local papers, seemed to have seen it.

By the end of the trip, they had agreed on a plan. Over the next two weeks, they would speak individually with members of their Township Committee and other local officials, and then start having discussions at public meetings.

At the Siting Board's regular monthly meeting the following Thursday, there was much excitement. While further public airing of the issue in Springfield was not to begin until the next Monday night, it was discussed openly and extensively in Trenton. Since the Siting Board had paid for the trip to Barnwell with public funds, there was no thought of not mentioning the name of the town or of talking only in a closed session.

Like most of the Board's meetings, however, this one was attended by fewer than a dozen people in addition to the members and staff. Although notice of each meeting was published in newspapers throughout the state and mailed to officials in every municipality and to anyone who had asked to be on the Board's mailing list, no reporters were present and everyone in the room was a supporter of the siting process.

When the meeting ended, a number of informal discussions about Springfield and strategy continued in the hall and in the Board's office. One of the participants was Marianne Kunz, part of the legislative and public affairs section of Jersey Central Power and Light Company (JCP&L), now known as GPU Energy. On behalf of her company, she attended every Board meeting, and had been helpful in critiquing a number of our draft publications and in arranging several speaking opportunities for Board representatives.

Kunz suggested that we contact Robert Singer, the state senator who represented Springfield to let him know what was happening. She pointed out that Singer was the vice-president of a hospital that generated radioactive waste, and added that she thought he would be both knowledgeable and helpful. Since she knew the senator better than I did, we agreed that she would call him. The theory behind talking with him was that people generally are more open to information if they are told about it early and directly, and that politicians, in particular, like to be forewarned.

Late that afternoon, Kunz called to warn me about the call I was going to receive. A minute later, Senator Singer was screaming at me on the phone. He wasn't talking about Springfield or radioactive waste, however. He was talking about his hometown of Lakewood and CAFRA, New Jersey's Coastal Area Facility Review Act.

In 1993, after years of discussion, the Legislature had amended CAFRA, a controversial law that requires state approval of all residential developments of 25 units or more proposed along the New Jersey Shore. Part of my job at the time as an assistant commissioner of the Department of Environmental Protection was responsibility for New Jersey's coastal management program, and I had been actively involved in the legislative debate.

Senator Singer had testified at the bill's final hearing before an Assembly committee. He asked that the language be changed to provide for less strict regulation in Lakewood. When CAFRA was originally drafted, eastern Lakewood, which is located 10 miles from the ocean, had been included in the area to be regulated because of its proximity to the tidal portion of the Metedeconk River. Most other areas located that far inland were deleted from the bill before it was enacted in 1973, but Lakewood had remained.

On a map of the CAFRA area, Lakewood sticks out as a large discrepant inland thumb, and Singer's request to reduce the extent of state regulation was not unreasonable. Had he raised it months or even weeks earlier, the change probably would have been included in the bill. But at that point, the end of the legislative session was only days away. Any change in the bill would render it different from the version that was before the Senate, and there would not have been time to send the revised bill back through the Senate before the session ended.

At the hearing, I told Senator Singer that the change he sought could be accomplished by the Department of Environmental Protection through regulation once the amendments were enacted. On the strength of my assurance, he withdrew his objection and later voted for the bill in the Senate.

This telephone call, three-and-a-half years later, was our next conversation on the subject. The DEP had not made the regulatory change I had promised,

I was a liar, and he was going to make sure I didn't get to build a radioactive waste facility.

I responded by trying to remind Singer that Governor Whitman had replaced Governor Florio soon after the amendments were enacted and that changes in governors always bring much of the bureaucracy to a standstill for a while. As a result, the DEP had not even proposed the necessary regulations by the time I left the department in September 1994, and even now, several years later, the regulations were still not proposed. I said that although I had had a lofty title, I was not a king, and that moving these regulations had been among the many goals I had not been able to accomplish in the DEP.

When he continued to yell, I tried to steer us back to low-level radioactive waste. Didn't he recognize that this was a serious problem facing the state? Didn't he know this from his own experience at the hospital? Was he really going to try to thwart the possible resolution of a state problem, a resolution supported by a governor from his own party, just because he was mad at me for something I allegedly did several years earlier? At that point, he said, "You're damn straight I am."

He concluded, "You're not going to put this in Springfield and you're not going to put it anywhere else in my district." When he asked if I understood what he was saying, I started to say, "Yes, but ...." But by the time I was finished with the "Yes," he had hung up.

Senator Singer never made a public statement about the issue, but by the time the Springfield Township Committee met four days later, the mayor and other local officials somehow all knew that even if they were brave enough to support studying the idea, the senator was going to do whatever he could to keep the disposal facility out of their town. That night, the Planning Board voted to end its pursuit of the possibility of siting, and the Township Council followed suit a week later. On March 15, the township clerk wrote to the Board, saying:

> *The Springfield Township Council would like to thank you for giving our Planning Board the recent opportunity to visit the low-level radioactive waste disposal facility located in Barnwell, South Carolina The Planning Board gained valuable information from the trip... However, the Township Council at its meeting on March 13, 1996 voted unanimously to advise you that Springfield Township has no further interest in discussing the matter with the Siting Board.*

Would New Jersey have been able to build a disposal facility in Springfield if the town had been represented by a different senator or if someone else had held my job? The proposal would have soon generated opposition even without Singer's involvement, and that might well have grown strong enough to quickly stifle the initiative. On the other hand, Springfield's visionary Planning Board members just might have been able to lead township residents to the same conclusion they had reached — that this could indeed be beneficial for their town.

CHAPTER 18

Lower Thumps Dump

D uring 1995, a new member of the Siting Board, Bob McKeever, had been suggesting to people in his town that they consider hosting the disposal facility. Among those who were receptive was an elected member of the Township Council.

Lower Township occupies the southern end of Cape May County. It is several miles south of Upper Township, with imaginatively named Middle Township located in between. In September, McKeever arranged a small meeting at the home of Mayor Robert Conroy. Conroy, who makes his living as a home builder, expressed skepticism about the possibility of finding a suitable location. The meeting concluded with Karl Muessig of the Geological Survey promising to review maps and other information with one of the town officials.

Since Cape May is best known as a tourist destination for bathers, bird-watchers, and fans of Victorian houses, Muessig and the local official were surprised to identify at least two areas on the Lower Township maps that were

Lower Township
Cape May County – 1996

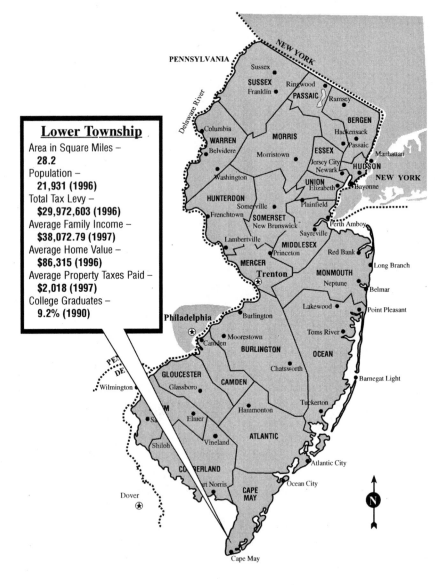

Lower Township

Area in Square Miles –
28.2
Population –
21,931 (1996)
Total Tax Levy –
$29,972,603 (1996)
Average Family Income –
$38,072.79 (1997)
Average Home Value –
$86,315 (1996)
Average Property Taxes Paid –
$2,018 (1997)
College Graduates –
9.2% (1990)

Sources: N.J. Department of Treasury, New Jersey Almanac, NJ Legislative District Databook

outside the 500-year floodplain, largely free of wetlands, and at least 150 acres in size.

By the beginning of 1996, discussion among local leaders had led to the idea of creating a citizen's committee to explore how to control development and taxes in Lower Township. The disposal facility could be one of the options the committee would weigh against other suggestions. Discussing radioactive waste in this context might lead to more informed and less emotional public meetings than other towns had experienced. If the committee concluded by recommending consideration of the disposal facility, it would present residents with their evaluation of the pros and cons of all the possibilities that had been reviewed.

In February, Mayor Conroy sent letters to about 20 Lower Township residents congratulating them on their appointment to the newly created Mayor's Advisory Board. At the same time, he wrote to the Siting Board saying he had informally polled the Council and that the "majority were receptive to this concept" of considering a low-level radioactive waste disposal facility for Lower Township. He asked that I give a presentation to the first meeting of the Mayor's Advisory Board.

Shortly after 7 p.m. on March 12, he opened the meeting by indicating that there had been thought of establishing such a group years ago, but that their attorney at the time had said it would be illegal. They now had a different attorney with a different opinion. He said the group was needed because Council meetings were generally disrupted by a small group who seemed intent on saying no to everything. He added that he had handpicked this Board to represent the people he considered the leaders of the community.

As the meeting began, 19 of its members were seated at one long table with local officials seated behind them at a similar table. Later, when one of the few audience members raised an issue concerning the possibility of turning a site he had identified into a public fishing area, he was added to the Board and moved to the head table.

The disposal facility was the first item on the agenda. Mayor Conroy introduced McKeever, noting that Bob had brought this idea to his attention and that he thought it was worth considering. McKeever then introduced me. I spoke and answered questions for an hour. Questions concerned truck accidents, what type of permits the facility would need, financing the closure of the facility, when benefits would start to flow to the municipality, what Greenpeace and other anti-nuclear groups advocated doing with the waste, why these facilities hadn't yet been sited in other states, and whether a facility in New Jersey could be forced to accept waste from other parts of the country.

As the meeting closed, the mayor established several committees. The committee to look into the siting process would be chaired by McKeever, one chaired by a woman from the Chamber of Commerce would focus on commercial development issues, and the newest member would chair a committee to look at fishing possibilities. Members were asked to volunteer for the committees of their choice, and by the end of the evening the disposal facility committee had five members.

The next day, the *Lower Township Gazette* gave the meeting front page billing with a banner headline:

TOWNSHIP WOULD GET $100 MILLION OVER 50 YEARS; LOWER CONSIDERS BECOMING RADIOACTIVE WASTE STORAGE SITE

The strategy of looking at the disposal facility as just one of several economic development possibilities had clearly been lost by making this issue the focus of the Advisory Board's first meeting, but the newspaper story was thorough and informative. At the bottom of the last column, the paper promised that, "Next week, the *Lower Township Gazette* will take a look at how a facility would operate, explain other safeguard precautions and explore future economic benefits for a municipality such as Lower Township."

The following day, members of the disposal facility committee drafted a "Dear Neighbor" letter that began, *"We are five residents of Lower Township concerned about our rising tax rates."* It concluded, *"Over the next weeks and months, we plan to learn as much about this as we can. We would welcome your help."*

The Township Council's next scheduled session occurred, however, before the committee members could meet to sign the letter. People arriving for the Council meeting were each handed a "clarification" on Township of Lower letterhead:

> *The Lower Township Council does not promote, encourage, support, condone, endorse or otherwise have a position on any proposed Low Level Radioactive Waste site. We have not discussed it, nor have we been formally approached about it.*
>
> *A request was made to make a presentation to the Mayor's*

*Advisory Board and after that nothing further has been done. If a
presentation had been made to the Rotary Club it would have the
same status as it now has with the Council.*

*No Low Level Radioactive Waste Facility will be permitted in
Lower Township without a public referendum. We personally will
not support such a facility without a public referendum.*

The promise of a referendum satisfied none of the people who came to the
meeting to speak against consideration of the disposal facility. One man said,
"We don't want $2 million that bad." A woman said, "If a dump comes here,
Cape May might as well forget about Victorian tours." And a man who said he
had helped coordinate the response to the accident at the Three Mile Island
nuclear plant in Pennsylvania noted that "the level of anxiety there was hor-
rendous" and asked that the disposal facility be considered no further.

The next day, the *Lower Township Gazette* reported:

LOWER THUMPS NUCLEAR WASTE DUMP

"Less than six calendar days after a bid to find a place to store waste
became public, Township Council on Monday declared its opposition," the
Gazette announced. Mayor Conroy told the reporter, Jack Smyth, that a quick
poll of the Council members showed they were opposed to the facility. When
asked how firm the opposition was, he said, "Obviously, it's not going to happen here."

A week later, the councilman who had been the initial supporter told me
that they responded so quickly because they didn't want this to be an issue in
the forthcoming election. November was actually nine months away, almost as
far off as an election can be, but apparently it was already very much on the
minds of those who would be running. The councilman said that comments on
the issue on the local call-in radio station had been about evenly divided, and
that he felt the Council's action left the door open to reconsider the issue at the
end of the year.

Not only was the issue not raised again, but the Mayor's Advisory Board
never held a second meeting.

CHAPTER 19

Somebody Wants It

T he New Jersey Department of Community Affairs (DCA) was created in March 1967 to help the state implement the Great Society then emanating from Washington. While its vision and expectations have since narrowed substantially, the DCA remains a dim beacon to New Jersey municipalities with dreams, needs, or crises.

One afternoon in March 1996, DCA Commissioner Harriet Derman had the type of meeting that was very familiar to her. Mayor Viola Thomas of Fairfield Township had driven two hours north to Trenton to say that her municipality desperately needed financial assistance. Fifteen percent of the town's more than 6,500 residents lived in poverty, property taxes were much too high, and the local government was about to adopt a budget that was barely balanced.

Derman was joined at the meeting by veteran Assistant Commissioner Chuck Richman. Mayor Thomas was accompanied by Charles Nathanson, a consulting planner who worked under contract for a variety of private clients and local governments, including Fairfield. At the end of an otherwise discouraging discussion, Richman suggested that Fairfield consider the disposal facility. He noted that the $2 million a year the Siting Board was promising equaled about half of Fairfield's annual budget.

Nathanson called the next day, and a week later he and Mayor Thomas were in the Siting Board's office asking questions and discussing the voluntary siting process. Seven weeks after that, on May 21, the Fairfield Township Committee voted to "gather and analyze information on the potential financial and other benefits of hosting such a facility, together with information on potential costs and risks therewith associated." They assigned Nathanson to this task and stipulated that his costs be paid by the Siting Board.

Newspaper reports of the Township Committee's action included a quote from Mayor Thomas: "Whether we decide to go with it or not, we get a tremendous education. If we do decide to go with it, we get a tremendous ratable."

While the town's financial need made the disposal facility attractive to local officials, Fairfield had a number of attributes that interested the Siting Board. The geography was rural and flat, with several large areas that seemed likely to be suitable. One of them consisted of 250 acres that were half wooded and half active farmland, adjacent to the Fairfield Federal Penitentiary.

The popular local perception was that the prison had been built by the federal government without meaningful consultation with anyone in Fairfield. As a result, the voluntary siting process might be seen as particularly refreshing. In addition, so far the government had run the jail as it had promised, causing none of the problems that opponents had feared. This might add some credibility to the pledges another government agency would make.

We also thought the school superintendent could be an early and influential supporter. He was the one who had become intrigued when the district's architect had heard me speak before the Trenton Rotary Club meeting a year earlier.

On June 18, at Mayor Thomas's request, Board staff spoke to a handpicked group of interested officials and residents. The town's consulting engineer, Dick Carter, asked most of the questions and seemed sufficiently satisfied that by the end of the meeting he was reassuring others in the room.

The mayor's guest list had also included the township solicitor, clerk, and emergency management coordinator, the Environmental Commission chair and vice-chair, one member of the Township Committee, the chiefs of both local fire companies, and Miles Jackson, a reporter for the *Bridgeton Evening News*, the major local daily paper.

We had worried that Mayor Thomas would be criticized for selecting the people to invite and our fears had only increased when we learned that a reporter was among them. But no one complained, and Jackson's article the next day provided a good description of the facility and the siting process under the headline:

TOXIC WASTE DUMP RIGHT FOR FAIRFIELD?

Two weeks later, the headline was:

COUNTY WANTS SAY IN NUKE DISPOSAL SITE

Jackson's article reported that several county freeholders had expressed strong concern about Fairfield's explorations. One said, "We need answers to many questions. This is not just a township issue." Another added, "It is unbelievable that a township with limited resources would get involved in something like this without county participation." A third said that the state government "should look toward sites that have already been ruined, rather than looking at what Cumberland County has protected." Finally, the county planning director expressed his opinion that, "This would be just one more stigma to overcome in attracting high-quality industry."

The county officials seemed peculiarly unaware of the limited extent of their powers. Some said that a countywide referendum might be necessary before a town could approve a development as significant as the disposal facility. One freeholder asked, "Is it possible for the state to locate something like this in a township and bypass the county?" In New Jersey, the answer is usually yes, because county governments have little power in land use matters. This is unfortunate because regional and statewide development patterns might have evolved more sensibly if they had been overseen by the 21 county governments instead of 566 municipalities. Still, it was a surprising question for a county official to ask.

The county opposition seemed to mobilize Fairfield officials, who had spent the previous three weeks painstakingly debating and planning strategy.

Fairfield Township
Cumberland County – 1996

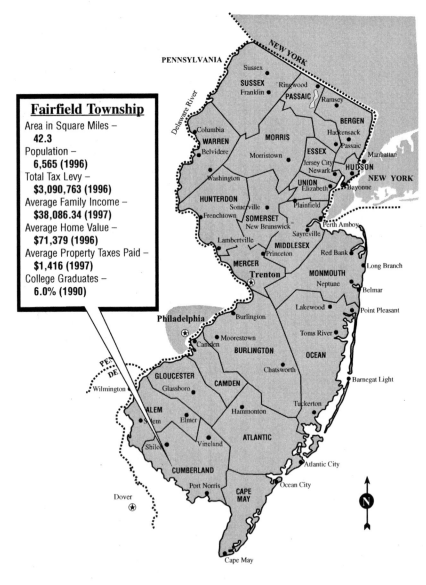

Fairfield Township
Area in Square Miles –
42.3
Population –
6,565 (1996)
Total Tax Levy –
$3,090,763 (1996)
Average Family Income –
$38,086.34 (1997)
Average Home Value –
$71,379 (1996)
Average Property Taxes Paid –
$1,416 (1997)
College Graduates –
6.0% (1990)

Sources: N.J. Department of Treasury, New Jersey Almanac, NJ Legislative District Databook

The Environmental Commission invited the Siting Board staff to its July meeting and devoted the evening to the subject.

Mayor Thomas expanded her informal working group and convened its second meeting. On July 17, she sent a letter to all township residents, advising them of the Township Committee's decision to study the issue:

> *We have many questions as you probably do too. What is low-level radioactive waste? Would this facility be safe? Would the benefits enable Fairfield to lower taxes and provide additional public services? Will this facility create a negative image problem? ... We want to assure you that NO decisions on this matter have been made, other than the decision to investigate the possibilities. We will keep you informed as the study goes on, and we welcome your comments, advice, and participation.*

With the signing of this letter, the step that Alloway had delayed too long, Fairfield became the town to have advanced the furthest in the voluntary process. It may have been like starting to root for a perfect game after a pitcher has a good first inning, but two months after the first article appeared, the process in Fairfield seemed to be starting to thrive. We began to focus more intently on the identification and analysis of specific locations. For the first time, we directed Foster Wheeler to join us in visiting potential sites. Two of Foster Wheeler's staff joined three of us from the Board for a four-hour Fairfield van tour conducted by Township Engineer Dick Carter.

The tour was a great success in that we all agreed Fairfield had two very promising sites. One was the site adjacent to the federal penitentiary, and the other was across the street. Several years earlier, a mining company had applied to use the first one. Although the variance it required had eventually been denied, the soil borings and site investigation undertaken at the time led Carter to be optimistic that the site would satisfy the criteria for the disposal facility.

Both sites were accessed by two-lane roads with very little traffic. It looked as if they might need, at most, some minimal improvements to accommodate the trucks bringing in waste. On two sides of one site, and three of the other, were farms. The disposal facility would have no impact on their operations, but if the owners preferred, the Siting Board would be able to buy them and manage them as part of a huge buffer or resell the farms in the future.

A new housing development was adjacent to one side of the first site.

While this was several hundred feet away, it seemed inevitable that unless the residents happened to all be physicists and geologists, some of them would be among the opponents.

The most interesting neighbor to consider for both sites was the prison. It was next to one site and across the street from the other. The buildings and yards were surrounded by a wide wooded buffer. Combined with the buffer of at least 300 feet required on the disposal site property, prison inmates and staff would be further than some residents of Barnwell from the concrete structures containing low-level radioactive waste.

Nevertheless, potential danger to the people close to any proposed site would be an understandable concern to anyone evaluating the idea, and an inevitable focal point for people arguing against it. Opponents to the facility proposed in Ward Valley, California, for example, were focusing on possible effects to the river located 22 miles from the disposal site, and opponents to the facility under consideration in Texas also were concerned about the health impact on people located miles away.

During the three weeks of seeming inactivity after the first news articles about Fairfield, consideration of the roles of local personalities had slowed the strategic planning. Two people apparently had each been told that they were to lead the public education and discussion effort. This confusion was resolved when one backed out due to an extremely busy schedule and his wife's discomfort with the issue.

Mayor Thomas then announced that the informal working group, to be called the Township Exploratory Committee, would be chaired by Cindy Zirkle, a well-known resident environmental activist who headed a group called Cohansey Area River Protection (CARP). She and her husband, Ernie, who was the state veterinarian, had fought against the incinerator proposed by the Hazardous Waste Siting Commission in nearby Maurice River Township a decade earlier. She had also been a leader of a more recent successful effort to oppose a commercial barging proposal.

After the county freeholders had met in early July, Freeholder Director Douglas Fisher had written to area environmental and civic groups, including CARP, asking their "reaction to the potential siting of a low-level radioactive waste facility in Cumberland County." Fisher went on:

> *As I look around Cumberland County today, I see many very positive things happening. The new state prison in Bridgeton . ... the possibility of a major hotel and downtown outlet mall ... the success*

*that Millville and Vineland are having with their Urban Enterprise
Zones ... the potential that ecotourism and tourism in general hold
... our farm industry ...*

*How can the siting of a low level radioactive waste facility com-
plement this agenda? How can it help turn the county's image
around? How does it fit in with our ability to promote a clean environment and good quality of life? Obviously, there are numerous
questions that need to be addressed as to the regional impact of a
project of this nature.*

In response, Cindy Zirkle was quoted in the *Bridgeton Evening News* as
saying: "It's sad in this day and age that some people are so hysterical they
can't accept the byproduct of modern medicine. This is a byproduct of aspects
of our society we're not willing to give up. It's great that Mayor Viola Thomas
is taking a look at the problem and attempting to be part of the solution."

When asked about the freeholders' concern with the county's image, Zirkle
said, "The four prisons in the county do more to hurt our image than a small,
unmarked nuclear disposal facility would have. For the county to oppose this
before looking into all of the facts about it is the height of irresponsibility."

Zirkle's comments appeared in an article headlined:

MIXED REACTION TO WASTE FACILITY

The article began, "If county officials are expecting unified opposition from
environmental groups to a low-level radioactive waste disposal facility in
Fairfield Township, they are in for a big surprise."

The area's two other environmental groups were more cautious, but not
hostile to the facility. The head of Citizens for Cumberland County said her
group had not yet taken a position. The leader of Citizens United to Protect the
Maurice River and Its Tributaries expressed concerns about whether the types
of incoming wastes could be limited, but also praised the Siting Board for the
voluntary siting process.

The county government continued to express its opposition, now by reviv-
ing a dormant Tourism Advisory Council. "One of the council's first official

acts," the *Atlantic City Press* reported, "was to pass a resolution opposing the siting of a low-level radioactive waste disposal facility in the county."

The Tourism Council's subsequent letter to me said that members were "appalled by the recent news that your board would consider us as a serious candidate for location of a facility which is so obviously incompatible with our long-held objective of marketing Cumberland County as a premier tourist destination."

When I called the group's chair, Robert Rose, we realized that we had spoken in the past — he in his role of running a wonderful folk festival in Bridgeton and mine as a disc jockey. He agreed to schedule me to speak to the group at its August meeting. He also acknowledged that they did not have any specific ideas for attracting tourists to the properties we were considering adjacent to the federal prison.

On July 25, the Township Exploratory Committee held its first meeting under its new name and chair. Zirkle said the major purpose of the meeting was to establish four working subcommittees. She asked the 50 people who showed up to consider volunteering to focus on local involvement, health and safety, community benefits, or legal issues.

Mayor Thomas then spoke and for the first time said that the Township Committee would hold a referendum before making any final decision about the facility. She noted that the vote would be at least two years away, after "we educate all the residents as to what the facility is about."

About an hour into the meeting, Freeholder Director Fisher rose to read a two-page statement that said in part: "The potential siting of a radioactive waste facility in Cumberland County is not about science. It's not about Fairfield. It's about fairness."

He referred back to the defeat of the hazardous waste facility and said that in the intervening ten years, "siting officials have gotten smart... They've rigged the siting process so that municipalities which find themselves behind the eight-ball financially are the ones most likely to ask for it. In the end, the result is the same. Areas of the state such as Cumberland County that have not met their full economic potential are often the ones targeted. It's the same old game... Pretty clever and pretty unfair! Cumberland County's citizens are being baited and manipulated. The process at work here is anything but fair and objective. It's biased, it's discriminatory, and it's targeted in ways that are unconscionable. That is wrong."

He concluded by saying, "Let's end this discussion tonight." One township committeeman, Robert Pierce, said he now agreed with Fisher: "I say stop it right now. The residents do not want it. I thought about the $2 million just like everyone else. But it doesn't do anything for me. I don't believe Fairfield needs it."

The next day, the banner headline in the *Bridgeton Evening News* read:

FISHER LEADS WALKOUT AT N-WASTE MEETING

The accompanying article said the county officials departed alone, and that the walkout followed Committeeman Pierce's remarks. Either way, about 25 people stayed and volunteered for the subcommittees.

Fisher's comments about discrimination were interesting. While Cumberland County is overwhelmingly white, Fairfield is unusually diverse. The 1990 census reported that the population included 2,800 blacks, 2,290 whites, 392 American Indians, 271 Hispanics, and 270 others. The black voters tend to vote Democratic and the whites Republican. Mayor Thomas and several, but not all, of the Township Committee members were black.

A number of people suggested that the county's attempted intervention was based on a lack of trust in Fairfield's government that would have been less pronounced if its elected leaders had all been white. This interpretation, as well as a feeling that the county had never done anything in the past to benefit the township, helped minimize the local impact of the county's opposition.

Actually, however, long before the possibility of locating the disposal facility in Fairfield had been raised, Cumberland County officials had made their feelings clear. In February 1990, after receiving the Siting Board's draft *Disposal Plan and Siting Criteria*, Cumberland's Board of Chosen Freeholders had passed a resolution advising the Board that it "should consider focusing attention on large industrial tracts which are homes for large radioactive waste generators at the outset of the process."

More recently, in 1995, Steve Kehs, the county's planning and development director, had responded to a letter the Siting Board had sent to the planning directors of all 21 counties. He had been the only one to say that he did not want to cooperate with the siting process. With words he would recycle a year later, Kehs wrote, "The siting of a low-level radioactive waste facility would be one more stigma that the county and its municipalities would have to overcome. Furthermore, it is possible that the siting of such a facility in this county or elsewhere would generate such a preponderance of similar land uses that it would become impossible to erase a growing reputation as New Jersey's low-level radioactive waste disposal area."

The July 25 meeting of the Township Exploratory Committee served as the kickoff for the active public involvement process in Fairfield. Two days later, on a sunny Saturday, four Siting Board staff spent the day at the Gouldtown Annual Firehouse Barbeque. We had been advised that this and a similar event to be held a month later by the other local fire company were among the major social events in Fairfield.

We set up the Board's traveling display next to one of the food trucks and spent the day informally answering questions. One of our handouts was a letter to Fairfield residents from Cindy Zirkle. She made clear that the Board had been invited by the fire company and the Township Exploratory Committee to "join us at today's barbeque to distribute informational material and answer questions." Throughout the day, she and several of the other residents who supported the siting process served as our hosts, warmly introducing us to many people.

The barbeque was a lovely low-key affair, with a horseshoe contest for four-person teams serving as the only real competition to good food and conversation. Mid-afternoon, when I was invited to fill a vacancy on one of the teams, I felt well on the way to building the familiarity and trust that would be essential for the years of serious dialogue and debate that lay ahead.

Monday's lead story in the *Bridgeton Evening News* was:

N-WASTE SITING CHIEF PICNICS WITH FAIRFIELD

It began: "John Weingart got clobbered at horseshoes at the Gouldtown Fire Department's picnic on Saturday, but it could have been worse. As executive director of the New Jersey Low-Level Radioactive Waste Disposal Facility Siting Board, Weingart could have faced much more hostility in the township, where emotions run high over the possibility of such a facility being built nearby."

Zirkle's letter had noted that Siting Board representatives would be at the municipal building every Monday evening and Thursday afternoon to answer questions, and that I would also be there the following Monday morning after I appeared on a popular local call-in radio show.

After the picnic Saturday, we moved the Board display to the lobby of the municipal building, where it was to remain indefinitely. There, people could look at a model of a disposal facility and pick up literature and the Board's video. Also available was a different letter from Zirkle, in which she thanked residents for taking the time to stop by, and gave her home phone number for anyone wanting to "know more about the committees established to study the issues."

My appearance on the radio Monday morning generated many calls with those that made it on the air divided about two-to-one against Fairfield's study. After the show, I spent the promised two hours in the municipal building, but found only a handful of people.

Before heading back to Trenton, I decided to make one more stop. I had wanted to talk with the freeholder director who obviously felt so strongly about the issue, but I had received no response to a letter I had sent him or to several phone messages. I knew he ran a local grocery and, as I left the radio station, I had asked for directions.

I got to the market shortly after noon and thought I would buy lunch, both because I was hungry and because I hoped the owner would be more receptive to a customer than to a mere bureaucrat. I took the fairly substantial meal I had collected to the cash register which was clearly operated by the man who ran the store. I stuck out my hand and said, "Hi, Mr. Fisher, I'm John Weingart." And he said, "You've got the wrong market. Doug's is three miles down the road."

So, 10 minutes later, I parked at Fisher's Market, bought a second lunch, and introduced myself to Freeholder Director Douglas Fisher. He said that going to the fire company picnic had been "a great idea," adding, "I'm sure we will be at odds at many times over the next year or two." Then he said that his concern was not about safety, but about process. He repeated that he had no doubt this facility would be safe, but felt that Cumberland County had not been treated properly by state government.

I told him I didn't want to take him away from his work and, since people were waiting to talk to him, we parted with a handshake and a smile. Later that day, I wrote him to ask, "If, in fact, this will be a safe, low-impact light industrial ratable, and a municipality in the county decides it is something they want, why should the benefits go to some other part of the state?" I also asked that he let me know convenient times to discuss this further.

This time, I did get a response. Fisher wrote thanking me for my letter "which follows up your impromptu visit with me." He added, "As I explained to you at that time, it is my firm conviction that the facility you propose is totally out of character with the direction Cumberland County must follow for a prosperous future for all our citizens." Then, in a marvelously clear demarcation of the end of a dialogue, he wrote, "Your request for a meeting on this topic is noted."

Over the next three weeks, the only scheduled event, other than the Siting Board's presence in the municipal building, was a Fairfield Planning Board meeting at which I spoke. The Board members wanted assurances that they would be able to keep waste from other states out of a Fairfield facility. While I said that the contract between the state and the town could preclude out-of-state waste, the *Bridgeton Evening News* reported the next day that Diane D'Arrigo of the Nuclear Information and Resource Center in Washington had responded that any such contract would be unenforceable.

The *Bridgeton Evening News* continued to give prominent and overwhelmingly accurate coverage to the issue. Miles Jackson wrote daily stories that portrayed some of the complexities of the issue. He continued to pursue the question of whether the state could limit the sources of waste. In one story, he quoted staff from the federal General Accounting Office saying, "The law allows states that have entered into compacts an exemption from the commerce clause. This allows them to exclude waste from other states."

Then he called Richard Sullivan, New Jersey's representative to the Northeast Compact, who told him, "There is no guarantee that out-of-state waste would not be accepted, but that's not the direction things have been going in."

Jackson's articles portrayed the issues surrounding low-level radioactive waste disposal to have at least two sides, but a "Reader Poll" his paper ran around that time seemed to locate only one of them. The paper asked whether a low-level radioactive waste storage facility should be established in Fairfield Township. The introduction to the article about the results said they were "expected" and that "it is interesting to see a variety of views expressed and it does offer a chance to have people affected by this issue respond." The Siting Board noted that many of the respondents did not live in Fairfield. Still, the 56 comments against and three in favor were not encouraging.

As even its introduction to the poll results made clear, the *Bridgeton Evening News* liked the voluntary siting process. A lead editorial proclaimed:

FAIRFIELD TWP. WILLING TO EXAMINE HOT ISSUE

The editorial took the position that the Fairfield Township Committee was "going about the touchy decision-making process of whether to agree to be host to a low-level radioactive waste site in the proper manner... While Fairfield residents obviously would love a break on their taxes and to see a host of civic improvements made to their community, they may not be attracted to the trade-off of having to serve as home to a low-level radioactive waste site. That's where the township committee is being wise. They have formed a committee to explore the idea. They are having the Siting Board visit community groups and provide information. They are determining for themselves whether the idea is practical."

Editorials also appeared in the *Millville Daily Journal* and the *Atlantic City Press*. Both papers considered Fairfield to be on the fringe of their coverage area and included occasional articles about the disposal facility discussions. *The Daily Journal* headline in its "Our View" column was:

WASTE FACILITY NOT WORTH THE COST

Its editors wrote that, "The planning and siting processes in New Jersey are such that acceptance of one facility greatly increases the chances that an area will be considered, willingly or not, for like facilities... And then there is the question of the adverse impact just one waste facility would have on the county. Quality of life and, closely related, an abundance of open space in Cumberland, are two selling points in local and county development plans. A glow-in-the-dark dump hardly fits in with ecotourism planning."

The *Atlantic City Press*, on the other hand, chided Cumberland County officials, commenting, "the least the county can do is let the process continue." Its editorial was headlined:

SOMEBODY WANTS IT

The editorial declared, "Freeholders want countywide public input on the proposal, especially from neighboring communities. That's a fine and necessary idea. But we hope that input does not amount to mere knee-jerk 'not in my backyard' grumbling."

While the disposal facility continued to generate front-page coverage, at least in the *Bridgeton Evening News*, and civic life succumbed to the general August slowing of activity, tension was growing among the local supporters of the study. As the Township Exploratory Committee started to plan a fall trip to South Carolina to visit Barnwell, the mayor and the committee chair each thought they had the authority to choose the participants.

The Township Committee's next regular meeting was the third Monday evening in August. This was to be the first meeting since Committeeman Pierce said he wanted to stop the process. That afternoon Mayor Thomas told me she had polled the committee and that if Pierce made a motion it would not receive a second. She also said there was no reason to invite supporters of the facility study to the meeting since she expected the issue to consume but five minutes in a meeting with a lengthy agenda.

It turned out that the mayor was wrong. About 150 people filled the municipal building's meeting room, and when Committeeman Donald Taylor asked if anyone in the crowd wanted to speak in favor of continuing the study, not one hand was raised. He then said, "I can see from the public input that the township residents do not want to proceed."

A township resident rose and said, "I stand before you and ask you to do the right thing for the people — your people — and vote this down, vote this down right now... Who is going to make a motion to put an end to this right now?"

Committeeman Pierce made the motion and not only found a second, but four affirmative votes, including Mayor Thomas, and only one against. The holdout, Thomas Munson, said, "Any clown that doesn't have the intelligence to look into something, I have no use for them."

As the meeting ended, Pierce said, "The one good thing to come from this is that the community came together, black, white, Hispanic, to defeat this thing. We proved the public is still the boss."

The next day, Mayor Thomas explained that once she saw the motion would carry, she decided to vote with the majority to increase her influence for raising the issue again in the future.

The headline over Miles Jackson's story in the *Bridgeton Evening News* the next day was:

WASTE NOT, WANT NOT IN FAIRFIELD

CHAPTER 20

Environmental Justice and Other Lessons from Fairfield

airfield's decision was particularly disappointing because it was not inevitable. If local supporters of the siting process had attended the final Township Committee meeting, the outcome probably would have been different. Unlike the quiet supporters the Siting Board had encountered in other towns, those in Fairfield would have stood up and made their views known. Their presence would likely have given members of the Township Committee the support and encouragement they needed to have voted down the motion to end the study.

Of course, if the Fairfield process had survived this hurdle, it would have faced many more. One of them would have been to demonstrate that placing a disposal facility in Fairfield was not environmentally unjust. The concern, which first gained widespread national articulation during the 1990s, is that poorer communities, often those with large minority populations, have had a disproportionate share of the nation's locally unwanted land uses imposed upon them.

These past practices are often categorized as "environmental racism" and confronting them is considered "environmental equity" or "environmental justice." President Clinton issued an executive order in 1994 designating the Environmental Protection Agency as the lead federal agency on the subject. By 1998, the EPA had received 48 administrative complaints concerning environmental justice. The complaints, all against federal, state and local agencies, generally alleged that a controversial proposed development would add to the environmental burdens already being endured by the population of an economically depressed area. Many of these projects were planned in cities that already were home to industries that pollute or pose potential environmental hazards.[50]

For the Siting Board, Fairfield represented the first candidate about which questions of environmental justice might be raised. The Nuclear Regulatory Commission's rules requiring at least 100 acres of open land excluded virtually all of New Jersey's cities, where the state's minority population remains concentrated. Even if that had not been the case, the $2 million the Siting Board was offering would have been a pittance to cities already receiving tens of millions of dollars in state aid based on complex need formulas.

Fairfield had little noxious industry, but it already had been saddled with a prison and had been one of the leading candidates for the state's hazardous waste disposal facility just 10 years before. The township is relatively poor, and its half black/half white racial make-up is distinctive in a state where the minority population is 30 percent. In the intense debate over something as controversial as the disposal facility, this would have been sufficient reason to consider whether the concept of environmental justice should have an impact on the final siting decision.

Perhaps of greatest significance, the nearest neighbors to the two most likely sites for the disposal facility in Fairfield were the inmates of the federal prison. Although the prison building was far outside of the safety zone required for a disposal facility, the potential for jeopardizing the prisoners' safety would certainly have been a subject of debate. Who would speak for the federal prisoners given that they would be ineligible to vote?

The consideration about environmental justice would have been further complicated by the rules of the voluntary siting process. If the voting residents of Fairfield chose to host the disposal facility, could they and the Siting Board still be accused of perpetrating an environmental injustice? Or would it really be inadvertent paternalism for objectors to say, in effect, that these people are so disadvantaged they don't see the danger before them?

This speculation became moot when Fairfield's Township Committee voted to end the siting process. But the process had been taking root and might have continued to grow. The fact that the voluntary process could have advanced much further — and perhaps succeeded — encouraged us to compile a list of 13 findings, including several which might have made a difference if they had been handled differently.

First, the primary topics of discussion in Fairfield were not safety-related, but concerned the ability of government at all levels to keep its promises. Most of the issues relating to federal and state agencies had been raised in other towns, such as whether waste from other states or other types of waste would be forced upon the town by the federal government, and whether the state might change the rules and give the town less than the promised $2 million a year once the facility was operating.

The new realization was that local government also was distrusted. We thought we were reassuring residents by telling them that their municipal government would be able to choose how to spend the revenue the town would receive. Yet, a surprising number in Fairfield said they wouldn't trust decisions their Township Committee made and feared the township would find a way to squander the money. This concern may have been particularly pronounced in Fairfield, where control of the local government shifts between parties every few years so that everyone is assured that the party they don't like will be in control at some point.

A response to this fear, we reasoned, might be to delineate from the beginning how the benefits could be used. We had thought that specifying $2 million a year, rather than citing an abstract formula, would better stimulate local imaginations on how the disposal facility could benefit a town. Perhaps, however, the benefits should be stated even more explicitly. It might have made a difference if the Fairfield Township Committee had said, for example, that if they eventually decided to host the facility, the benefits would be used exclusively to lower taxes, provide college scholarships for every local high school graduate, build a new municipal building and library, and/or protect open space.

To be so specific up front flies in the face of the theory that community decision-making processes should be as open and unformed as possible to provide for maximum citizen input. But if people aren't really going to trust the process no matter how it is structured, there may be benefit in more fully articulating the initial proposal before raising it to public scrutiny.

The second problem was that benefits would not have begun to flow to

Fairfield until the disposal facility was in operation. This was at least four or five years in the future, and more likely six or eight years. This is a long time to wait for something you are skeptical of ever receiving. Politically, it meant that Township Committee members knew that when they ran for reelection they would have to defend the study without having in hand any tangible benefits or proof of the state government's sincerity or trustworthiness.

In constructing the voluntary siting program, the Siting Board had scoffed at Connecticut's promise to give a community $250,000 just for expressing interest and an additional $250,000 six months later if it continued to participate in studying the possibility of hosting the facility. The Fairfield experience suggested we should revisit this issue for New Jersey.

Money was also at the root of the third problem. A cardinal principle for the Siting Board had always been that a municipality should incur no costs for exploring the possibility of volunteering. Support for this idea had been so universal that the Siting Board, its Advisory Committee, and its staff had never fully thought through the total cost or mechanics for implementing it.

The Board emphasized its willingness to provide local assistance three times in the *Voluntary Siting Plan*.[51] The most specific reference occurs in a table listing the "Resources To Support Community Consideration" that the Board would make available to residents of interested communities:

- *Local coordinator or liaison between the community and the Siting Board;*
- *Technical consultants to help a community evaluate the suitability of a potential site;*
- *Independent facilitators to help area residents consider all facets of the issue;*
- *Fees and travel costs for outside speakers on topics such as facility design options and safety considerations;*
- *Brochures, newsletters and/or other materials to be developed by a local community to help keep residents informed;*
- *Written, audio and video information on issues related to low-level radioactive waste;*
- *Trips for community representatives to currently operating waste disposal facilities to view operations and talk with local officials and residents; and*

- *Meetings of representatives of all communities participating in the voluntary siting process, so that they can share information and concerns.*

All of these fine phrases did not prepare us for the letter Fairfield sent the Board on June 13 requesting $7,405 to reimburse their planning consultants for the work they had conducted to date. The enclosed bill told us only that they had spent 73 hours working on the "Low-Level project," 25 of them at a rate of $85 per hour and 48 at $110 per hour.

While the Board was very happy with Fairfield's progress to that point, we had no idea how most of these 73 professional hours had contributed to the process. Many, we learned, had been time the consultants had spent familiarizing themselves with the issue. This made sense, but we had required no work plan or budget in advance to let the Board, or the local officials, know if there were limits or controls on the amount of money to be devoted to the study.

Implicitly, we had expected to provide reimbursement for time devoted to the issue by municipal employees and for consultants hired for specified tasks. We had not considered that most small towns would be relying on pre-existing and more open-ended contracts with planning, engineering, and legal consultants.

The result was that the Board did not pay the bill until after Fairfield had ended the process. Instead, throughout the summer, we haggled with the consultants, asking them to propose an overall budget so that we could consider the $7,405 in a context.

Four months after the process had ended, the Board received an additional bill from the township for $5,664. This included reimbursement for 11 items, including additional planning costs, as well as engineering and legal expenses, custodial services at the municipal building, and $1,500 for "use of building." When we asked the meaning of that last item, the township eventually submitted a revised bill with that charge deleted, and the Board paid it.

Fairfield Township and its consulting planner received a total of $11,569. In the context of the Siting Board's mission, this was money well spent and larger sums might also have been perfectly reasonable. Clearly, however, greater advance agreement about the types and extent of allowable costs would be helpful to build and maintain good will and trust between the Board and other municipalities it worked with in the future.

The fourth Fairfield lesson derived from the oft-stated concern expressed about New Jersey's ability to keep waste from other states out of any new facility. The questions asked were more pointed than those we had previously asked ourselves.

First, the federal law granting Congressional consent to the Northeast Compact gave the Compact Commission the ability to exclude waste from outside the compact region, but it did not obligate it to do so. The Northeast Compact Commission was now composed of one member from New Jersey and one from Connecticut. Once either state opened a disposal facility, the host state would get a second representative. If New Jersey was successful, the state's two commissioners, presumably, would not approve the acceptance of waste from other states if the host community objected, but that was not an ironclad guarantee. Perhaps the Board would need to explore developing a provision that could be added to the host community's contract with the Board to allow it to shut down the facility if the Compact Commission took action to which they objected.

The other related issue was about waste from Connecticut. While the two states had agreed that each would build its own disposal facility, waste volumes were continuing to shrink, and it had begun to seem increasingly likely that whichever state opened a facility first would be petitioned by the other for some type of partnership.

We had assumed that this eventuality could be addressed when it arose. Once a community had gone through all the analysis and soul-searching to agree to host the facility, it would surely understand that accepting Connecticut's waste would only increase the economic benefits it would receive, without increasing its risk or significantly adding to the new traffic it would face.

But the discussions in Fairfield made clear that virtually any change from initial promises or understandings would significantly undermine the Siting Board's credibility and could threaten an agreement. On this issue in Fairfield, at least one of the local officials supporting the study had promised his constituents and neighbors that no waste from outside New Jersey would be accepted. Subtleties and footnotes to the Board's statements would be irrelevant. If the rules appeared to change, some support would evaporate.

One revision this suggested was for the Board to consider telling communities from the beginning that the facility might include waste from Connecticut. This would provide more flexibility for the final stages of negotiating a community agreement and would help shore up one of the Board's weakest answers. But such a policy change needed to be discussed with the Siting Board, Connecticut's Hazardous Waste Management Agency, the Northeast Compact commissioners, and the governors of both states. A consensus would not be reached overnight.

The fifth lesson also concerned the necessity for the Board to stick with all of its initial promises. No matter what other costs the state might incur, the municipality was going to have to receive at least $2 million a year. In Fairfield, it might have proved desirable or necessary to provide financial benefits to the county and perhaps to adjacent municipalities. This money could not have been drawn from the $2 million without the Board and local supporters risking a serious loss of credibility.

Sixth, Fairfield would have welcomed and used more explicit guidance from the Siting Board regarding how the consideration process should proceed. While the philosophy of the voluntary process was to be flexible and not tell people how or what to think, it would have been helpful to have had a fairly specific course of events to suggest. In new towns considering hosting the facility, such a schedule might be ignored or significantly altered, but it would at least offer a template from which to work.

The seventh lesson was that the role of the Office of Dispute Settlement was highly valued. While it would have been easy to view Fran Snyder as the eighth member of the Siting Board staff, Fairfield officials seemed to find her exclusive focus on community participation and process uniquely valuable to them. As a result, Mayor Thomas did not invite Board staff to the strategy sessions she held in her kitchen, but she did include Fran.

Eighth, groups from outside Fairfield and Cumberland County played only a minimal role in the local deliberations. Diane D'Arrigo of the Nuclear Information Resource Service was occasionally quoted in the newspaper, and the two leaders of New Jersey's fledgling group CHORD wrote a long letter to the *Bridgeton Evening News*, but their comments did not seem to have much impact. This lack of outside influence was surprising, given the experience of communities in other states in which disposal facilities have been proposed, and even considering New Jersey's experience in Roosevelt. It may have been because Fairfield is outside major media markets, so the three-month process was not picked up by the *Associated Press* or reported in the *Philadelphia Inquirer, The Times of Trenton* or any other papers likely to be read outside of New Jersey's southernmost counties.

Also, at least in 1996, the Internet had not yet made many inroads into Fairfield. No one mentioned information or contacts they had found on the web. In fact, more houses in Fairfield seemed to lack telephones than to have computers.

Ninth, some local environmental activists and groups were helpful and openly supportive of the process. Having Cindy and Ernie Zirkle as strong, confident, and outspoken leaders was a significant help, and they were not alone. Members of the township Environmental Commission and other residents who identified themselves as environmentalists were at least open to considering volunteering, and several organizations resisted the readily available opportunity to jump on a bandwagon of opposition. This environmentalist support was very important. It prevented the press, and also in this case the county, from characterizing the debate as good against evil.

Tenth, the institutions that generate waste played no role. While the nearby nuclear power plants are New Jersey's largest producer of low-level radioactive waste, no one connected with its operator, Public Service Electric & Gas, spoke at a public meeting, wrote a letter to a paper, or otherwise contributed to the discussion. The same was true for the three other waste generators in the area.

The eleventh finding was that the press provided accurate, informative, and extensive coverage. This replicated the Board's experience in other parts of the state. In Fairfield, the dedication and skill of one reporter, Miles Jackson of the *Bridgeton Evening News*, made a great difference.

Twelfth, the end of the process took the Board completely by surprise. As had happened in Alloway, we were unaware of the extent to which local opposition was building and were unprepared to respond to it. The lesson seemed to be that local supporters needed to be every bit as dogged as opponents and attend every municipal meeting during which the subject could possibly arise.

Finally, Fairfield demonstrated that the Board had a message to which some people would listen. There were clearly residents who viewed the video and read the literature, attended meetings and talked with their neighbors, and concluded that the disposal facility might be something they could live with.

To come away from a town knowing what we would do differently if given the opportunity to do it again felt like a great accomplishment. The Board members and staff now believed more than ever that the voluntary siting plan was the right approach and that, with some adjustments, it would enable New Jersey to solve its low-level radioactive waste disposal problem.

CHAPTER 21

Falling Dominos

W hile Fairfield was the Siting Board's major hope and focus during 1996, people in other towns were also assessing whether or not they should volunteer to host the disposal facility. We now knew to expect less of these inquiries than we had a year earlier. Hackettstown in the north, and Oldmans and Pine Hill in the south were among the municipalities that became the subject of thin folders in our filing cabinets.

In National Park, an oddly named borough of 3,400 people, the *Gloucester County Times* reported that the mayor had suggested that the town consider volunteering to be the host site. The article noted that he raised this idea as he informed the Borough Council that they had just been denied $100,000 in state aid, and that the municipal budget they needed to approve totaled just $1.4 million.

The headline announcing this possibility was:

▬▬▬▬▬▬

Hot temptation — Radioactive waste could bail out National Park

▬▬▬▬▬▬

While the front-page story produced no local outcry or opposition, the area's high water table compelled the Board to tell officials that their town was ineligible.

A professor at Rutgers University told us that one of his students was engaged in promising conversations with officials in his hometown. Periodically, the professor would ask us to send him multiple copies of the Board's literature and videotape that he could pass on to the student, who would then pass them on to the local officials, perhaps in plain wrappers during the dead of the night.

In two other towns, interesting discussions that seemed to have more potential occurred concurrent with Fairfield's deliberations. The Board learned about one of them from a newspaper article; the other never received any public attention.

Commercial Township

The first was Commercial Township, also located in Cumberland County, just two towns down Delaware Bay from Fairfield. The same page on which the *Bridgeton Evening News* first reported the Fairfield Township Committee's vote to enter the voluntary siting process had included another article headlined:

COMMERCIAL CHECKING WASTE SITE

This story reported that the Commercial Township Committee had voted "to engage in an 'exchange of information' with the Low Level Radioactive Waste Siting Board." It explained that the Township Committee had asked its Environmental Commission to look into the issue, and that the Commission had recommended that the town proceed. "Based on a review of the technical considerations and requirements," the Commission had written in its short report, "it appears that Commercial Township might have some locations that could be suitable for the facility described in the literature."

The article appearing on May 24, 1996, quoted a member of the Township

Commercial Township
Cumberland County – 1996

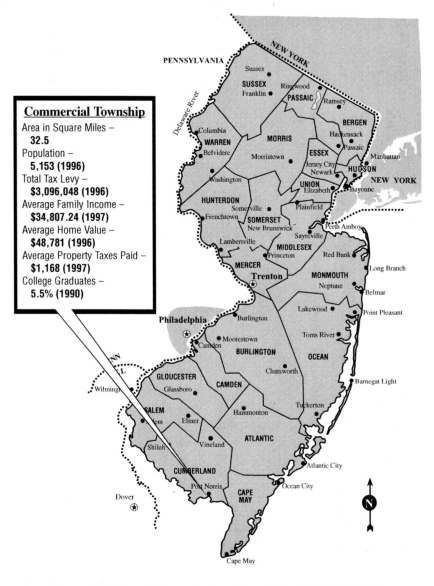

Commercial Township
Area in Square Miles –
32.5
Population –
5,153 (1996)
Total Tax Levy –
$3,096,048 (1996)
Average Family Income –
$34,807.24 (1997)
Average Home Value –
$48,781 (1996)
Average Property Taxes Paid –
$1,168 (1997)
College Graduates –
5.5% (1990)

Sources: N.J. Department of Treasury, New Jersey Almanac, NJ Legislative District Databook

Committee who said: "I think we should look into it. We can always say 'no.'" The reporter concluded the story by writing, "With $2 million a year for 50 years at stake, the committee agreed it couldn't hurt."

Unlike Fairfield's mayor, however, Commercial officials showed no interest in talking with anyone from the Siting Board for several months. Their mayor and I had our first substantive conversation when I called in response to a July 26 front page article headlined:

COMMERCIAL TWP. COOL TO LOW-LEVEL WASTE

The article began, "Mayor George Garrison says the township has received no information from the state about the possibility of a low-level radioactive waste facility here, but he doubts that there is a suitable site available."

After we talked, the mayor asked the Board to review available maps to determine if there might be acceptable sites. He followed up with a short, somewhat indirect request for assistance: "The Commercial Township Environmental Commission has recommended to the Township Committee that further investigation take place to see if the Township of Commercial has a site that would meet the criteria for a low-level radioactive waste disposal."

Karl Muessig of the Geological Survey did find potentially acceptable sites, but the mayor soon called to say that the county's opposition in Fairfield had discouraged them from proceeding further at this time. On August 29, just nine days after the Fairfield Township Committee voted to stop, Commercial followed suit. Mayor Garrison's letter said, in part:

As you are aware, it was upon recommendation of the Environmental Commission that we requested the state to look in our direction as a nominee, but upon further investigation by the Environmental Commission they altered their original findings and submitted a written statement to the Governing Body that Commercial Township withdraw. Also, the Committee had discussed the proposal of an application for a Low-Level Radioactive Disposal Facility in our community and this was not well received by the public in attendance... Thank you for your consideration in this matter.

A Town With No Name

The other town the Board worked with during the summer of 1996 chose a different approach. This town, about an hour's drive from Fairfield and in a different county, had the largest population of any municipality to seriously consider hosting the disposal facility. The impact of significant growth and rising tax rates in recent years had become major local issues in this town.

As soon as the voluntary siting plan was adopted in February 1995, the town's consulting planner, mayor, and a few other officials had started considering the revenue a disposal facility would bring. Although $2 million a year would have been a less dramatic addition to the annual budget here than it would have been in Fairfield and other smaller towns, the group felt it could help them address a number of problems, and possibly be an asset.

In July 1996, staff from the Siting Board and the Geological Survey met with the town's planner and toured the specific area they had in mind, a closed landfill owned by the municipality. It seemed potentially suitable. The analysis and monitoring would be more complex, but a preliminary review of records held by the town and the Department of Environmental Protection indicated that the landfill had never accepted any radioactive materials. This meant that identifying a leak from the disposal facility here might be no more difficult than at any other location.

Early that evening, we met in a conference room in the town hall with the mayor and several other local officials. The next day, the planner wrote to the town administrator, confirming that his firm could "begin immediately to assemble information to evaluate the potential for a waste disposal facility in the township." Over a three-month period, he would "attend meetings, keep minutes, perform research and coordinate activities on behalf of the Township."

For this work, the planner would charge $7,500. It was understood that he would be paid by the municipality, since the Siting Board would not have made such a payment without making it public and the town officials did not want anyone else to know about their exploration until they learned enough to decide whether this was a viable proposal they could recommend. We did not learn how the town's payment to the planner would be kept secret. Perhaps they assumed it would be such a small expense in the public record that no one would be likely to notice.

On August 21, the planner called to say that town officials were ready to sign his small contract, but just wanted to make sure Fairfield was not already so far ahead that they didn't really have a chance. I explained wistfully that Fairfield had bowed out the night before.

About the same time, the mayor was driving to a meeting that happened to be taking place in Cumberland County. Listening to local news on the car radio, the mayor heard analysis suggesting that a major factor in the Fairfield debate was that the town and county governments were controlled by opposite parties. The mayor, who represented a town with a similar political dynamic, called me from the car to withdraw from the siting process.

Thus, Fairfield, the Board's greatest success story to date, led directly to the withdrawal of two other potentially interested towns. Both Commercial Township and the town with no name were dominos knocked over when Fairfield said "waste not, want not."

CHAPTER 22

The Poll

W hile Fairfield's failure led two towns to lose interest, we had hoped that the relative success of the process would stimulate others. After all, there had been three months of interesting, civil public discussion with no hostile shouting matches or name-calling. Many residents who met with Siting Board representatives or reviewed the material seemed to have decided that the disposal facility could be safe and beneficial, and the end had apparently occurred largely because of a local personality conflict other towns could seek to avoid.

Ironically, however, the lack of statewide publicity that may have helped prolong the Fairfield process also kept its relative success from being widely recognized. Just as the Board and staff had created a support statement by doctors, scientists, and educators to try to demonstrate the general professional consensus regarding safety, we needed a way to show people throughout New Jersey that most of the residents of Fairfield, after three months of public discussion, had been open to hosting the facility. We needed to demonstrate that the people who had packed the final Township Committee meeting had not represented the majority, or at least not an overwhelming majority, of the local population. We decided it was time to take a poll.

Advocates of low-level radioactive waste disposal, and of risk communication in general, have initiated many polls over the years. The results always seem to confirm that the general public is in surprising agreement with the sponsoring organization. In this case, that means that the public would rather have one well-designed disposal facility than have waste left on site in many scattered locations, and that people would prefer a decision-making process that included them from the beginning to one they felt took them and their concerns for granted.

Ever since the voluntary siting process had been on the horizon, the Board had spurned suggestions for polls, feeling that the results would change neither the opinions of the general public nor the approach of the Board and its supporters. But proof that the siting process and perhaps eventually the disposal facility itself could have enjoyed continued popular support in Fairfield would attract attention and perhaps lead to opening some doors. There was even a possibility that the poll could help revive the process in Fairfield, perhaps after the 1997 election.

The Eagleton Poll is the most well-known public opinion survey in New Jersey. It is run by the Center for Public Interest Polling at Rutgers University's Eagleton Institute of Politics. The surveys often receive extensive press coverage and are highly regarded. In addition, Eagleton offered the same advantage that had helped attract the Siting Board to *New Jersey Network* to produce its video and to the Geological Survey to perform preliminary site analyses. Because it is part of state government — in this case, a state university — the Board was able to negotiate a contract at a good price without enduring a competitive bidding process that would have delayed the project for months.[52]

Over several evenings in November, interviewers hired by Eagleton had telephone conversations with 348 Fairfield residents. They asked a set of questions that the director of the poll, Janice Ballou, had developed with the Siting Board staff.

Unfortunately, the results were not what we had expected. Although 43 percent agreed that the Township Committee had been right to begin the exploration process, 71 percent supported the study's termination and only 27 percent said it should have continued. While occasional political candidates are able to overcome early polling deficits of this magnitude, it is a rare event. It is hard to imagine local officials becoming motivated by these numbers to begin a siting process in their towns.

The other results of the poll, while interesting, were also not encouraging. To begin with, discussion of low-level radioactive waste really had dominated life in Cumberland County all summer. Only 19 percent of the respondents said they didn't know much about it, while 42 percent knew something about it, and 40 percent knew a great deal.

Forty-eight percent of the people said they had first heard about the disposal facility study from a newspaper, and 28 percent mentioned word of mouth. At the same time, 80 percent received some information by word of mouth, 72 percent from newspaper articles, 53 percent from newspaper editorials, 41 percent from letters to the editor, 36 percent from the mayor's letter to every household, 27 percent from the township meetings, 25 percent from seeing the Siting Board display at the municipal building or firehouse, 19 percent from the radio, 18 percent from attending meetings with the Siting Board, and 15 percent from viewing the Board's video.

It was discouraging to learn that 35 percent of the respondents did not know the town could receive $2,000,000 a year. Also, 68 percent said it was not at all believable or not very believable that the disposal facility could be safe. Only 31 percent believed it was very possible or somewhat possible that any site with radioactive materials could be safe.

The reasons people most often cited for opposing the disposal facility were:

- The state would not keep its promise to keep out other types of waste, such as high-level nuclear waste (79%).
- The facility would have waste from nuclear power plants (77%).
- The state would not keep its promise to keep out waste from other states (74%).
- The facility would not be safe (71%).
- The facility would hurt property values (70%).
- The facility would hurt Cumberland County's ecotourism industry (50%).
- The township would not spend the $2,000,000 wisely (42%).
- The state would not keep its promise to provide that amount of benefits (40%).

The poll itself became a minor local controversy. Some of the Fairfield residents contacted by the poll-takers called Cumberland County officials. These officials then worried publicly that the poll might mean the Siting Board was not really abandoning Fairfield. We responded that the results would be made available two months later to anyone who asked for them. But two months later, no one remembered, or cared enough, to make the request.

CHAPTER 23

More Planning for the Election

O nce the Siting Board decided to shift to a voluntary process, the quest to find a municipality that would eventually agree to host New Jersey's disposal facility for low-level radioactive waste always occupied center stage. But the Board simultaneously continued to prepare for the steps that would be necessary once the process succeeded and representatives of a municipality said, "OK, let's go."

One part of this effort, begun in 1995, was working with Foster Wheeler to fully articulate the site analysis and characterization processes. Other elements that came to the fore in 1996 included designing the facility, developing a financing plan, and continuing to examine the New Jersey Siting Act to determine the changes needed to fully implement the voluntary process.

Facility Design

The cover of the Siting Board's *Voluntary Siting Plan* includes a sketch of a disposal facility. A large piece of land has been cleared. About half the site is

New Jersey's Voluntary Plan
for Siting a
Low-Level Radioactive Waste Disposal Facility

New Jersey Low-Level Radioactive Waste Disposal Facility Siting Board
New Jersey Radioactive Waste Advisory Committee

STORAGE FACILITY ENVISIONED: The Siting Board's vision of what New Jersey's low-level radioactive waste storage facility might look like was pictured on the cover of its Voluntary Siting Plan.

vacant and the other half is occupied by four structures that look like extra-long closed-in greenhouses or self-storage lockers. One of them is in the process of receiving waste, via a crane, from a truck nearby.

A similar, more detailed vision of the New Jersey disposal facility was provided by the three-dimensional model Board staff carried to most public meetings. This gave a better sense of the layout and scope of the entire site, and its proximity to surrounding farms and houses. A viewer could see trucks stopped at an inspection station near the entrance.

The Board's videotape also provided some images, but we tried to make clear that New Jersey's disposal facility had not yet been designed. We wrote in the *Introduction to the Issues* booklet, for example: "The design of the facility, including whether it will be above or below ground, will be determined after a thorough evaluation of local geology and hydrology, and the preferences of the host community."

If, however, the Board followed this to extremes and took no steps to design the facility until a site was "thoroughly evaluated," many more months could be added to the time that would elapse before the Nuclear Regulatory Commission (NRC) license application would be submitted and before the facility could open. It seemed clear that the design process should not wait.

Moving ahead also was consistent with the comments we were hearing both in towns in which volunteering was considered and from more general meetings and conversations. People wanted to know what this thing would look like. Being told that the answer depended in part on what they and their neighbors wanted did not inspire their confidence.

While such community participation has led to wonderful, imaginative configurations for playgrounds, the concept did not seem to translate as well for creating structures needed to protect the environment and multiple generations of people from potentially dangerous exposures to radiation. One of the lessons we had learned in Fairfield Township was that this is one of the areas in which the prevailing theories about risk communication may be in need of revision.

The general consensus is that when a controversial idea is unveiled, it should be far from fully formed. People will respond more openly if they feel they are being admitted to the process early and that they are not being presented with a take-it-or-leave-it proposition.

The flip side, however, is that greater fears can be raised by a concept if it is too sketchy. It is easy to imagine the worst, or accept the scariest rumor, if a proposal is not sufficiently detailed to dispel it. Moreover, the prevailing distrust of government does not lead people to have confidence that even the most objectionable options will automatically be eliminated during the design process.

As the voluntary process proceeded, lessening the uncertainties in the disposal facility design came to seem increasingly important. By the end of the Fairfield episode, we were encouraging the initiators in other towns to spell out which locations should be considered for the disposal facility and how the money received in benefits would be used. For our part, we were trying to develop a better response when asked what the facility would look like.

From the beginning, the siting process in New Jersey had been structured so that the Siting Board would use competitive bidding to select one private company that would design, build, and operate the facility. That company would rely on the site characterization work performed by Foster Wheeler, which was prohibited by its contract with the Board from bidding on the subsequent work.

Late in 1995, the Board had hired Edward Truskowski as a seventh member of the staff and asked him to prepare the Request For Proposals (RFP). Truskowski had worked on radioactive waste disposal issues for a number of private companies and, like Jeanette Eng, was a health physicist. We now had two scientists in the office.

Writing the RFP proved to be, as former New Mexico Governor Bruce King once said of something entirely different, like opening a box of Pandoras. It required articulating in precise detail what would happen between the day a municipality signed an agreement with the Siting Board and the day about 60 years later when the facility would be full and entering a period of long-term monitoring.

Trying to outline a work plan for that period identified a host of issues and questions that neither the Legislature nor the Siting Board had considered. Most of them dealt with how the facility would be financed both while it was being constructed and during its operation.

Because of the uncertainty and duration of the approval process and the bonds that must be posted against any type of problem, huge amounts of money are required to participate in this business. The work is considered very specialized, and no more than five or six corporations were expected to consider submitting bids. We also had heard reasons why all but one of them might choose not to participate at all. The most likely contestant would be Chem-Nuclear Systems, which not only operated the Barnwell facility, but also had signed contracts to build facilities in Illinois, North Carolina, and Pennsylvania. That, in itself, raised the question of how the Board should evaluate its bid. Wouldn't Chem-Nuclear have conflicting interests if it decided, for example, that the market would not support facilities in both Pennsylvania and New Jersey?

As we struggled with the RFP, there were few places to turn to for help. Representatives of the two entities of state government that issued contracts, the Division of Purchase and Property and the Division of Building and Construction, each made clear that they felt the other one was best suited to assist with a complicated 50-year contract that would involve the state, a municipality, hazardous materials, and controversy.

The two agencies did advise us of a way we might solicit advice from the potential bidders. As long as the Board advertised widely and asked the same questions of everyone, we could seek information without violating the integrity of the competitive bidding process. Following that suggestion, the Board issued an 18-page Request For Information in July. We mailed copies to the companies we thought might be interested and placed advertisements in several national trade journals.

We received two replies. One, from a consortium of US Ecology and Bechtel National, provided five pages of reflections and suggestions. The second, not surprisingly, was from Chem-Nuclear Systems. Chem-Nuclear said it would "continue to assist you and your staff as the New Jersey Siting Board continues their efforts to successfully site a LLRW Disposal Facility in the State of New Jersey." The second paragraph of the reply, however, stated: "It is our position that our responses to these questions will constitute an integral part of Chem-Nuclear's future proposal and, as such, a premature release of this information may have an adverse effect on our competitive position."

These responses provided two important pieces of information. First, the prevailing wisdom had been proven slightly suspect, because it had not included US Ecology and Bechtel, either separately or together, as likely bidders. The second was that the Board still needed to figure out how this project could be financed.

Financing

The New Jersey Siting Act fashioned by the Legislature in 1987 appropriated $500,000 for its implementation. It also devised a formula and procedures for the distribution of benefits to the host community, but it offered little direction about the financing of the disposal facility.

By 1991, the Board had received $2.7 million from annual appropriations in the state budget. This was more than sufficient to pay the Board's start-up expenses, including engaging consultants to conduct the statewide screening for acceptable sites and to design the public outreach program. Each year's appropriation, however, was preceded by suspense as the Board lobbied and waited to see how its little agency would fare in the debate and bartering that accompany the passage of the state's annual budget.

The uncertainty of the annual budget process is a challenge that accompanies most government programs and many in the private sector as well. Vice President Al Gore's 1993 *Report on Reinventing Government* notes that "annual budgets consume an enormous amount of management time" and create "an enormous amount of busywork." He noted that 20 states have adopted two-year budgets and recommended that Congress do the same.[53]

Moving to biennial budgeting in New Jersey would, as Gore suggested, give the governor and Legislature "much more time to evaluate programs and develop longer-term plans." It would create an opportunity for great improvements in the operation and productivity of many government programs. It would not, however, have made a significant difference in helping the Siting Board make the multi-year financial arrangements necessary to plan and build a disposal facility expected to operate for 50 years.

The constitutional inability of legislatures to make binding long-term financial commitments is one of the reasons why many have created relatively independent authorities to oversee major public construction projects. These agencies generally have their own bidding, contracting, and bonding powers. The Siting Board had been given the power to award contracts, but initially it had no authority to raise funds.

In June 1991, the Legislature amended the Siting Act to enable the Board to collect fees from generators of low-level radioactive waste in New Jersey "sufficient to meet all expenses incurred by the board and the department in implementing the provisions of this act." This change had been sought by the Board and was supported by the waste generators, including the nuclear utilities and the Chemical Industry Council. It was touted at the time as a rare instance of an industry asking that it be taxed.

Within a year, the Board had adopted fee assessment regulations and used them to assess the generators $12.7 million. This was the amount the Board estimated it would need to complete the statewide screening process, characterize several sites, and designate one of them for preparation of the application for an NRC license.

Once the license was received, the Board assumed that all subsequent costs would be borne by a combination of additional fee assessments and by the private firm it would hire to build and operate the facility. In 1991, the Board did not feel it needed to resolve how those costs would be divided. Ultimately, the generators would pay them, either through additional assessments or through disposal fees charged by the facility operator.

Now, in order to write a Request For Proposals from potential facility operators, this issue had to be addressed. Any bidder would need to know what up-front costs it would be expected to incur.

The situation had become much more complicated over the years since the Siting Act was passed. In 1985, California had chosen US Ecology, Inc. as its "license designee." Its agreement required the company to shoulder all the planning and construction costs for the proposed disposal facility in Ward Valley. As the planning and regulatory process for that facility had entered its second decade, US Ecology had suffered financially to such an extent that its bankruptcy was frequently rumored. Any company bidding in New Jersey was going to expect the state, through either its treasury or its generators, to demand much less risk than US Ecology had accepted in California.

At the same time, New Jersey generators were no longer nearly as quick to support calls for substantial expenditures. Watching the events in California and other states unfold had made them more wary. Many wondered if a new facility would ever be built. In addition, at least for the moment they now had two out-of-state options that made them feel less desperate and less willing to support a facility at any price.

Not only had Barnwell reopened to most of the nation, but a private company called Envirocare was now accepting some high volume/low radioactivity waste from around the country at a huge site it owned in the West Desert, 80 miles west of Salt Lake City, Utah. Its new facility had the physical capacity to accept all of the nation's waste for many decades, but it was unclear whether the State of Utah had the financial need and political will to periodically renew and perhaps expand Envirocare's operating license.

In the course of preparing the first fee assessment in 1992, the Board had estimated the entire cost of the facility, from planning through construction and operation to closure, would be $80 million. Late in 1996, the Board decided it needed to be able to work with an updated and more precise cost figure.

The Board arranged for the National Low-Level Waste Management Program to analyze the likely cost. This program, funded by the U.S. Department of Energy, is part of the Idaho National Engineering and Environmental Laboratory. Through competitive bidding every few years, the DOE hires a private firm to run the lab. Lockheed Martin Idaho Technologies Company was the current operator.[54]

While its pedigree is complex, the National Low-Level Waste Management Program was refreshingly helpful and informal. It functions as an agricultural extension service for low-level radioactive waste. A small staff of extremely knowledgeable people is available upon request to provide free assistance to states and regional compacts. In this case, the national program was able to begin a comprehensive financial analysis within weeks of the Board's request and to complete it less than three months later. Their work was not only fast, but also directly responsive to the questions the Siting Board asked.

The national program was able to start so quickly because its staff and consultants had performed a similar analysis for Massachusetts in 1994. That work, in turn, had been based on a preliminary feasibility design prepared for the Illinois Low-Level Radioactive Waste Disposal Facility. Two of the people who had worked on the Massachusetts study, Karl L. Sorman and Paul R. Smith, were assigned to work with New Jersey.

The Board requested cost estimates for a "small" and a "large" disposal facility. Small was defined as having a capacity of 1,000,000 cubic feet, and large as 3,000,000 cubic feet. The latter was the Board's estimate of the amount of waste New Jersey would generate in 50 years if all the contaminated parts of the state's four nuclear power plants were taken to the disposal facility when they were shut down. The lower figure assumed that arrangements were made for the nuclear plants to delay decommissioning for at least 50 years.

The Board also asked that two different time frames for the operating life of the disposal facility be examined. The Board planned to design the facility to accept the waste it projected New Jersey would generate over 50 years. Our experience in Fairfield, however, had made at least some of us more open to the possibility that the Board and the host community might end up agreeing to accept waste from Connecticut as well. In that case, we estimated a facility of the planned size could be filled in 30 years.

The report the national program prepared for the board, *Preliminary Cost Estimates for Building, Operating, and Closing a Low-Level Radioactive*

Waste Disposal Facility in the State of New Jersey, estimated that the total cost, from planning in the late 1980s through monitoring the closed facility well into the 21st Century, would be between $421 million and $1.021 billion. The lower figure was for the smaller facility accepting 750,000 cubic feet of waste over 30 years; the larger was for a facility accepting 3,000,000 cubic feet over 50 years.

For the Board and for the waste generators, two other numbers were more important: the amount of money that would need to be spent before the facility could begin operating, and the fees a facility operator would likely have to charge to recover costs and make a reasonable profit.

The figures for the pre-operation phase were not too far removed from the $80 million estimate the Board had been using. The National Low-Level Waste Management Program projected that by the time a 1,000,000-cubic-foot facility opened in New Jersey, $83,610,000 would have been spent by some combination of the Legislature, waste generators and the selected facility operator. To open a facility large enough to accept the decontamination and decommissioning waste from the nuclear plants would cost $112,535,000.

The other figure of note was the fee generators could expect to pay to dispose of waste. The invoice price to customers using the Barnwell facility was based on a variety of factors, including a surcharge for the level of radioactivity and special handling charges. But the amount charged per cubic foot of waste was the part of the cost the New Jersey Siting Board relied upon to compare options.

The National Low-Level Waste Management Program estimated charges ranging from $324 per cubic foot for the larger facility operating for 30 years to $562 for the smaller facility also open for 30 years. These figures, higher than the $315 generators were then paying to use Barnwell, would have been inconceivable when Barnwell opened in 1971. At that time, it charged 95 cents per cubic foot; nine years later, the cost escalated to $6 per cubic foot. Even in 1990, four years after South Carolina had first added a surcharge of at least $40 per cubic foot, the fee had jumped only to $79. But compared to the fees projected for the other new facilities under consideration around the country, these numbers were not out of line.

The preliminary cost estimates served as a sanity check. It showed the Board and the waste generators that a facility built just for New Jersey waste might be economically competitive. It could be feasible even though the costs had multiplied and the waste volume had decreased since 1991, when the Legislature had last revisited the Siting Act.

New Legislation

At the end of 1995, the Siting Board and staff in the Governor's Office had anticipated that amendments to the Siting Act could be debated and perhaps enacted within six months. The amendments would add explicit recognition of the voluntary siting process to the act and facilitate its implementation.

The changes were relatively minor and would further the Legislature's goals of choosing a site for a disposal facility with a process that provided enhanced public involvement. Although it was possible that objections might be raised by the few people who actively campaigned against having a disposal facility anywhere, our primary fear was that indifference would lead to inaction.

Instead, what happened in 1996 was that several representatives of waste generators told the Board they would not support the proposed amendments in their present form. In particular, the Chemical Industry Council, which represents several pharmaceutical companies, asked for Board consideration of a set of modifications it proposed. These made it clear that the Chemical Industry Council and ostensibly the companies it represented had become uncomfortable with the open-ended siting process and perhaps with the mission of trying to build a disposal facility in New Jersey.

Their concern was that New Jersey might follow North Carolina, California and other states, which had by now spent tens of millions of dollars with no end, or opening date, in sight. As a result, the industry group wanted the following sentence added to the Siting Act: "In order to make assessments of more than $1,000,000, an affirmative vote of at least two of the three authorized membership of the board who represent industries that generate low-level radioactive waste shall be required."

The Chemical Industry Council had also become worried about the Siting Board's commitment to fund information-gathering by potentially interested local governments. Its fear was that a local request might allocate money to a group with beliefs and policies fundamentally opposed to the Board's mission. The Board thought it had addressed this issue by wording the proposed amendment to say that one of its new powers would be "to contract with any person, under such terms and conditions as the board deems appropriate, in order to meet the goals and objectives of this act."

The industry group, however, wanted to remove any ambiguity and add that the "person" had to be "engaged in activities that advance the siting of a low-level radioactive waste disposal facility." These revisions and most of the others it suggested would not have changed the Siting Board's behavior. The

Board was not going to undertake any major action by a six-to-five vote. It generally operated by consensus and spent many anguished hours when even one member was not supportive of a proposed action.

Similarly, the Board was not going to award funds to groups with a clear political agenda opposing, or not believing in, the safe disposal of low-level radioactive waste. Municipal officials who were considering volunteering, as well as the Board, were no more likely to want to make a grant to Greenpeace or the Nuclear Information Resource Service than they were to the Nuclear Energy Institute or the Chemical Industry Council itself.

The Chemical Industry Council responded that its concerns did not arise from the current members of the Siting Board, whom they praised both publicly and privately. They worried about subsequent appointees who might have different views and operate less collegially.

While incorporating the industry group's requests might have had little substantive impact, it would have made the proposed amendments a magnet for other opposition. If two-thirds of the industry representatives had to support major decisions, others would argue that two of the three representatives of civic and environmental groups should also be in the majority. If limits were placed on the types of organizations to which the Board could make grants, additional limits would be requested by others.

On the other hand, the Board felt that it couldn't seek legislative support for the amendments if there was going to be opposition from the Chemical Industry Council or any other significant group. The only legislator who seemed to have an active interest in the entire subject of radioactive waste disposal was Senate Majority Leader John Bennett, a Monmouth County Republican who had sponsored the Siting Act when he was an assemblyman. He might have been able to move the amendments as one item on a list of minor actions, but he was unlikely to try if the Chemical Industry Council showed up to testify against them.

While it was never clear whether more than a half-dozen of the 100 or so waste generators in New Jersey were even aware of these discussions, the reservations expressed by the Chemical Industry Council derailed the Board's plans to seek legislative changes. Throughout 1996, the Board's legislative subcommittee and Chemical Industry Council representatives met occasionally to seek common ground, but never quite found it. The Board never abandoned the amendments. As 1996 ended, the Board felt they would probably still be necessary, and perhaps the Legislature could consider them in a year or two when a volunteer municipality might be ready to join in supporting them.

CHAPTER 24

O Little Town of
Bethlehem

For several months after Fairfield ended its siting process and Commercial Township followed suit, no towns were publicly considering hosting the disposal facility. The Siting Board continued to hear from people who were still thinking about raising the possibility in their area, and we received occasional new inquiries, but none of the callers seemed to have included more than one or two other people in their explorations.

Then, two days before Thanksgiving, a reporter from the *Hunterdon County Democrat* called to ask for a comment about the previous night's Township Committee meeting in Bethlehem. It seems that the Committee had asked the chair of the town's Environmental Commission to look into the possibility of hosting the disposal facility.

In a front page article the next day, the reporter, Ken Serrano, wrote that, "With property taxes continually rising and municipal budgets under increasing strain ... resistance to such a dump might be ebbing." He quoted the mayor, Walter Baumgarten, as saying approvingly, "This is a big ratable," and

Bethlehem Township
Hunterdon County – 1996

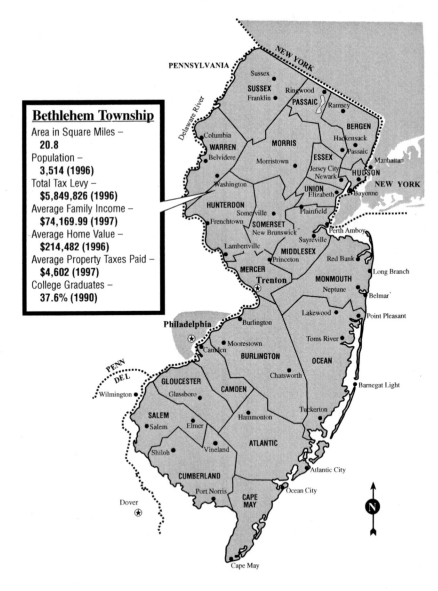

Bethlehem Township

Area in Square Miles –
20.8
Population –
3,514 (1996)
Total Tax Levy –
$5,849,826 (1996)
Average Family Income –
$74,169.99 (1997)
Average Home Value –
$214,482 (1996)
Average Property Taxes Paid –
$4,602 (1997)
College Graduates –
37.6% (1990)

Sources: N.J. Department of Treasury, New Jersey Almanac, NJ Legislative District Databook

210

a resident agreeing that if they hosted the facility, "You could buy up all the farmland in the valley." The article headline was:

$2 MILLION A YEAR HAS TOWN MULLING RADIOACTIVE DUMP

Bethlehem Township had not been on the Board's radar screen or in our files. We quickly learned that it is a rural municipality in Hunterdon County with about 3,200 residents. Once the article appeared, I called the Environmental Commission chairwoman, Roberta Morganstern, who said she was surprised at the positive reception the idea had received at the meeting. Two of the three Township Committee members had been clearly supportive, so the mayor had charged her with gathering more information. I described the Board's publications and videotape, and said I would mail multiple copies of each. The one item I faxed immediately was the support statement from New Jersey doctors, scientists, and educators, hoping that it could help her frame preliminary replies to some of the concerns about safety that people might express to her over the next few days.

Morganstern noted that her interest in this topic had first been raised when she read an article in the Spring 1995 issue of the Association of New Jersey Environmental Commissions (ANJEC) Newsletter. At that time, she had just been appointed to the Bethlehem Environmental Commission and felt too new to suggest volunteering. Earlier this month, however, when Bethlehem voters had approved a referendum for open space funding, she had been emboldened to write to both the Township Committee and Environmental Commission.

When I spoke with Mayor Baumgarten, he was unreservedly enthusiastic. He noted that he worked for a pharmaceutical company and was knowledge-able about radioactive materials. He also said that while Morganstern had talked of devoting the revenue from the disposal facility to open space, that was not necessarily the community consensus. He thought at least a major part of it should go to tax relief.

The second time I spoke to Morganstern, she told me she planned to talk with each member of the Environmental Commission before their next meeting. She said she also now thought the $2 million should be split between open space preservation and tax relief. With surprising speed and ease, the first potentially divisive issue appeared close to resolution.

The article in the *Democrat,* Hunterdon County's dominant weekly paper, had also quoted William Neil, a township resident who worked as assistant director of the Audubon Society. He speculated that the area might be disqualified because it was rich in limestone. He was the other person I called, both because he might have useful information and because he could have a significant impact on public opinion, particularly if he decided to support the Environmental Commission's study.

The following week, Karl Muessig of the Geological Survey reported that limestone with possible caves and sinkholes comprised the northern quarter of the town. This type of topography, known as Karst by geologists, would make it difficult to fully analyze a site because extensive drilling would be required to assure that no closed depressions or underground drainages might underlie the disposal cells. The rest of the town, however, was crystalline rock, which might be feasible for study and ultimately for siting. He suggested we meet with township officials to discuss the maps and sites they had in mind.

On December 12, the Environmental Commission held its first meeting to discuss the disposal facility and voted unanimously to explore it. The *Democrat* appeared that day with the headline:

AG BOARD LIAISON RESIGNS POST OVER WASTE DUMP ISSUE

The article explained that a local attorney had resigned in protest as Bethlehem's representative to the county Agricultural Development Board. It quoted her saying that officials were "panicking the public," and that, "You might as well bring the plague into the township."

The story went on to note, "Meanwhile, the township has passed an early checkpoint in its inquiry" and described the initial review performed by the Geological Survey. An article in the adjacent column was headlined:

WHAT WOULD RADIOACTIVE SITE HOLD?

The sidebar provided clear, accurate, basic information about the waste and the disposal facility. It also listed Hunterdon Medical Center and three other locations in the county where low-level radioactive waste is generated.

Also in this issue of the *Democrat* were the first letters on the subject. The theme was set unambiguously by township resident Hugh Shannon Jr. who wrote:

> *I read with great alarm that Bethlehem Township is looking into the possibility of locating a low-level radioactive waste dump within the township. Instead of mulling this over, dreaming of $2 million for the township coffers annually, the township's Environmental Commission ought to be saying, No way! ...*
>
> *If this happens in Bethlehem Township, and I pray that it doesn't, it will not only affect Bethlehem ... Any community that is on a roadway that this hazardous material is transported over will be affected...*
>
> *Late last year ...the Planning Board entertained the idea of K. Hovnanian building a few thousand homes on farmland that was to be sold for development. We are still tangled up in a Council on Affordable Housing nightmare as a result.*
>
> *Now the Environmental Commission wants to consider putting a radioactive waste dump in the township. What is wrong with these people?*

The following Monday afternoon, Greta Kiernan and I, along with Fran Snyder from the Office of Dispute Settlement, met for the first time with Mayor Baumgarten and Environmental Commission Chair Morganstern. Also attending the meeting in the municipal building were the township clerk, a member of the Planning Board, and the town resident who had spoken up in support at the Township Committee meeting. Bill Neil, the assistant director of the New Jersey Audubon Society, had also been invited with a voice mail message, but was not there.

As the meeting began, *Democrat* reporter Ken Serrano walked in, saying he had received an anonymous phone call informing him of the meeting. Serrano regularly covers Bethlehem and had, therefore, gained custody of this story. After some awkwardness, Baumgarten said Serrano could stay, even though he did not think working meetings like this had to be open to the press.

Despite this accommodation, the Planning Board member, who was also a former mayor, announced that he was leaving because he thought the entire community should have been told of the meeting. He added that he was opposed to meetings that were "by invitation only."

The *Democrat's* article the following Thursday was headlined:

OFFICIALS SEEK RATIONAL TALK ABOUT DUMP

The story was remarkably kind. While he eventually mentioned the anonymous phone call and the one-man walkout, Serrano began by writing:

> *With the controversy just gaining momentum, state and township officials and a township resident cautiously set a course this week for presenting the idea of accepting a low-level radioactive waste disposal facility into the township.*
>
> *The group met Monday afternoon to discuss the thorny issue, which some opponents want to quickly nip. The meeting – not announced to the public or the press – focused in part on alternatives to large, open gatherings, which prove more emotional than informative, state officials said.*

The paper also had letters from three more Bethlehem residents. Louis J. Judice wrote:

> *With towns all over America struggling with the problems of toxic and radioactive waste in their midst, it is appalling to even consider bringing them here. There is no scientist who can possibly claim that such a site is 100 percent safe. And in a township with narrow, winding roads, the possibility of a disastrous accident while transporting material to the site is even more likely. We don't need discussion ... We need to say no – right now – to this outrageous idea.*

The other two letters, however, supported the study. One was from Morganstern, who responded to concerns that the disposal facility would reduce property values. She wrote that if the town had "$1 million a year of tax relief and $1 million a year of open space preservation, then in a relatively short time Bethlehem Township would have very low taxes, very good facilities and its physical beauty and rural environment would remain intact. This would be very good for property values."

The third writer, Styra Eisinger, said:

> *It is really too bad that the mention of a low-level radioactive waste site in Bethlehem Township has brought forth so much automatic anxiety instead of a willingness to look further into the matter before making a decision. When I see comments ...that no one would ever again eat any food from farms in the township if the material is stored here, I can't help but wonder if that person drinks French wine – from a country where 80 percent of the power is generated from nuclear power plants.*

The final item concerning the disposal facility in this issue of the *Democrat* was the Siting Board's first display ad. We had thought of taking out ads in other towns, but the local officials with whom we talked had felt they would be premature. This time, we decided unilaterally, and just informed the mayor and Environmental Commission chair. The half-page ad began:

> *Last week, the* Democrat *reported that Bethlehem Township may consider whether to volunteer to host the disposal facility New Jersey needs for its low-level radioactive waste.*
>
> *You may have questions and concerns about whether such a facility would be safe and beneficial for the area. The New Jersey Siting Board will be happy to speak with you and send you literature and a videotape ...*
>
> *You might be interested to learn that many New Jersey Doctors, Scientists and Educators Agree that, "A disposal facility for low-level radioactive waste would be a safe neighbor."*

This was followed by the rest of the text from the support statement by doctors, scientists, and educators, and the names and affiliations of the approximately 200 current signers.

The goal of the advertisement was to place this information into some local discussions before too many positions hardened. We hoped that awareness of the views and credentials of the signers would give people who might be open to the study support, courage, and information they could share with others.

Another reason to place the ad immediately was that it would appear in the issue of the paper many people would be reading as the Township Committee met that Thursday evening. This was the Township Committee's first meeting since opposition to the disposal facility had begun to surface and we worried that the governing body would be beseeched, in the words of one of the letter writers, to "say no right now to this outrageous idea."

Morganstern planned to present the Township Committee with a letter from the Environmental Commission which recommended that the study proceed, and that the next steps include two open house forums for township residents and the formation of a study group. She also had commitments from several supporters of the study to attend the meeting. In addition, she was prepared to respond to any petition objecting to the study by requesting time to circulate one in support.

This preparation, and a surprisingly small turnout, prevented the meeting from being dominated by the opposition. About 25 people attended, only two of whom spoke against the study. One was the president of the Builders Association of Hunterdon and Warren counties, who said the town could gain just as much revenue by building the 700 proposed new homes, and that that would create a better community.

The Township Committee voted to set up the two open houses, take out legal ads for volunteers for the study group, and invite the Siting Board to address a subsequent meeting. It also authorized the township engineer to meet with representatives of the Siting Board to discuss the suitability of sites in Bethlehem for the facility.

In an unrelated matter, one member of the Township Committee who had just been reelected in November announced that he was resigning. He worked as an engineer and had been informed by the township attorney that he would face a conflict of interest if he participated in any projects that required inspection by township officials. Rather than forsake a percentage of his income, he reluctantly quit.

This Township Committee meeting was reported not only by the *Democrat*, but also by a daily paper, the *Express-Times* of nearby Easton, Pennsylvania. Its first headline on the disposal facility, appearing five days before Christmas, was:

NUCLEAR DUMP'S BOARD COMING TO BETHLEHEM TWP

After a holiday lull, organized discussion resumed. When one of the area's environmental groups, the Musconetcong Watershed Association, held its reg-

Stop the Dump Now!

**Call Your Township Officials TODAY and
Tell Them to STOP THE DUMP IMMEDIATELY.**

**Mayor Frank Kehle
Walter Baumgarten
Committeeman
Roberta Morganstern
E.C. Chairperson**

**Call The State Radioactive Dump Site Board
And Tell Them We Don't Want a Nuclear Dump:
John Weingart, Director:**

**Spread The Word To Your Neighbors and Residents
In Surrounding Towns. Remember, Radioactive Waste
Will Be Trucked Through Union, Clinton, Hampton!
Groundwater Leaks Could Threaten Them Too.**

Write The Hunterdon Democrat or Courier News.

Call Radio Station NJ 101.5

Put Up A Sign In Your Driveway.

OPPOSITION FLYERS: Anonymous leaflets like this one from Bethlehem Township often were the first signs of coalescing local opposition.

ular monthly meeting. Bill Neil from the Audubon Society and Beverly Graczyk, the woman who had resigned in protest from the county Agricultural Development Board, led a discussion concerning the disposal facility. The group's minutes reported that they were both "outraged by the action of the Environmental Commission. They believe the information being disseminated is one-sided, favoring the safety of the disposal site." Later, the minutes note, "after a lengthy discussion, the members decided to ask the Siting Board to present information at their February meeting."

On Tuesday, Morganstern and a member of the township engineer's firm met with Board staff in Trenton. With Karl Muessig of the Geological Survey and staff from Foster Wheeler, we reviewed maps and eventually concluded that six areas of the township merited further review as possible locations for the disposal facility.

On Wednesday evening, Jeanette Eng, Greta Kiernan and I attended the Environmental Commission's regular meeting. Although our presence had been publicized in advance, only about 30 people attended. At least half of them, though, were prepared with tough, generally hostile, and sometimes sarcastic questions. Eng's quiet expertise and competence, however, seemed to win at least grudging respect from almost all of them, and the meeting turned into a focused and civil two-hour introductory seminar on low-level radioactive waste.

The Township Committee was scheduled to meet the following night. While we had been pleasantly surprised by the tone and small turnouts at each of the previous meetings in the town, opposition was clearly becoming more visible and perhaps growing as well.

Hugh Shannon, one of the first people to write a letter to the *Democrat*, had started a web page on the subject. In early 1996, web pages were a relatively new phenomenon, but the contrast with Fairfield was impressive. There, six months earlier, we had difficulty contacting some people because they could not afford telephone service. While this web page made an attempt to provide information from all viewpoints and did include all the material the Siting Board submitted, Shannon also made clear that he was opposed to having the facility in Bethlehem.

Louis Judice, another early letter writer, now wrote a much longer letter to the *Democrat* that began, "A futile desire to thwart development has led Bethlehem Township into what I believe is its most misguided misadventure: consideration of hosting a nuclear waste storage facility."

Describing the town as "a strange mixture of ex-urban professionals and longtime residents, some of whom continue to pursue farming and other agricultural work," he addressed the arguments of "some well-meaning individuals" that the disposal facility would enable Bethlehem to avoid the 700 proposed new homes.

Judice wrote, "There are just a few problems with this scenario. No one wants to buy a home in a town with a radioactive dump." While acknowledging that "the state has a plan to protect property values," he worried about the impact during the "two-year protracted debate over whether to put the dump here ... Two years is plenty of time to discuss the issue and should also provide enough time for the word to get out to the real estate

GUARANTEE:

If New Jersey's disposal facility for low-level radioactive waste is built in your community, and if you attempt to sell your property and find that its value has been reduced by either the operation or the perception of the facility . . .

. . . the Siting Board will arrange for the purchase of your property *or* will pay you the difference between what your property sold for and what certified real estate appraisers calculate it would have been worth had the disposal facility not been sited in your community.

The New Jersey Low-Level Radioactive Waste Disposal Facility Siting Board

PROPERTY VALUE PROMISE: To alleviate fears that a low-level radioactive waste disposal facility would lower property values, the Siting Board promised to compensate homeowners for any loss of resale value.

buying public that there's a serious problem brewing in the Musconetcong River valley."

In conclusion, Judice asked, "Why are we so afraid of the inevitable development that eventually will occur that we would actually hold ourselves hostage by accepting poisons in our midst?" At the same time, he acknowledged that the facility might be safe: "Technically, you can probably prove to some level of statistical confidence that a low-level radioactive waste dump is safe to a degree. But undeniably, having any nuclear waste in your neighborhood is still more nuclear waste than most of us would prefer to live near with our children."

Beverly Graczyk, after resigning from the county Agricultural

Development Board and then helping lead the discussion at the Watershed Association meeting, seemed to be expressing her opposition in every available forum. Her second letter had also appeared in the *Democrat*, and she had sent two lengthy essays to Roberta Morganstern critiquing Siting Board publications.

In one of her letters, Graczyk analyzed the ANJEC newsletter that had first attracted Morganstern's attention. She wrote that "it was obviously prepared by a professional marketing or 'communications' firm to get the desired message to the desired audience." Though this was not true, she was correct that the article, which was in the form of an interview with me, had not been based on an actual interview. That had been the original plan, but due to scheduling problems, the ANJEC staff had asked me to make a list of questions I had heard environmental activists ask and then write answers to them.

The format Graczyk chose for her analysis was to repeat each question and then write a short sentence followed by a supplementary paragraph. Each of her sentences started with the words, "The answer" and then varied from "is superficial" and "is entirely misleading" to "doesn't address the question."

The final question I had asked was, "Will I glow in the dark?" My response had been to recommend that readers seek out information from a variety of sources and decide for themselves about the health, safety and environmental issues.

This answer seemed particularly infuriating to Graczyk. "The very phrasing of the question is insulting," she wrote. "Is this some failed attempt at humor? Or an attempt to ridicule real and true concerns for safety and health which most responsible persons would have?"

With the opposition apparently mounting, we were worried that this time the Township Committee would face much more significant pressure to cut the study process short. If we could forestall its next meeting until after the open houses, perhaps enough support would build to encourage the Committee to keep the process on track.

Miraculously, our wish was granted: Mayor Baumgarten came down with chicken pox and was out of commission. Since a replacement had not yet been named for the Township Committee member who had resigned, the Committee, which required two members for a quorum, couldn't meet.

If a disposal facility was ever to open in Bethlehem, public policy makers would analyze why the voluntary process had been able to succeed here. Although the mayor really did have spots, cynics and conspiracy theorists would probably speculate that the illness was faked.

The two open houses were scheduled to take place in the gymnasium of Bethlehem's Conley School. With one on Wednesday, January 15, from 3:45 to 9:00 p.m. and the other the following Saturday from 11:00 a.m. to 3:00 p.m., we hoped most interested people would be able to find a convenient time to stop by. Open houses are a technique for attempting to reinvigorate public participation in decision-making about controversial issues. The now-traditional public hearings in which all eyes and ears and any available television cameras are drawn to the loudest and most emotional presentations have become much more of a tactic than a discussion. Those whose opinions are already set dominate the event, and other attendees can only watch the sometimes vicious ping-pong match that ensues. There is little opportunity for people who are just curious to ask questions or for anyone to search for compromise or common ground.

In its letter of invitation to all Bethlehem residents, the Township Committee described the open houses as "informal forums where residents can stop by for a few minutes or hours, collect information, see displays, and ask questions of knowledgeable experts." It then listed eight issues residents might want to consider:

- *Bethlehem Township is under enormous developmental pressure. Hundreds of houses are under construction with two thousand more in the planning stage. The township does not have the schools, roads, police, fire protection, sewage treatment or public water supply to support this population. Township committee members estimate that if this development occurs, taxes will double or triple to pay for these services.*
- *Open land, farms and the rural quality of our township are threatened.*
- *Low-level radioactive waste storage offers the township $2 million per year for 50 years, jobs, and preferential use of local merchants and services.*
- *Bethlehem Township might decide to use the $2 million yearly to offset taxes, preserve open space and farms, and pay the legal expenses of developers' lawsuits.*
- *There is reason to believe that the advantages provided by the storage site would increase property values, but this is controversial and in need of exploration.*
- *The safety of the storage site is, of course, a major considera-*

tion. There seems to be a general consensus that the radioactive medical waste is relatively safe but concerns have been expressed about the safety of low level radioactive material from nuclear power plants. (Fuel rods, uranium, waste material from nuclear weapons and other high level radioactive waste would not be stored in the township.)

- *Three storage sites in other parts of the country were closed because they had design problems and leaked. The siting board says that the lessons learned will result in improved and safe design in New Jersey. Two facilities, in Barnwell, South Carolina and Richland, Washington, have been successfully operated since 1969 and 1965. Residents of those communities are said to be pleased with results in their communities.*

- *Concerns have been raised about the transportation of the material: 100 truck deliveries a year would occur. Ideally, the 100-acre site would be accessed from a major thoroughfare. No liquid material would be transported or stored. Material is prepackaged under strict federal guidelines.*

If nothing else, the flyer showed that at least a small group of Bethlehem residents had made substantial progress in framing the issues the town would need to evaluate. The Siting Board bought a second ad in the *Democrat*. With a tone similar to the earlier ad reprinting the support statement, we wrote:

If you've read the past few issues of the Democrat, *you know that Bethlehem Township is in the initial stages of exploring what a disposal facility for low-level radioactive waste might mean for the community.*

To help inform residents about the issues and to answer their questions, the Siting Board, which is responsible for finding a site to ensure the safe, long-term disposal of the low-level radioactive waste generated in New Jersey, will participate in an Open House on ....

Newspaper articles also publicized the scheduled events. Not only did the *Democrat* continue its thorough weekly reporting, but the open houses sparked coverage from the *Courier-News* of Bridgewater and the *Star-Ledger*. Now, with the *Express-Times*, three daily newspapers were paying attention to Bethlehem.

Our choice of the word "participate" in the ad in the *Democrat* was deliberate. While the Board would have a display set up in the gym, township officials had tried to make clear that any other group that asked would be allotted an equal amount of space. All participants would be treated equally, and the Board would bring refreshments and pay the minor costs associated with using the school facility.

By the time the first open house started on Wednesday afternoon, the Health Physics Society and the League of Women Voters had each set up a table, as had the Siting Board, the Department of Environmental Protection's Geological Survey and its Radiation Program, and the Office of Dispute Settlement. The only opponent who had responded in advance to the invitation

ON THE SPOT: John Weingart, executive director of New Jersey's Low-Level Radioactive Waste Siting Board, answering a reporter's question at the first open house in Bethlehem Township.

was Hugh Shannon, who had arranged for a table to demonstrate his web site.

Soon after the doors opened a small group of people arrived and "demanded" space for the so far unnamed group of opponents they said they represented. We had anticipated this possibility and quickly set up additional tables.

Over the next five hours, and the four hours that followed on Saturday, about 475 people came to the gym. While some were from neighboring towns, the events clearly attracted a significant percentage of Bethlehem's 3,200 residents.

As people walked in, they were greeted by a person from the town and another from the Siting Board staff sitting at a sign-in table. They were offered a short written description of each of the organizations which had prearranged to have tables, and then left to wander around the gym.

The Geological Survey, with its radiation meters and rock samples as well

---

Indoor Radon in Bethlehem Township compared with LLRW Disposal Facility

► The average indoor Radon concentration in the United States is estimated to be 1.08 pCi/l, leading to a dose of 200 mrem/yr.

- Of that 200 mrem/yr, about 60 mrem/yr is from indoor exposure away from home; 140 mrem/yr is received at home.

► Assuming you spend 62% of your time inside your home, you will receive about 130 mrem/yr for every pCi/l of Radon in your home (in addition to the 60 mrem/yr from indoor Radon exposure away from home).

► The average Radon screening test in Bethlehem Township is 7.5 pCi/l. This correlates to 1,035 mrem/yr, including the 60 mrem/yr received away from home.

- 46% of the 173 Radon screening tests in Bethlehem Township are above 4.0 pCi/l (current action guideline).
- The maximum screening test was 240 pCi/l - if unmitigated, this would have lead to a dose of 31,200 mrem/yr!

► The maximally exposed individual may not ever receive more than 25 mrem/yr from a LLRW disposal facility.

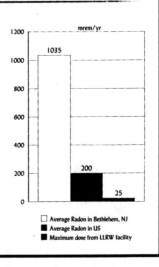

☐ Average Radon in Bethlehem, NJ
■ Average Radon in US
■ Maximum dose from LLRW facility

ENVIRONMENTAL IMPACT STATEMENT: State environmental officials issued a handout showing that the maximum exposure to radiation from a low-level radioactive waste facility would be less than 4 percent of the level already present in the average Bethlehem home due to naturally occurring radon gas.

as maps of the entire town, was the most popular stop. People would pore over maps and talk with Karl Muessig about why particular locations in the area would — or wouldn't — be good to explore for the disposal facility. Some residents took advantage of the opportunity to ask questions about local soils for farming and groundwater for drinking.

The DEP table offered various publications, including information about radon levels in Bethlehem Township, which were among the highest in the state. One handout that was only in draft form for the open houses estimated that "radon remediation of three homes in your town could easily reduce the total radiation dose to the people of your town more than they would be allowed to receive from a radioactive waste disposal facility."

The opponents had decided to cluster their tables together, so that under one of the basketball nets people could view Hugh Shannon's web site, collect literature from the Nuclear Information Resource Service and several locally produced flyers, and sign a petition opposing the facility. The petition drive had been stimulated by Mayor Baumgarten's apparently spontaneous promise at the last Township Committee meeting that he would stop the siting process if he received a petition signed by 1,000 people.

The leader of the local builder's association had written two flyers. One offered "The 'Other' Side of the 'Low Level Radioactive Waste Dump Story.' " The second one began:

Hunterdon County:
- *Rural Lifestyle*
- *Excellent Schools*
- *Radioactive Waste Dump?*

Another resident had written a three-page "open letter" asking, "How can managing a nuclear waste site in our township protect, enhance and develop our natural resources?" He gave his name and address, and identified himself as a "Friend of the Earth."

An unattributed handout told residents that:

- *Today's event was bought and paid for by the nuclear waste generators – mainly PSE&G and GPU.*
- *The State Siting Board has paid professional marketing firms and polling companies to help them "SELL" this idea to us. Even the League of Women Voters has received grants ... They are here as paid lobbyists.*

The Siting Board area included the three-panel display, the model of the facility, a barrel filled with examples of the types of materials that would be disposed at the facility, and a table of literature and videotapes. The facility model attracted the most attention and became the focal point for many of the news photographs that subsequently appeared.

All of the articles reporting on the open houses gave accurate interesting accounts of the hour or two each reporter had spent in the gym. Even the headlines reflected the purpose and tone of the events:

OPEN HOUSE EXPLAINS NUKE DUMP
- Courier-News

FRIENDS, FOES OF DUMP STATE CASE
- Express-Times

LOW-LEVEL RADIOACTIVE SITE TOUTED AS SAFE
- Star-Ledger

The newspapers, however, couldn't fully capture the energy in the gym each day. The room looked like an indoor county fair focused on low-level radioactive waste and felt like a model of democracy in action. Some residents who arrived with little knowledge about the subject could approach each table by asking, "What's this all about?" Others, including the surprising number of physicists and geologists who seemed to live in the Bethlehem area, could begin their conversations with much more detailed questions.

Some people walked straight to the opposition tables and never visited the others, but most made it around the gym. A few conversations were heated and unfriendly, but most were quiet exchanges of information and opinions.

Many years earlier, Bill Neil of the Audubon Society had taken a course taught by Gerry Nicholls, who was at the Department of Environmental Protection's table in his role as director of the Division of Environmental Safety, Health and Analytical Programs. Neil and Nicholls pulled chairs into an unpopulated area of the gym and argued their opposite points of view for over an hour. Other active supporters and opponents also chatted together as they met at the refreshment table.

The open houses were exhilarating and exhausting. By the end of the second one, the Siting Board members and staff felt public discussion in Bethlehem would now be much better informed than it had been, and more civil than it might otherwise have become.

Most township officials had showed up on one or both days, including Robert Housedorf, who had just been selected as the replacement member of the Township Committee. After talking with him on Saturday about where the facility might be located, we agreed that Karl Muessig and Greta Kiernan would meet him on Monday to drive around the town. They would stop at the six sites that had been listed at our meeting with the township engineer two weeks earlier, as well as others that people had suggested while visiting the Geological Survey table.

The Bethlehem tour took place as planned, and on Tuesday morning, Muessig and the Siting Board staff met to discuss the findings. He reported that the only site that might be suitable was the one already committed to housing proposed by the Hovnanian Corporation. Each of the other five that had looked promising on paper contained unmapped wetlands or steep slopes, or had been partially developed since the maps were prepared so that too little land would be available for the disposal facility. And all of the sites that had been suggested at the open houses were too small.

After all this activity, the Board was going to have to disqualify Bethlehem! Later that morning, our weekend exhilaration faded even further when we received calls indicating that support on the Township Committee for continuing the study was eroding rapidly. Apparently, the experiment in democracy looked less exciting to some residents who were imagining these four- or five-hour sessions expanding to fill the next two years.

Given this confluence of bad news, we decided to reject Bethlehem before the town could reject us. On Thursday, I faxed a letter to Mayor Baumgarten informing him "that it appears unlikely that there are suitable sites" for a disposal facility. I explained that the results of Monday's inspection were only preliminary and that we would be happy to review any conflicting material they might want to submit. "Unless and until we receive such a request from you," I wrote, "we are now discontinuing our exploration of Bethlehem Township as a possible host for New Jersey's disposal facility."

The next issue of the *Democrat,* appearing Thursday morning before my letter was signed, had a Page One article headlined:

WASTE SITE IDEA DUMPED

The paper reported that two of the three Township Committee members had announced plans "to bury the study of the dump at a special meeting set for Saturday."

The next day, the daily papers, whose reporters' connections in Bethlehem were not as extensive as the *Democrat's* Ken Serrano, had a different emphasis. All three led with news of the Siting Board's letter and were less definitive about the intent of the Township Committee meeting scheduled for Saturday. The *Courier-News*, for example, wrote that Mayor Baumgarten "plans to discuss the proposal one more time with residents and find out if there is any support for it." Its headline was:

NUCLEAR DUMP MAY BE MOOT: NO ROOM, OFFICIALS SAY

The *Express-Times* and the *Star-Ledger* interpreted this set of news similarly. The *Express-Times* declared:

LAND UNSUITABLE FOR NUKE DUMP

The *Star-Ledger* wrote:

STATE 'TRASHES' SITING DUMP IN BETHLEHEM TWP.; LAND FOUND UNSUITABLE FOR RADIOACTIVE WASTE

On Saturday, the Township Committee met and voted to end their exploration of the disposal facility. In a letter dated that day, Mayor Baumgarten wrote to the Board:

> *Thank you for all of your time and patience displayed during your efforts to inform the residents of our township of the siting procedure.*
>
> *It has been made clear to the Township Committee that the sentiment of our community is that there is virtually no support for the continuation of the siting process, and the residents of our township have spoken out without question to halt this process now....*
>
> *The problem of disposal that you are faced with is certainly a serious issue indeed, and needs a solution at some point in time. I would like to take this opportunity to wish you the best of fortune towards that solution, and hope that you will be able to keep your faith during this most difficult task.*

The *Courier-News* chose to report this final action with an odd headline, given that they had reported three days earlier that the matter might be moot:

BETHLEHEM TO FIGHT DUMP

But the other reports, in the *Hunterdon County Democrat* and *Star-Ledger* respectively, were sensible and straightforward:

RED-HOT DUMP STUDY IS ENDED

BETHLEHEM TWP. KILLS NUCLEAR WASTE PLANS

CHAPTER 25

'An Appalling Lack of Ignorance'

F or Bethlehem residents, the saga did not end when the Siting Board and the Township Committee each said, in effect, "You can't fire me, I quit." It continued to be a factor in local politics.

The issue of the *Democrat* that announced "Waste Site Idea Dumped" had another article asking:

WAS THE DUMP PROPOSED TO FEND OFF HOVNANIAN?

An accompanying editorial noted: "There were rumors that the proposal was just a ploy by township officials to scare off a big housing development that threatens to double the township's population and boost school costs."

In fact, over the next two years, the town eventually was able to fend off the housing developer by buying the land for $5.4 million. It received a Green Acres grant from the New Jersey Department of Environmental Protection covering about half the cost and financed the rest locally.

Discussions of this victory for open space have alternately accused and credited Mayor Baumgarten with pursuing a deliberate strategy that was, depending on one's perspective, either devious or brilliant. When his term on the Township Committee ended at the end of 1998, the *Democrat* ran an editorial entitled "Here's To Baumgarten," which observed:

> *His most notable achievement is his key role in the buy-out of the massive Hovnanian housing development, which saved one of western New Jersey's finest scenic vistas and headed off a costly population explosion in the township.*
>
> *But he also lost some of his constituents' trust by using oddball and ethically doubtful tactics, such as inviting the state to make its pitch for siting a radioactive-waste dump in the township in order to dampen Hovnanian's home-building ardor.*

To write history from just this and other similar newspaper stories would give a mistaken impression. It was Roberta Morganstern, chair of the township's Environmental Commission, who first suggested studying the disposal facility. For two months, Morganstern, Mayor Baumgarten, and the other members of the Environmental Commission and the Township Committee thought the disposal facility might be a workable alternative to the housing development. For better or worse, no one was clever enough to anticipate that the town would find a way to purchase the site, and that it would schedule multiple weeks of quality time with the Siting Board as the means to achieve it.

The disposal facility debate also created and enhanced local political careers. It attracted resident Lou Judice to Bethlehem politics, and the next year he was elected to the Township Committee. He defeated the appointed incumbent, Robert Housedorf, who had been the last member to continue supporting the siting study. When Baumgarten stepped down, Judice became mayor, but then was defeated when he ran for a second term. In January 2001, another active opponent of the disposal facility, Beverly Graczyk, who had just been elected to the Township Committee two months earlier, was selected as mayor.

The week after the Bethlehem process ended, the *Democrat's* lead editorial was headed:

THE DUMPED DUMP

After mentioning the response the Siting Board provided to "fears of radioactivity in general, government in general, and the prospect that the project would lower their property values," the paper went on to say:

> *This is probably true, but today who believes government representatives, even those as affable and knowledgeable as Mr. Weingart, a scientist and longtime environmental official who is local (he lives in Delaware Township) and a radio (not radioactive) deejay, to boot. For that matter, people are suspicious of 'safety' claims from some industries, too.*
>
> *But if they're concerned about radioactivity, how many people in Bethlehem have tested their air and water for radon?*

The following week, the *Democrat* printed a letter in response from one of the leading opponents of the facility in Bethlehem. He missed the chance to point out that I am not a scientist and even agreed that I was affable and knowledgeable. His mission was to object to the paper's reference to testing for radon, but in opening his letter, he got a little carried away. He wrote, "Your incredible lack of ignorance on this topic is appalling."

PART 4

"Working For The Government Is Cool"

CHAPTER 26

A Bureaucrat's Journal
From April 28,1997
to February 12,1998

Early in 1997, the Siting Board staff was discouraged. The demise of Fairfield, coupled with the dismal results of the Board's subsequent poll, had been particularly unsettling. The subsequent Bethlehem experience had been interesting and even exhilarating, but residents had raised lingering thought-provoking questions about whether a disposal facility could be accepted in any largely residential town. While the Siting Board seemed quite comfortable with having the voluntary process continue, several of the members and staff were having quiet discussions about how we would know if, or when, it was time to give up. We had now helped nurture a number of siting initiatives, only to see each die in infancy.

Taking stock of the situation, the Board found that the likelihood of securing a permanent disposal option outside New Jersey continued to be unclear, and that we knew of three more New Jersey municipalities that might be interested in volunteering. Small groups in each of these towns seemed to be preparing to initiate a wide public discussion about hosting the disposal facility.

The voluntary siting process would continue, but to me it now seemed finite. We would continue to welcome inquiries from people in any part of the state, but if each of these three towns lost interest, I thought it probably would be time to throw in the towel.

When I started to work for the Siting Board, the two former directors of New Jersey's Hazardous Waste Siting Commission, Sue Boyle and Rich Gimello, advised me to keep notes. They expressed regret that they had not chronicled their long, dramatic, and ultimately unsuccessful crusade, and said they thought the story of what was involved in trying to site unwanted land uses should be told.

During my first two-and-a-half years at the Board, I had written down occasional thoughts and comments. Now, I became more systematic. In April 1997, I began a journal explaining that three towns were possibly on the verge of stepping forward to volunteer as host communities. I envisioned the journal ending when one of the towns and the Siting Board signed an agreement to build the disposal facility or when the third town said "No, thank you."

I wrote the journal hoping that others would someday read it. Whether or not the voluntary process succeeded, I hoped the Siting Board's experience would offer lessons that could help improve other governmental efforts. At first, I did not mention the name of two of the three towns because their leaders had not yet made a commitment to publicly consider volunteering to host the facility. Thus, at the beginning, I refer to "Town Number Two" and "Town Number Three." If the few people quietly talking with Board representatives had decided to abandon the idea of volunteering, I would have honored the Board's commitment to treat those conversations in confidence, and the town names would never have appeared.

In its immediacy, I hope retaining the journal format offers an added vantage point for viewing the siting process, the mechanics of government, and the experience of being part of it.

April 28, 1997

I am not as optimistic as I was when I began this job, but three towns are talking to us, three towns in which at least a few people are thinking that volunteering to host the disposal facility New Jersey needs for low-level radioactive waste might be good for their community.

To an extent, my early optimism has been rewarded. Local officials and others in a dozen towns have publicly suggested that hosting the disposal facility might be worth exploring, and representatives of many more have spoken with us quietly and say they may bring it up in their area "when the time is right." That is far more interest than many people expected to see. But we had to eliminate three of those towns because the sites they suggested were unsuitable. In the other nine, vehement local opposition led their governing bodies to pass resolutions expressing no further interest. As a result, my optimism has waned. Nevertheless, here we are, two years after formally adopting New Jersey's first voluntary siting process, with people from three towns talking seriously with us.

One is the town I live in, Delaware Township in Hunterdon County. This small rural municipality is located less than three miles from the Delaware River and is the proud home of New Jersey's only remaining covered bridge.

The Planning Board invited me to address a meeting last month. When I called Mayor Ken Johnson to suggest that it might be better to first quietly explore whether our town has any land that could be suitable for the disposal facility, he said, "Oh, don't worry, nobody's serious about this thing." But he was wrong, and the Planning Board members, himself included, asked a series of thoughtful questions and then listened to public opinion. That night, it was split, with two people speaking in favor of looking into this possibility further and one person wanting any consideration to end at once. The Board then thanked my colleagues and me, and went on to other business.

This low-key, deliberative discussion among about 30 people, however, was immediately front-page news in the *Hunterdon Democrat*, our local weekly. The articles led to lots of discussions around the post office, general store, and soccer fields, and prompted letters to the editor generally questioning the sanity of anyone who would entertain the idea.

I contacted each person who had written to the newspaper. I wrote to the person I know whose letter asked, "Do you believe the government has ever given the truth to the people about radioactivity?" But I decided to call the person I didn't know who wrote, "Mr. Weingart's wrong-headed type of logic needs to be recognized as a local manifestation of right-wing conservative logic."

Despite the fact that this second letter began with a reference to our township's "own resident Dr. Strangelove," I thought that if I talked with this man and explained that I was still proud to have worked in George McGovern's presidential campaign, he might at least question his premise and consider

Delaware Township
Hunterdon County – 1997

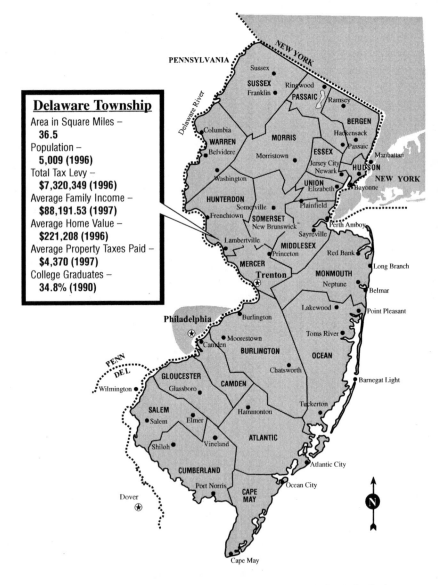

Delaware Township
Area in Square Miles –
36.5
Population –
5,009 (1996)
Total Tax Levy –
$7,320,349 (1996)
Average Family Income –
$88,191.53 (1997)
Average Home Value –
$221,208 (1996)
Average Property Taxes Paid –
$4,370 (1997)
College Graduates –
34.8% (1990)

Sources: N.J. Department of Treasury, New Jersey Almanac, NJ Legislative District Databook

240

exploring the issue a bit further. The phone call was a mistake. As soon as I introduced myself, he started to yell at me, with a woman who I assume was his wife yelling louder in the background. When I meekly said I thought he might want some information, he said, "I don't have any information, and I don't need any information."

Last week, Planning Board members announced that if any governing body should consider this issue further, it is the Township Committee, and not them. This decision, too, was front-page news. What, if anything, will happen next is unclear.

In any town, the people who find this idea to be of interest are attracted by the economic benefits the host municipality will receive. In my town, people are very concerned about the number of large farms that may be sold for housing developments. With one 100-acre or 200-acre site dedicated to the disposal facility, the town could spend all or part of the $2 million each year to preserve a large amount of farmland and open space. Proponents of the idea think this option might provide the optimal approach for maintaining the town's character, which we all consider charming and important.

Town Number Two also is residential and rural. People there seem attracted to the disposal facility by the opportunity to lower taxes and also by some sense that as residents of the smallest municipality in their county, no one has taken their interests into account, and it is time that someone did.

If the disposal facility ends up being located there, it will give the League of Women Voters reason to be proud because this was the only municipality to send representatives to both of the statewide workshops the LWV convened in New Jersey on low-level radioactive waste disposal. These events, funded by the Siting Board, were intended to provide a safe place for local officials and others to learn more about the possibility before deciding whether they thought the idea had sufficient merit to raise it for local discussion. The first workshop, held a year ago, attracted about 140 people, but only about 25 attended the second one this past December; six of them, however, were from Town Number Two.

The six people apparently have been laying the groundwork for a community-wide debate. During the winter, they spoke to 30 others, all of whom, they say, agreed to serve on a study committee if one is established. They have spoken frequently, and now met, with Fran Snyder of the state Office of Dispute Settlement. She has been advising them on how to design a process for local consideration of this issue that will have a reasonable chance of leading to wide public involvement, productive discussion, communication, learning, and ultimately decision-making.

The town is now poised to go public. The Township Committee is planning to send a letter to the 30 people inviting them to the first meeting of the study group next month. A few days later, they plan to send a slightly different letter to every resident, informing them that hosting this disposal facility looks like it could be good for the town. The letter will add that they, as the town's elected leaders, have formed a study group to look into the possibility of hosting a facility and will report back to residents by December. The town leaders understand that anything involving radioactive materials is controversial, and that local groups against "the dump" spring up wherever this idea is broached. They think any opposition that arises will come from surrounding towns and not their own.

The third town has a different perspective. It already has significant industry and is interested in the disposal facility as the anchor development for a potential research park. Supporters think that such a park might attract some of the pharmaceutical and other industries and laboratories that generate low-level radioactive waste. The town successfully sited a cogeneration plant several years ago, and many of its residents now have firsthand appreciation for the intensity that can be reached in public debate on developments when some people feel their health may be threatened.

This town, too, is taking its time. Its Township Committee invited me to speak to one of its meetings well over a year ago. Perhaps because they saved my talk for the end of the meeting and took a 45-minute recess immediately before calling on me, it received no press attention. In fact, we have never formally heard another word of interest from them. We have, however, been contacted from time to time by two people who seem to be taking the lead in putting together a process the town could use to consider volunteering. They recently told us that they will soon be submitting a draft plan for our comments. The idea seems to be that the Siting Board would pay one or both of them to coordinate the effort and act as a broker between the town and the state.

There are other towns that also might decide to consider hosting the disposal facility. The long gestation period that seems to precede public expressions of interest means that we could hear from another town any time. The contact may come because residents heard a speech or read an article or publication months or even years ago. And there are several specific towns we hope to hear from, towns that already have some industry, available and perhaps contaminated land, savvy political leaders, and serious financial problems.

I fear, however, that we are approaching the end of this experiment, and the actions of these three towns may determine if it succeeds or fails. From the 12 other towns in which some interest in volunteering was publicly expressed, a depressing pattern is emerging.

Like many issues, this one gets little media attention as a national or state problem in need of resolution until a specific site or town is mentioned. As a result, the first time most residents of these 12 towns heard of low-level radioactive waste was immediately after their local governing body passed a resolution or publicly discussed whether the disposal facility might be safe and beneficial. The ensuing public outcry then led them each to quickly say they had learned enough and had no further interest. The three-month process in Fairfield was our greatest triumph so far, the longest period of time we were able to withstand public fear.

We do, however, take pride that the process ended in three of the 12 because we determined that they did not appear to have suitable sites. We mention this often when critics allege that a voluntary siting process means environmental and health standards would be compromised immediately at any local expressions of interest.

I had believed that the hardest part of this endeavor would be finding the first one or two towns that would explore volunteering. After the ice was broken, I thought, others would be more likely to follow, but the effect may be working the other way. Some local officials are clearly being discouraged by the stories they hear of the negative reactions evoked in other areas. Earlier this year, when Bethlehem Township publicly considering hosting the facility, opponents asked why they should take something that all these other towns have had the good sense to turn down.

As we enter what may be our — or at least my — last campaign in the arena of low-level radioactive waste, I am left with two alternate images of our situation. When the optimism shines through, I feel we are like Bill Murray in the movie *Groundhog Day*. We keep getting to replay the day until we finally do it right and get the girl. Ultimately, the residents of one of these three towns will explore the issue and decide that this would indeed be a safe development that would fit into their community. The Siting Board will conclude that the town has a site that meets all the federal and state requirements. Representatives of the town and the Board will then negotiate a mutually agreeable contract. This success will offer lessons for resolving other controversial land-use issues, it may put a small dent in the walls of cynicism that greet almost all efforts of government to solve problems, and it will show that foolish optimism can sometimes be rewarded.

The other image that haunts me is that the Siting Board members and staff are the Washington Generals, the team sent in to play the Harlem Globetrotters night after night in one town after another.[55] We get praise for our ability and tenacity, but everyone knows the Globetrotters have better theatrics and will be the eventual winners.

If all three towns decide they don't want the disposal facility — whether it is because they think it unsafe, because they don't trust government to deliver on promises, or just because they don't want the controversy — then we will probably conclude that this interesting experiment has failed. Theoretically, the voluntary process could continue until it had been tried in many more parts of the state. After all, New Jersey has 566 municipalities. But even Harold Stassen stopped sounding optimistic in his later presidential campaigns.[56]

May 13, 1997

Town Number Two is about to get a name. We have emphasized that people can talk to us in confidence, and that an individual can request information about radioactive waste or the voluntary siting process without finding his or her name in the newspaper, even if the person is a local official.

As a result, when reporters ask whether any towns in their area are considering volunteering, I respond that individuals talk with us all the time, but that I will not say who they are or where they are from because they are simply asking for information from a government agency. I explain that there will be about two years of public discussion in any town that decides to pursue this possibility. So far, this response has been accepted without comment.

Other states have taken different approaches to this issue, with the siting agency feeling obligated to tell anyone who asks the names of people they have met and spoken with, and even of people who have requested that literature be sent to them. Some siting agencies have, therefore, given grants to nonprofit organizations to conduct their local outreach with the hope that local officials will talk with them more freely than they would with the government. This is one of the areas in which the rights of the public to open government can collide with the ability of government to accomplish its assigned tasks with something resembling planning and efficiency.

But last week, on May 5, the five members of the South Harrison Township Committee sent letters to 35 residents asking them to be part of a study group to explore whether their township should host New Jersey's disposal facility for low-level radioactive waste. The Township Committee letter notes that they have been "exploring ways to preserve farmland, manage growth, stabilize property taxes and maintain the rural character of our community," and that they have "taken a preliminary look into an option that might offer us a way to achieve all of these goals." The letter invites them to a meeting next Monday evening, at which the Board "will be invited to discuss the voluntary siting process, the state's need for such a facility, the benefits the host community will receive, and to answer your questions."

The Committee is sending a slightly different letter to all township residents that is to be in the mail today or tomorrow. This letter will tell them of the formation of the study group and invite them to an "open house" with the Siting Board at the municipal building the following week on September 29 from 4 to 9 p.m. A third letter is being sent by the Township Committee to their Congressman, state legislators, county freeholders, and the mayors of adjacent towns.

South Harrison is in Gloucester County, about an hour south of Trenton. It is right on the border of Salem County and has a population of 2,100 people living in 15.8 square miles. The town is a mix of farms and residential areas. Its residents paid $2.8 million in property taxes in 1997 and received $1.4 million in state aid. A new development paying the town $2 million would make a significant contribution to increasing municipal services and/or lowering its tax rate.

In our office, we are now nervous. Will reporters start to call before next Monday's meeting, and will there be opposition at the meeting? Is there anything we can do now in this short remaining lull? I sent a note to the Governor's Office so they will know about this if the governor happens to be in that area in the next few weeks. Should we be contacting the three Democrats competing in the primary to run against her so they are less likely to make some off-the-cuff comment that confuses the situation or creates more opposition?

Rowan University is not far from South Harrison, and it has a new Center for South Jersey Affairs. Can we find ways to enlist the university in this process? In Bethlehem Township, we had circulated a flyer from the state Department of Environmental Protection noting that remediating that area's high radon levels in just three homes would cut radiation exposure more than the amount that would be added by the disposal facility. We checked and found that radon levels in South Harrison are much lower than in Bethlehem, but is there something similar we can do? Can we anticipate questions people are likely to ask and generate answers now?

And, finally, a state bureaucracy question. It is mid-May and we are entering the annual "blackout" period, a time in which the Department of Treasury tries not to process requests for expenditures so it can close out the books for the fiscal year that is ending on June 30 and begin the new one with a clean slate. So, what funds do we anticipate needing for events in South Harrison between now and mid-August when spending under the new fiscal year budget will be flowing more smoothly? Mostly, we just wait and meet every day or two as a staff to see if anyone has any new ideas.

Meanwhile, we lost the town I live in and may have gained another. When I took in my recycling on Saturday morning two weeks ago, I ran into several members of the Planning Board. They apologized that the Board had decided to "duck the issue" and pass it on to the Township Committee. They said they had no idea what would happen next. But two days later, the Township Committee met and voted to send me a letter saying they had "no further interest." The vote was 4-0, with one member absent. I learned of the vote from a friend who was at the meeting for unrelated business, and then from a newspaper article.

Although I still haven't received the formal notice, I sent a letter to the *Hunterdon Democrat* which was printed last week, praising the town government for considering "a possible mechanism for saving significant amounts of farmland and open space, even though they knew the idea would be controversial." I said, "Almost everyone I spoke with, no matter how skeptical they may have been, was courteous and willing to listen and consider information that may have been new to them." This was a slight overstatement, but I wanted to follow the management advice of "catching people doing something right."

I will miss having my town as a candidate. I didn't really think we had a viable site, but it seemed to help the credibility of the process and certainly to get people's attention for my backyard to be in the running. My wife has no regrets, however. Knowing that this issue generates both passion and fairly convoluted conspiracy theories, she had worried that if Delaware Township had eventually agreed to host the facility, critics would have complained that I acted improperly by participating in a decision that would lower my taxes or otherwise improve the town in which I live.

While Delaware Township is now off the list, yesterday we received a nibble from another town. The mayor called and said, "I've certainly been getting enough mail from you people." He and a member of his Council have science backgrounds and are confident the disposal facility would be safe. He has a specific site in mind, which he described to me. I told him we would look into

South Harrison
Gloucester County – 1997

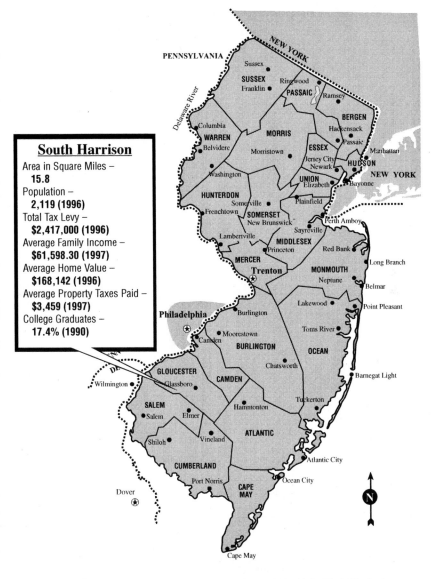

South Harrison

Area in Square Miles –
15.8
Population –
2,119 (1996)
Total Tax Levy –
$2,417,000 (1996)
Average Family Income –
$61,598.30 (1997)
Average Home Value –
$168,142 (1996)
Average Property Taxes Paid –
$3,459 (1997)
College Graduates –
17.4% (1990)

Sources: N.J. Department of Treasury, New Jersey Almanac, NJ Legislative District Databook

OPPOSITION HUMOR: A hand-made cartoon from South Harrison Township.

it and get back to him. He acknowledged that it is either on the border of or straddles one or two other towns, and we both agreed the towns would have to work together. That would be a very exciting and challenging possibility because few of New Jersey's 566 towns ever work together on anything involving land use.

May 22, 1997

The first meeting of the South Harrison Study Group was a success. Twenty of the 35 invited residents showed up. The only other person who attended sat in the back without saying a word. Our fears proved unfounded that the meeting would be overrun by people who had become concerned after receiving the Township Committee's letter or after reading a small article that had appeared in Sunday's *Philadelphia Inquirer*.

The town clerk, Jean Riggs, opened the meeting and asked everyone to spend a minute introducing themselves and giving their background. Then she introduced us. Jeanette Eng and I spoke for about 30 minutes and then took questions.

After another 40 minutes or so, Riggs cut off the questions and moved on to logistical issues. Calls had come to her office complaining that the open house conflicted with the school play. Some alleged that this was a deliberate ploy to limit public discussion. The group decided to go ahead with the meeting as scheduled, but to add two more sessions. They made sure that all three would be on different days of the week to minimize the number of people who would be unable to attend at least one.

They also decided to call themselves the "Study Group" and to participate, themselves, in the open houses. They will wear nametags that identify them as members of the group and have their own table. They willingly put their names in the blanks on a chart prepared by the clerk ahead of time so that at least two of them will be present at all times throughout the five-hour events.

We had been concerned before the meeting that the study group had no designated chair who could run meetings effectively and be a spokesperson to other residents and the press. The group decided to pick one at their next meeting and concluded that anyone interested should give their name to the clerk. They then picked a date for their next meeting, putting it between the second and third open house, and adjourned at 10:00 p.m., three hours after they had begun.

One member had suggested that the study group have a newsletter and had offered to write the first draft. He has now faxed it to us, asking for comments, and we have both faxed and telephoned suggestions. Most were factual corrections to his summary of "What the State People Said," since we tried to stay away from second-guessing his sense of strategy. His draft, for example, asks others interested in joining the study group to submit their names, although the Township Committee had not wanted to increase the size. This is a problem for them to resolve, not us. He wants to get it in the mail by tomorrow so people have it before the first open house. He is also working on a questionnaire he plans to have available at the open houses.

The next day, we received a call from a woman living more than 100 miles away in Hamburg in northern Sussex County. She had helped lead or create the opposition two years ago when a landowner had suggested that her town consider hosting the disposal facility, and then helped form the group CHORD.

The woman said she was calling to check on a rumor that the Siting Board had started site characterization in South Harrison and had already drilled several holes. I told her this was not true, and that what had happened was that a study group had had its first meeting and set up some events for wider public discussion throughout the town. Since she didn't ask me when those events were, I assume she already knew. I think she also knows enough about the Board's process to know that we couldn't possibly have started site characterization. So her phone call, despite a pleasant enough demeanor, felt less than honest to me, but I can't guess what she hoped to accomplish that couldn't have occurred in a more direct conversation.

She also said she had heard the town I lived in was considering volunteering. Her manner encouraged me to ask her about a letter she had written to a newspaper a while back saying she had a videotape that showed the Siting Board's "manipulative tactics." I had written at the time asking to borrow the tape and had never received a reply. When I asked this time, she said she had returned the tape to the other cofounder of CHORD. I assume the tape is of the public meeting held in Roosevelt in March 1995.

The most notable aspect to South Harrison's consideration so far has been the role of Fran Snyder of the Office of Dispute Settlement. Originally formed as part of New Jersey's unusual Department of the Public Advocate, a state agency created in 1974 that had the power to sue other state agencies in the public interest, Fran's office was moved to the Public Defender's Office when Governor Whitman abolished the Department 20 years later. The Siting Board had contracted with it in 1992 specifically to gain the services of Fran as it began to seriously consider designing a voluntary siting process. Each year, the contract has been renewed, first by my predecessor and then by me.

Fran's role in this siting process is both unusual and a little confusing. A critic, or even a casual observer, would be likely to conclude that Fran is just another member of the Siting Board staff. But in Fairfield, Bethlehem, and now South Harrison, local officials and residents have found a distinction much larger than I thought possible. Over most of the last year, Roger Samartino, the South Harrison Township Committeeman who has taken the lead in pushing for a study of the disposal facility, has talked with her at least once or twice a month, trading ideas about public processes and strategies for gaining fair treatment of a controversial issue. And in the weeks leading up to the first study group meeting, Fran and the town clerk spent hours together by telephone and in person.

May 28, 1997

As recently as two weeks ago, Township Committeeman Samartino was saying that he had talked to no one opposed and couldn't imagine where opposition would come from. Now a newly formed local group called the Citizen's Coalition Against Nuclear Waste is being quoted in the news stories that are appearing with increasing prominence.

Ten days ago, the *Philadelphia Inquirer* ran a story in its "Neighbors" section that was hard to find; today, South Harrison's flirtation with the disposal facility is the lead story on the front page of its "South Jersey" section. In between, this past Sunday, the *Gloucester County Times* ran a front page banner headline:

RADIOACTIVE WASTE SITE SPARKS ALARM

Tonight, the Citizen's Coalition has scheduled a meeting and Samartino plans to stop by. He called to ask for my responses to some of the claims being put forward by critics about the amount of waste coming from nuclear power plants and other issues. He is clearly surprised by the amount of opposition and personally hurt by the comments coming from people he has known a long time who are not only opposed, but also questioning his motives and honesty. I haven't talked with him too much before, but today was the first time I heard him express concern for two people running for the Township Committee this November and his hope that this issue would not hurt them. He told me that his main interest is seeing the town's land-use ordinance changed, and his pursuit of the disposal facility is motivated as a means towards that end.

Clearly, the next few weeks will be critical. I tried to remind Samartino, and myself, that all the opposition will be apparent soon and that most area residents are watching the two sides like a ping-pong match. They will listen to what we each say and how we respond to determine who is most credible. Such perspective is hard to maintain from Trenton, and I know it is even harder for the local residents on the study group and others associated with this idea.

South Harrison is looking into the
possibility of hosting New Jersey's disposal facility
for low-level radioactive waste.

Just

We believe this will be a safe, low-impact,
light-industrial development. Some say it won't be.

Say

Decide for yourself.

Maybe.

Today's 'Open House' is one of
several opportunities you will have
to learn about the issues.

Decide for yourself if hosting this facility
might be right for South Harrison.
Talk to people who work with
radioactive materials. Speak with
opponents of nuclear power.

Then make up your own mind.

For more information, call the Siting Board
toll-free at 1 (888) 777-1558.

A CALL FOR OPEN MINDS: In a takeoff on Nancy Reagan's famous "Just Say No" anti-drug slogan, the Siting Board urges South Harrison residents to "Just Say Maybe" to the possibility of serving as host municipality for the state's low-level radioactive waste disposal facility.

June 2, 1997

New Jersey's low-level radioactive waste disposal facility is not going to be in South Harrison. The end came amazingly quickly. The meeting called to oppose the facility attracted about 50 people, who walked *en masse* to the ongoing Township Committee meeting, and asked that the study be ended. The Committee members apparently made clear that they were committed to this process, though they were far from deciding about the facility itself, and asked that the objectors participate.

The open house the next day attracted more than 400 people. The town clerk had been receiving calls earlier that afternoon from people asking directions to South Harrison, which led to visions of a large convergence of opponents from Philadelphia and northern New Jersey. Messages from the Green Party of New Jersey on the Internet that day had asked "concerned environmental groups and citizens" to attend the open house, which would begin at 4 p.m., and a "press conference and informational picket" to be held at 6 p.m. In response, the town had arranged for plainclothes police to be present.

The open house, however, was very civil and non-threatening. The only opponents who traveled any distance seemed to be three or four from the CHORD group. As people arrived, we asked them to sign in and almost everyone included their address. About half the attendees were from South Harrison with almost all the rest from surrounding towns.

As in Bethlehem, the open house was a wonderful thing to behold, an exercise in democracy with a sizable percentage of the local population moving from table to table collecting literature, asking questions, and expressing opinions, all about low-level radioactive waste. The Siting Board had its exhibit, as did the League of Women Voters, the Health Physics Society, and the New Jersey Geological Survey.

Interspersed were other tables from the week-old South Harrison Study Group and the even newer local Citizens Coalition Against Nuclear Waste. In addition, CHORD, as the would-be statewide opposition group, and the national opponents, the Nuclear Information Resource Service, each had a table. Thus, three of the seven tables were staffed by people opposed to building, or studying the possibility of building, the facility. Someone had printed up expensive-looking posters saying, *Shut The Siting Board Now, Say No To Nuclear Waste In South Harrison Township.*

While the five hours were lively and exhilarating in many ways, they were also worrisome. Several members of the study group did a great job soliciting their neighbors' concerns and fears while also defending the plan to study the idea, but the prevailing comments seemed more negative than they had in Bethlehem. A reporter asked me if I had spoken to anyone in favor, and when I pointed to the people on the study group, he said they didn't count. Apparently, because these local residents had decided the week before that this study was a good idea, they were now not worth quoting in a news article.

In truth, I had only talked to one other person at that point who had expressed support and I couldn't find him. Later, when another person said he thought this would be good for South Harrison, I led him over to the reporter, triumphantly saying, "I found one." This comment led a woman staffing the local opposition table to chastise me for trying to bias the press.

It turned out that what was occurring at the school musical next door was more important than anything going on in the municipal building. We had thought the coincidental scheduling of the two events in neighboring buildings was fortuitous because more people might consider dropping by the open house on their way to the musical. But in the school auditorium, members of the Township Committee were feeling besieged by people who thought studying the possibility of hosting the disposal facility was a terrible idea. One Committee member later said that the concerns were all expressed by women, most of whom felt this was going to harm their children's health.

I had arranged to speak the next night at the monthly dinner of the Association of Mayors of Gloucester County, where South Harrison is located. This timing too seemed fortuitous. It could be the perfect opportunity to talk with officials of the surrounding towns at the moment when they would be most focused on the issue. South Harrison's flirtation with the Siting Board had just become a lead story in all the local papers, so they would know about it but might not yet have had a chance to publicly say they were unalterably opposed. Word of this meeting's agenda had spread so that there were a few pickets at the banquet hall before the meeting.

The association is a wonderful informal group. It conducts business in a way that cynics of the political process would believe never happens, and that reformers who advocate no government behind closed doors would probably like to outlaw.

One Friday evening each month, the mayors of the 24 towns in Gloucester County, Democrats and Republicans, get together with any guests they want to bring to socialize and discuss common issues. All but two of the mayors show up regularly. It seemed to me that genuine friendships and relationships had been formed that would make it easy for any of them to pick up a phone and call each other as potential conflicts or opportunities for cooperation arise.

My talk seemed to be well-received. Perhaps because there is no press at these dinners, the mayors asked thoughtful questions and felt no need to make speeches. I drove home feeling great, except I was puzzled that Edmund Crispin, the mayor of South Harrison, and his two guests had left at the end of the meeting without saying goodbye.

As it turned out, they knew their Township Committee was having an emergency meeting the next morning, Saturday, at 9 a.m. to vote to end the process. On Sunday, the vote was on the front page of the *Gloucester County Times*. Mayor Crispin was quoted saying that the amount of money the Board was offering was not worth tearing the community apart. He referred to his constituents and added, "It's their township, not ours, and we don't want what they don't want."

The end of the process in South Harrison generated far more press coverage than had occurred in any other town, including Fairfield, where the public debate had lasted three months as opposed to 10 days. I gather this was because South Harrison is in the coverage area of the *Philadelphia Inquirer* and, once it covered the story, the articles were picked up by the *Associated Press*.

The slant the news articles all put on South Harrison's withdrawal was that the Siting Board was now zero for 14. This referred to the number of towns that had publicly considered hosting the disposal facility. One newspaper ran an editorial comparing the Siting Board's record with the Philadelphia Phillies and suggesting, without any specifics of course, that we take a "new tack."

Though I used the word "disappointing" when I responded to reporters' inquiries, "depressing" would have been more accurate. The only positive aspect I could find was that Governor Whitman was reportedly asked about "the controversy in South Harrison" on a Philadelphia radio show, and responded with understanding and support for the voluntary siting process.

Representatives of waste generators often ask me about the governor's position on the siting process. I now had semi-tangible evidence to back up my response that she is behind it. They clearly wish she would use her pulpit to encourage towns to host the facility, rather than following the path of most governors, which is to avoid expressing opinions about controversial matters as long as possible. Politicians usually hope the controversial issue will go away, and believe that if it doesn't, the very end of the process often is the time when their intervention can have the most impact.

South Harrison had had so much going for it — suitable land and people very concerned about saving open space. The county landfill had been imposed upon them, giving the town residents some experience with which to compare advance fears and promises with actual operation. A dedicated town clerk shouldered a tremendous amount of additional difficult and delicate work, and other local people had devoted a lot of time to thinking about how to present this topic to this town. Thirty-five people had agreed to participate in a study group, including at least one engineer, farmer, fire official, ambulance worker, real estate agent, physician, construction worker, attorney, and accountant.

So, what went wrong, and could it have been prevented? Once again, town officials underestimated the extent of opposition. Once again, opposition focused on the process, asking how local people could have talked about this for a year without letting everyone know and questioning who picked the members of the study group. Property values quickly became a major issue when a local couple widely distributed a letter describing how their plans to sell their property had fallen through because of the buyer's concern about the disposal facility.

And, as in Bethlehem, the Internet became a factor. Local opponents had quickly gathered negative information so that their signs and flyers immediately reflected arguments and charges that had been raised elsewhere. Others who approached the issue with open minds and were just seeking information also found the Nuclear Information Resource Service web page and based their questions on what it alleged. More than a few seemed to regard material found on the web as inherently valid. They held pages they had downloaded as if they were from the gospel and deserving of far more trust than the Siting Board's literature and responses.

Could this outcome have been different? I don't think so, but two growing theories in our office were corroborated. First, we should give up working with communities that are overwhelmingly residential. The attraction and the reason they contact us is that their budgets tend to be smaller so the $2 million represents a significant sum. But with residents' opposition to truck traffic and industry of any kind, fear of traffic accidents, and doubt that their local officials could negotiate a good deal, I don't think we would ultimately have been successful in South Harrison if we had been trying to get the town to volunteer to host a Ben and Jerry's plant.

The second conclusion is more generic. Not only is all politics local, but local politics is intensely personal. The Township Committee members in

South Harrison didn't abandon the voluntary process so quickly for "political" reasons, at least in any negative connotation of the word. They may have worried about the impact of this issue on their next campaign, but in conversations during and after the open house, it was evident that far more significant was the concern and often hostility from people they considered friends. They felt they were constantly defending their support for this process every time they ran into anybody anywhere, and realized that this debate might define most of their social interactions, and those of their family members, for the next two years.

June 17, 1997

I spoke to a class on "Science, Technology and Public Decision-Making" at Rutgers University's Camden campus. The other speaker was David Ryan, who works for Penn State University in a program focused entirely on educating the general public about radiation. The program, called Pellrad, is financed with voluntary contributions from Pennsylvania utilities.

It was helpful to take several hours to reflect with Ryan and the course's professor, John Gagliardi, about the difficulty of communicating about risk. Ryan quoted studies showing that women between the ages of 35 and 65 are the most important members of a community in terms of decision-making.

I noted that many of the people I have encountered seem to demand that government officials promise no risk and perform with absolute perfection. At the same time, they expect to find total incompetence and corruption. When the reality is almost always in between, their alienation only increases.

June 20, 1997

We almost gained a new town from the mayor who had called last month. But after reviewing the maps he sent us, we had to tell him that the land that was available had steep slopes. This would make the site very expensive, and perhaps impossible, to adequately evaluate.

We continue to hope to hear from others, and we continue to receive occasional queries and visits from people connected with Hamilton Township, a large municipality adjacent to Trenton with more than 86,000 residents. But I still think New Jersey's voluntary siting process will culminate with the three towns that were in the running when I began this journal in April. Two have now rolled over, and only one is left. Either we will site the facility in the remaining municipality, or we will end the voluntary siting process and scale back the Siting Board's efforts to keeping track of waste generated in New Jersey and monitoring the national scene. And I and most of the other staff members will have to find new jobs.

The third town still has no name, because it has not publicly expressed interest, but it is very different from any of the others we have worked with in ways that bode well for the siting process.

First, the town is not entirely residential. It has a significant industrial base. Second, it is home to a recently approved and constructed cogeneration plant. As a result, the residents have lived through controversy almost as intense as will be generated by the disposal facility proposal. Third, the town is located near three of the state's four nuclear power plants, which means that a fair number of people in the area should have some work experience with radioactive materials. Also, the idea of placing the disposal facility near the plants where much of the waste is generated has been appealing to people we have met throughout the state. And fourth, it has a municipal agency focused on economic development that seems willing to take an active role in the siting process.

The other difference presented by this town is that its approach to us is being orchestrated by an entrepreneur who wants to be paid to try to broker an agreement between the Siting Board and the town. He is an attorney based in Newark, but with local family roots. Last summer, the township attorney for Fairfield, who had grown up with this fellow, contacted him as someone the town might hire to help evaluate the desirability of volunteering. The Fairfield process ended before any hiring could occur, but the lawyer's interest was aroused, and he has formed a corporation called LowRad Inc. specifically to enable him to work with a town considering the disposal facility.

This entrepreneur is part of a five-person local team working to design and advance this proposal. It also includes one elected member of the Township Committee, the chair and executive director of the local economic development agency, and one area resident who had served on our Siting Board's Advisory Committee when the voluntary process was being considered, developed, and adopted. He resigned from the Advisory Committee during the winter so that he could work on this proposal without any conflict of interest.

We met with the group of five in our office last week. They presented us with a process suggested by the entrepreneur that is quite different from what the Board has been envisioning. Under this proposal, the Township Committee would pass a resolution, a draft of which is among the half-dozen documents we have now been given. The resolution would authorize the municipality's economic agency to hire LowRad Inc. to coordinate all tasks related to exploring whether this town should host the disposal facility.

The major problem presented by this proposal is its cost. The Siting Board has talked of reimbursing a town for up to $50,000 in related expenses it incurs during the year of preliminary site investigation. This proposal calls for $1.3 million over the same period. The money would be used to pay for the time of the entrepreneur and the former member of the Advisory Committee, and to enable the town to hire additional staff for the economic agency. In addition, almost every other municipal agency would receive a small grant.

The grants to local agencies are immediately suspect. They are suggested on the theory that each local agency will intensively study the impact of the disposal facility on their operations during the year. They appear, however, to really be unencumbered grants intended to encourage some local residents to feel grateful that the town is considering the disposal facility. I made clear, to begin with, that we only had legal authority to award funds for tangible products or services.

It is perhaps a sign of how unusual the voluntary siting process is, however, that we are taking this proposal, though not the specific dollar amount, very seriously. If it succeeds, even the full request of $1.3 million would be a small percentage of the overall cost of siting and construction. And if it fails, it will probably end quickly and cost only a prorated share of the amount agreed to for a full year.

The members of the Siting Board with whom I have discussed this arrangement are supportive of going ahead. Rick McGoey, who represents GPU Energy which owns one of New Jersey's four nuclear power plants and by itself pays more than 30 percent of the Board's budget, said that the voluntary process was set up to be flexible and allow for local creativity. He added, "We knew all along that the hardest part would be giving up control to a town."

June 27, 1997

Tom Kerr from the National Low-Level Waste Management Program came east for workshops in Connecticut on Tuesday and Friday, and offered in advance to spend the middle days in New Jersey. Originally, I had thought he could spend time in South Harrison, perhaps giving an introductory workshop on radiation to the local study group, but that was no longer an option. Instead, we arranged for him to meet with the group of five from Town Number Three. Tom Kerr is a hero to fans of voluntary siting because of his work with the Illinois Department of Nuclear Safety. While there, he had been the one to successfully negotiate an agreement with the Town of Martinsville to host a low-level radioactive waste disposal facility for the two states, Illinois and Kentucky, then included in the Central Midwest Compact. The agreement by the town officials was endorsed by local voters in two referendums, and the facility would be operating today except that it was disapproved by the Illinois Low-Level Radioactive Waste Disposal Facility Siting Commission. There are those who contend this final required approval would have been granted were it not under consideration in the midst of a gubernatorial campaign. Regardless of the final outcome, however, Kerr showed that it was possible for a town to voluntarily agree to host a facility like this.

Now, Kerr is the embodiment of what the federal government could be. He works for the federal National Low-Level Radioactive Waste Program, which is located in Idaho. He and several colleagues are paid to provide assistance to states and compacts. His office will prepare reports and analyses upon request, either doing the work themselves or subcontracting to knowledgeable consultants. In addition, he will come to a state on a moment's notice to conduct a workshop to introduce a group of people to basic radioactive science, to discuss his experience with the voluntary siting process in Illinois, or to help with any other issues related to radioactive waste disposal. His office is financed entirely by the U.S. Department of Energy. He can come into town and sincerely say, "I'm from the federal government, and I'm here to help."

The meeting took place yesterday in the office of the local economic agency, located in town at the local community college. We had been told earlier that the year-old commission had deliberately chosen an office outside of the municipal building to provide some distance from day-to-day governmental functions. Seated around the proudly displayed new conference table which we were the first to use, were Kerr and the same people from the town and the state who had met two weeks earlier in Trenton.

Kerr's description of the Martinsville, Illinois process provided an inspiring template for the discussion. If the facility is built in this town, I will look back on the accidental scheduling of this meeting as a lucky and important break. The group of five was clearly fascinated, as we all were, to hear from someone who had done it, who had actually signed an agreement with a town's governing body to host a disposal facility for low-level radioactive waste. They had so many questions that the hour-long meeting that began at 10:30 a.m. didn't end until almost 2 p.m.

In some ways, the issues we discussed were familiar from our experiences in other New Jersey towns, but having an entrepreneur who wants to make money from the beginning of the process makes it more complicated. There could be advantages in having someone involved who is trusted by the town government and has a lot of basic knowledge on this issue. This would be useful when residents are first notified that a group of their neighbors believe the disposal facility is worth exploring further. He could be an immediate and reasonably informed coordinator, collector of questions and issues, and arranger of meetings, something we've never really had in any of the other towns. At the same time, he is putting time into this now and feels he should be paid for it. After all, we have said we will reimburse a town for all reasonable expenses it incurs in considering the issues.

Because the Board can issue checks only to local government agencies without going through a lengthy competitive bidding process, we reached general agreement yesterday that we could reimburse the economic agency if it chooses to hire this fellow and he submits his expenses to them. How they would hire him without their own local competitive bidding process, I don't know, but that is largely their problem.

One of two related questions is when do the costs start? The entrepreneur feels he has already put more than $30,000 of his time into this project. Since he is a lawyer in private practice, his time costs more than the local coordinators we had been envisioning. I don't want to start reimbursing him until we have some sense of an overall budget and game plan on which we agree. Since we can't agree to the amount of money in his first proposal, we need a new budget, which he is now preparing.

The other question is what the community reaction will be if the town decides to go forward and people then learn that a lawyer from Newark has already received a substantial amount of money to look into a project they are just hearing about for the first time. Will he have any credibility to serve as the honest broker he wishes to be?

During my years in state government, I have been repeatedly frustrated by all the rules limiting flexibility in many matters, including the expenditure of money. Like many other people, I have thought that some of these rules must waste money. For example, they have sometimes prevented state agencies from purchasing needed supplies at a discount store because it is not on a preapproved list of vendors. And I continue to believe that the greater than usual flexibility the Siting Board enjoys is essential to our ultimate success. But in these discussions, as with Fairfield last year, I do wish we at least had written guidelines to help frame what types and amounts of payment are acceptable.

At the meeting, we also discussed beginning to examine specific sites in the town. The Siting Board's contract with Foster Wheeler enables us to ask the environmental consulting firm to promptly undertake analysis of any site or sites in New Jersey. This arrangement provides us with additional flexibility that is unusual for a government agency and very welcome. Each year, the contract is continued into the next, so that when we get interest from people in a town, we do not have to say, "We'll just put out an RFP for a consultant and be back to you in a year or so to have the work done.'"

I offered to ask Foster Wheeler to begin the preliminary site investigation process now, while we continue to discuss the other issues. This is work that we know will have to be done sometime. Doing it soon could give us all more information about how many sites, if any, seem promising, and could lessen the amount of work that needs to be done after a decision is made to proceed and begin public discussion. At first, we talked of having Foster Wheeler contact the economic agency for any local information they needed. By the end of the meeting, however, the group of five was a little more nervous. We decided to have Foster Wheeler just collect the mapped data that can be found without making any local contacts.

The other issue that went back and forth during the meeting was whether I should give local talks. At first, the director of the economic agency was going to get me invited to the Rotary and Exchange clubs to which she belongs. I liked this idea because it would help more people in town know something about the issue should it be considered further. It might also identify people who would consider helping in one way or another. But I think the Township Committeeman's attitude by the end of the meeting put that on hold, a decision my colleagues in the office agree with and I don't.

One passing comment that may or may not have future relevance was the Township Committeeman saying of his fellow Township Committee members, "We're not getting along so well at the moment."

One nice phrase from the meeting was the chair of the economic agency referring to "the rubric I want to hang my hat on." After the meeting, we found a good local diner that is open 24 hours a day.

September 4, 1997

After the flurry of activity in June, the pace of discussions with Town Number Three slowed over the summer. The entrepreneur has sent us revisions to his proposal, but we have reached no agreement. The group of five has still not made a commitment to actually pursue the possibility of siting the disposal facility.

This period of calm has brought into focus the major downside of the Siting Board's relative independence. I believe we are much more likely to be successful than if the low-level radioactive waste siting program had been housed, for example, within the Department of Environmental Protection or the Department of Health, rather than as a guest on a floor of the Department of Personnel building whose staff refers to us as "the glowworms." We are able to focus resources on one mission, and neither our staff nor our funds are pulled away to help with other emerging pressing problems. We also have greater flexibility in hiring and spending and can make quicker decisions.

Some of this flexibility is a product of the high level of mutual trust and respect within the Board and between the Board and the staff that can more easily mature in a smaller organization. The only problem is that when work on this one subject is slow or dead, there are no other tasks to which we can reassign ourselves until things pick up.

A more significant and newly pressing issue for us is accomplishing our public purpose while still complying with the laws and spirit of open, non-secret government. It is ironic that this is a problem for us because the decisions to be made under the voluntary siting process will be so unusually accessible and participatory. And this problem is being caused by just one person.

We have learned from our experiences in various towns that formidable opposition rises literally overnight, but support for considering the disposal facility takes time to develop and must be nurtured. Our conversations with Town Number Three have attempted to build from that lesson.

Next week, I expect to sign an agreement with the local economic agency committing up to $200,000 as reimbursement for expenses it incurs while considering the disposal facility over the next four months. The agency will use about three-fourths of the money to hire the entrepreneur to prepare an analysis of the pros and cons of the disposal facility, and to recommend how to conduct the voluntary process if the town decides to proceed further. The rest of the money will be spent directly by the agency to double its staff size to two. This will allow it and, by extension, Town Number Three to better explore the feasibility of attracting other industries to a complex centered around the disposal facility.

The local agency will study this concept for four months. If its members are convinced that the disposal facility could be good for the town, they propose to then make a presentation to the Township Committee. They would recommend that the township formally enter the voluntary siting process, and they would propose a specific plan of action to encourage full and productive public exploration, culminating in a township-wide referendum to be held about a year later. They would also negotiate with the Siting Board at that point for additional money to continue to support their efforts.

At the Siting Board office, we are excited for a number of reasons. First, geographically, the town itself continues to seem promising. It has several open sites that appear viable, proximity to the Salem and Hope Creek nuclear power stations, and the recent experience with the cogeneration facility. More significantly, however, several years ago town officials and residents had appeared to support hosting an incinerator when the Hazardous Waste Siting Commission had proposed building one adjacent to the town's major industrial plant. Although the incinerator was never built there or anywhere else, we have learned that this was the only town considered in which substantial local opposition did not arise.

In addition, because Route 295 and the New Jersey Turnpike run through the town, it is fair to say that most of the low-level radioactive waste generated in New Jersey and the states to the north already spends short amounts of time passing through here. Also, the town is in Salem County, rather than Cumberland County, so the official county opposition we faced in Fairfield is not inevitable. Finally, and significantly, the town has a distinct municipal agency ready to take responsibility for this exploration.

Our discussions with the town so far have been encouraging. The group of five that we have now met with several times seems smart, enthusiastic, and willing to both listen and compromise. Initially, they envisioned building the facility on municipally owned waterfront land. Once we told them the land was too wet, they decided to continue even though this meant the facility couldn't be sold to residents as the answer to quite as many municipal problems. And, through our discussions over the summer, they have agreed to move ahead with far less money for reimbursement of expenses, and none of the money they had requested as outright grants to the other municipal agencies.

There are two novel aspects to this proposal that have revived my faith that

New Jersey may someday have a disposal facility. First, the local agency has no intention of proposing to restrict the facility to New Jersey waste. Rather, members expect to speak only of the amount and type of waste they would accept. This is important, for not only does it dramatically increase the possibility that the facility could be economically viable, but it may also lessen the controversy raised in other towns over a state or compact's authority to deny access to out-of-state or out-of-region waste.

The second innovation is to modify New Jersey's voluntary siting process so that negotiation between the town and the Board over a binding facility development agreement occurs before, rather than after, the expensive and time-consuming site characterization work. The local group of five proposes this schedule because they believe all the information necessary for a decision will be known at that point, and because they believe an earlier agreement could enable the economic benefits to start flowing sooner.

We like this schedule because it means the Board would not have to commit the millions of dollars needed to analyze and evaluate the site until the municipality was committed to hosting the facility. This revision would actually make New Jersey more consistent with the plans adopted by all the other states with voluntary siting programs.

Yet, when the Siting Board held its regular monthly meeting today, I said none of this. How could I not tell the Board I work for that we have reason to be encouraged about our work and that I am agreeing to spend $200,000, more than ten times the amount we have paid to any other municipality?

I faced a choice. To avoid even the appearance of secrecy, I could have outlined this entire proposal at the Board meeting today, and perhaps even had a representative of the local group make a presentation. It would have been a good, quiet discussion. While Board meetings are open to the public, and in fact are advertised with expensive legal notices each month, almost no one ever attends who is not actively supportive of the siting process. And reporters, who receive individually mailed invitations, are rarely present.

The discussion would have been noted in the minutes of the meetings. The minutes would have been approved by the Board at its meeting the following month and then forwarded to the governor. Ten days later, unless the governor raised an objection, the minutes would have been sent to the mailing list of about 50 people who over the years have asked to receive our minutes.

A year ago, I would have made that presentation to the Board. The local group would still have had the time they feel they need to quietly learn whether the disposal facility was something they wanted to recommend. They would have had time to then develop a plan to promote civil civic discourse about the possibility.

But recently, one of the members of CHORD asked to be added to the list of recipients for copies of the Board's minutes, and she has already written to request every document referenced in the first few sets she received.

So, had I talked about this town this morning, in roughly six weeks a woman committed to keeping a disposal facility out of New Jersey would have learned about it. Although there is a possibility that she would have just added the minutes to a pile and not read them for months, there is probably a better chance that within days she would have made phone calls, put out a press release, and posted a notice on the web mobilizing the people on her mailing, phone, and e-mail lists to stop the process before it could really begin. Based on our experience to date, it seemed entirely likely that she could succeed.

So I didn't talk about the town's plans at the Board meeting this morning. Instead, over the last few weeks, I spoke with most Board members by telephone, and encouraged several to attend this afternoon's meeting of the finance subcommittee. At that meeting, which does not need to be open to the public as long as a quorum of the Board is not present, we did discuss in detail the town and its proposal. While no votes were taken, individually every member present, from the representatives of nuclear utilities to the representative from the League of Women Voters, supported signing the agreement and concurred with my decision to keep the discussion out of the formal Board meeting.

One of the purposes of New Jersey's Open Public Meetings Act is to prevent a government agency or official from spending significant sums of money out of public view. Yet it can also have the effect of preventing the members of a Board from meeting in private to plan strategy, discuss different options, or just get to know each other.

This act and similar laws and regulations were adopted around the country in the wake of the Watergate scandal in an effort to check the arrogance and secrecy that historically defined some government officials and agencies. These rules, which were a needed reform to real abuses, now need to be reformed themselves, but it is hard to see support emerging for a movement to restore just a little bit of secrecy to government. Yet some such changes may be necessary, as in this situation, to enable citizen volunteers on a state commission to meet behind closed doors to consider how best to accomplish their assigned public mission.

September 25, 1997

I had told the individual members of the Siting Board and the local group of five that I would sign a contract with the mystery town by September 12. Two weeks later, it is still unsigned. Issues are being raised by Susan Roop, the state's deputy attorney general in charge of contracts, and language is being traded back and forth. Some of the changes she suggests will make the document clearer and more responsive to potential situations that could arise. Others, she says, are necessary just to make it legal. I think we will eventually resolve these issues and be able to sign the agreement.

It has disturbed me a bit that so many of her comments are new information to the entrepreneur, who is also a lawyer, but I know that contract law is not an area in which he has specialized. More troubling is that Roop clearly thinks the arrangement proposed between the Siting Board, the town, and the entrepreneur is very peculiar and could be subject to great criticism. Although her concerns are not about legality and therefore are not really part of her assignment here, she is a smart, public-spirited person, and if this makes her suspicious, how will others receive it?

The Siting Board can explain that spending $200,000 to try out a different approach is a good investment in the context of a program that is now spending more than a million dollars a year and so far, one could say, has nothing to show for it. Town officials can explain that they were approached by an entrepreneur with an interesting idea that he would explore for them without cost, because he could get all the necessary funds from the state.

But what will be the public perception of one lawyer with no real experience in this subject earning $190 an hour, and up to $150,000 for four month's work? Will this agreement survive the red-face test? Will criticizing this arrangement be an easy way out for political leaders, editorial writers, and others who may be intellectually open to the possibility that the facility would be safe, but are still emotionally uncomfortable with the idea?

We take as a given that opponents of the disposal facility in any town will criticize the process that has brought it to public attention. Despite all the thought that has gone into designing the voluntary process, no one has ever said about this or probably about any other public issue that though they disagree with the goal, they think it is being considered fairly. In the face of that reality, it is tempting to just move ahead, worrying increasingly less about the nuances of the process. But there are points at which potentially open-minded people will say that, regardless of whether the idea has merit, the process stinks. I think we risk that here, but I can't think of any alternative.

The rate and amount of money going to the entrepreneur has made all of us uncomfortable from the beginning, but he eventually convinced us that his plan was worth trying and that he would only do it if it was in his financial interest. We focused then on dramatically reducing the total preliminary commitment he had been seeking for himself. This has now dropped from the original $1.3 million to the present $200,000. Now I am sorry I didn't push harder to lower his fee, though even cutting it in half to $95 an hour would probably do little to mitigate a likely public perception problem and a potential public relations disaster.

The entrepreneur has been our central point of contact in these discussions, but this afternoon I called the Township Committeeman who has been in several of our group discussions to express my concerns. He is the one elected politician in the mix so far, and I wanted to know what he thought. Like the Cumberland County freeholder director I had approached during the Fairfield discussions, he too owns a local food market. When I reached him there by phone, he immediately said that he understood my concern. "You're asking a question that only God can answer in the end," he said. "We both know that we don't have a lot of control. These things don't happen according to plan so much as the way they just happen."

In the end, he said, continuing down the path we have started is the only way to go and does have some chance of success. I had half-hoped he would suggest that the town and the Siting Board proceed by working together directly and eliminating the entrepreneur, but he didn't, and that would have lost us the talents and enthusiasm that the entrepreneur offers.

The Township Committeeman and I also discussed our recent discovery that there is another economic possibility under consideration for the town. Proceeding almost as secretly as the Siting Board, the Delaware River Bay and Bridge Authority is eyeing a waterfront site in the town — a site that would not be acceptable for the disposal facility — for a major industrial development. The committeeman said that "from a political standpoint, it's probably competition," and that the authority has "a deeper pocket than you do." He also felt that there might be possibilities for "synergistic effects" that could make the two developments mutually supportive and perhaps interdependent.

The Committeeman's assessment of the likely public reaction seems quite realistic. He noted that most people have "a built-up innate distrust of anything they can't assimilate into their values." His planned approach is to never miss an opportunity to talk about the benefits and to say, if asked, "You've got to be nuts not to look into this possibility." He is up for reelection this November, so it is unusually courageous of him to be even quietly pushing this issue. When I asked about the reaction he anticipates when this becomes public in December or January, he said, "I hate to say this, but that's so far in the future."

October 11, 1997

The week that just ended provided a surprising set of events that reminded me why I enjoy working in government. Explaining why may take some time.

Several years ago, I started noticing billboards that read, "Working For The Government Is Cool!" The subject wasn't immediately apparent to me, but with the logo of a television network in the corner, I thought that maybe some-one had created the program I had often imagined.

This show would be like *LA Law* or *E.R.* and various other hour-long weekly programs now almost always focused on law or medicine. There would be an office full of interesting characters each working on different projects, any one or two of which could become the focus for a particular episode. Each show would include subplots involving romance and other intrigue in the per-sonal lives of the staff.

In my show, the office would be a government agency much like the New Jersey Department of Environmental Protection where I had worked. The show would be entertaining with pathos, romance, and humor, but it would also pro-vide insights into environmental issues of various magnitudes, as well as how bureaucrats think and work. One week, for example, a lead actor might be focused on an investigation into a possible cancer cluster that has area residents alarmed, while another might be equally occupied by a homeowner picketing the Statehouse because the agency had stopped him from filling wetlands to widen his driveway. Meanwhile, staff might be comparing rumors about the next departmental reorganization and a colleague's impending divorce.

The show would be so compelling and well-written that viewers would get hooked by the characters and subplots, and inadvertently pick up a better understanding and appreciation of government. Several years later, observers would wonder how much credit the show deserved for the new public pressure for sensible reform measures, and the increasing number of young people gravitating toward careers in government.[57]

The billboards, though, were advertising *The X Files*, the soon-to-be wild-ly popular show about two federal agents, one driven and one assigned to investigate the possibility of aliens as the cause of otherwise inexplicable and generally horrible sets of events. As entertaining as it is, the show is almost the opposite of what I had envisioned. It contends that government, and everyday life for that matter, is in the hands of murky conspiracies and evil individuals, rather than being something we have both the responsibility and the capability to improve. When we meet the enemy in *The X Files*, unlike in the comic strip *Pogo*, it is never "us." To add irony to insult, surprisingly often the culprit seems to involve radiation.

While the network billboards for *The X Files* were meant to be ironic, my work this week definitely felt "cool" to me. Over the previous weeks, it had begun to seem that we were going to be unable to find a mutually satisfactory mechanism to provide funding for the local agency's preliminary exploration. But on Monday morning, the deputy attorney general sent me a set of contracts with language acceptable to her, the entrepreneur, and the town's lawyer. The entrepreneur and I congratulated each other on the phone and agreed that he would ask the chair of the economic agency to sign first, and I would then have copies to sign by Thursday or Friday.

The regular monthly meeting of the Siting Board was going to be on Thursday and, because the chairman was going to be unavailable, I called vice-chair David Steidley on Tuesday to discuss Thursday's agenda. Steidley is chief physicist in the Department of Radiation Oncology at St. Barnabas Medical Center in Livingston and has been a member of the Siting Board since its inception.

Steidley cordially asked what was new? I described our interactions with the mystery town and my delight that I was going to be able to sign a contract committing up to $200,000 to reimburse them for expenses related to the siting process. He did not share my delight.

Instead, he was surprised and alarmed that such a major action could take place without first having been discussed at a Board meeting. As I explained that the idea had been discussed and supported in subcommittee meetings over the previous two months, I realized that he was one of the two Siting Board members who had missed all those discussions. To make matters worse, as he talked I became far from certain that he was wrong. He said, "You tell us about every Kiwanis Club you speak to: How could you not tell us about this?"

The rest of Tuesday was spent trying to think of ways that the Board could formally be informed without taking away the town's ability to explore the issue before deciding whether the possibility had sufficient merit to raise publicly. At Thursday's meeting, I could describe our interactions and proposed actions with the town without mentioning its name. In addition, perhaps the Board should vote to approve the action. Or, the Board could pass a more general resolution empowering the executive director to sign payments of this scope and magnitude. Any of these approaches would provide a possibility for open discussion and evidence of that discussion in the meeting minutes and transcript.

Roger Haas, the deputy attorney general who regularly works with the board, had already determined that I had the authority to sign this contract, but he too would be unavailable for this month's Board meeting. His supervisor perhaps would attend.

Tuesday afternoon, I flew to Chicago to attend a workshop for representatives of states and compacts that had been arranged by Chem-Nuclear Systems, the operator of the Barnwell, South Carolina. disposal facility. Chem-Nuclear was facing a serious funding problem at Barnwell and wanted to get input on various options it was considering.

The problem dated back to 1995 when the company helped convince the South Carolina Legislature to reopen the facility to the rest of the country. On the strength of the company's predictions that it would receive 650,000 cubic feet of waste per year, the Legislature imposed a surcharge that was expected to give South Carolina $137 million annually for college scholarships. Chem-Nuclear's predictions proved wrong, and the State Scholarship Fund was going to receive just $77 million in 1997 because only 240,000 cubic feet of waste, instead of the 650,000 projected, was being shipped to Barnwell.

While $77 million is a lot of money, Barnwell was a sufficiently controversial statewide issue that the reduced revenue was portrayed by some as an example of the untrustworthiness of anything positive said about the disposal facility.

As a result, the South Carolina Legislature had added language to the budget for the fiscal year ending this past June 30 requiring Chem-Nuclear to provide specified payments into the scholarship fund, regardless of how much waste it received. The immediate problem was that Chem-Nuclear expected to be about $7 million short on its $24 million bill.

The company chose a surprisingly touchy-feely approach to solving the problem. It scheduled three workshops to be led by an independent facilitator who had worked for the Conservation Foundation and other environmental organizations. The first two were held in late summer with representatives of waste generators. They had been described as interesting open discussions where all sorts of ideas were solicited, written out, and distributed later to the participants.

Once our session began, it became clear that this one would be different. After describing the problem, no one asked for our ideas. Instead, a Chem-

Nuclear official unveiled his company's proposed solution. He explained that options were limited. They could not raise prices because they had found that the amount of material sent to their facility had become extremely price-sensitive. While Barnwell used to have a captive market, generators now can send much of their waste that is lower in radioactivity, but higher in volume, to Envirocare in Utah. They can then store their remaining high-radioactivity waste on their own property if they feel the Barnwell fee is too high.

Chem-Nuclear's new proposal was to raise the funds it needed by selling *in advance* much of the remaining capacity at Barnwell. Anyone with money — a generator of waste, a speculator, or a government agency — could buy "shares" that would cost $235 per cubic foot. Before purchasing the space, they would be told the prescribed handling fee they would be charged when they brought waste material to the facility. All together, they would be paying less for disposal than the current charges at Barnwell.

The major attraction to purchasers, more than the lower price, would be a commitment by the South Carolina Legislature to keep the facility open for at least 20 more years. Generators of waste, therefore, could estimate the amount of space they would need for waste over that time period and then purchase that amount or more. If they found they had purchased too much, they could sell what they didn't need.

When access to Barnwell was reopened to almost the entire country in 1995, the nation's low-level radioactive waste disposal problem might have appeared to have been solved, except for one major obstacle: no legislature can legally bind a future one. Just as the current South Carolina governor and Legislature had reopened the facility that the previous administrations had devoted 15 years to shutting down, so the next election could result in a return to a much more restrictive policy. Since public and legislative opinions on Barnwell were reported to be close to evenly divided, even supporters of Barnwell could not be sure of its future.

Chem-Nuclear thought it had found a way to effectively address this problem and guarantee long-term access. The $235 paid up front for each share would be placed in a trust fund. For the plan to work, at least 5 million of the 7.8 million cubic feet remaining in the disposal facility would have to be sold, resulting in the fund starting with $1 billion. Chem-Nuclear would hold the trust, paying $84 million it would earn in interest each year to South Carolina that the state could use for higher education. These payments would continue each year that the facility remained open, much like the current system but with a guaranteed revenue stream.

If, however, the facility remained open for at least 20 years, the principal of the trust — $1 billion — would then be transferred to the state. If it closed before then, the principal would revert to those generators and others who had purchased the shares.

Although the current governor and Legislature would be asked to agree to this plan, their successors would still have the power to close Barnwell at any time. But, in so doing, they would have to explain why such a policy change was worth handing back $1 billion that would otherwise forever help higher education in South Carolina.

Chem-Nuclear proposes to try to implement this plan with speed. On November 1, less than a month from now, they will offer shares to the current users of Barnwell. Commitments to purchase with an initial deposit of $3 per share will be due by mid-January. This money will enable the company to fill the gap in the $24 million payment it owes South Carolina for the surcharge for the last fiscal year. Assuming it has received "down payments" on at least 5 million cubic feet, Chem-Nuclear will then work to have the Legislature approve the plan during the session that begins in January and ends in June. If it passes, purchasers of shares will have to provide full payment by the end of November. In less than 13 months, a plan no one had heard of an hour earlier could be fully funded and implemented. It could potentially resolve a national problem that has been festering for almost a quarter of a century.

Chem-Nuclear's proposed solution was creative and stunning. Our room full of jaded, somewhat burned-out bureaucrats struggled to avoid admitting that we were hearing an actual good idea, one that held great promise and had previously occurred to none of us.

We found concerns to raise. How would access be guaranteed for hospitals, colleges, research labs and other small (i.e., non-nuclear utility) generators which probably wouldn't be able to afford to buy disposal space in advance of their need? Would generators and others really decide so quickly to commit so much money? Was it reasonable to think this could be enacted when South Carolina's governor and Legislature were up for reelection? And the most intriguing and probably most ignored question was raised by Richard Sullivan, New Jersey's representative to the Northeast Compact Commission: Since all of us in our home states brag about how much more sophisticated than Barnwell the disposal facilities we seek to build will be, would placing so much more waste in Barnwell be the best option for the environment?

The questions were interesting, but no major concerns or seemingly insurmountable obstacles were apparent to me. Even though everything in waste management seems uncertain and old hands always point out that apparent answers to this problem have been just around the corner before, I know that I was not the only one to leave Chicago thinking that this may be the long-sought solution, at least for the next 20 years.

The Siting Board met as scheduled the next day and began as usual with a report from the executive director. I described the Chem-Nuclear plan, using overheads both to try to make the numbers and dates less confusing and to emphasize that this was big news. While everyone recognized the significance of this proposal, there was little to discuss beyond a few clarifying questions. We agreed that this obviously merited close scrutiny and monitoring, and decided we would invite a representative of Chem-Nuclear to next month's Board meeting.

At the end of my report, the vice-chair called us into executive session. That meant sending the 15 people, including some of our own staff, our environmental consultant, and a few generators of waste down three floors to our office, where they would be called when the meeting resumed. The only people to remain with the Board members were John Renella, the deputy attorney general covering for our regular lawyer, the representative of the Governor's Office who is assigned to monitor this agency, and me. Had any members of our Advisory Committee been present that day, they, too, would have had to leave.

This closed meeting is allowed because it fell into the category of "discussing matters of attorney-client privilege." Specifically, Renella questioned whether my signing a contract with a municipal agency for $200,000, without public discussion or vote by the Siting Board, would violate New Jersey's Open Public Meetings Act.

After an hour's discussion, it became clear that two to four of the nine Board members present were now uncomfortable with the way I had planned to proceed. The Board seemed relieved when Renella said the legality of the plan was "in a gray area, leaning toward black." They asked him to do more research on the subject.

While I would have been very disappointed with this outcome a few days earlier, now I almost welcomed it because it offered time to better absorb the Chem-Nuclear proposal. In addition, I couldn't quarrel with the concerns being expressed. Very rarely can one public official commit $200,000 with no explicit permission or approval from anyone else.

During the closed session, several Board members asked me to list all the other contracts I had signed without Board approval. They meant to be supportive, thinking I would have a long list into which this contract could fit, but in reality, there were none. Despite having broad authority as executive director, I haven't had the need or the occasion to use it. Certainly, I approve payments of all kinds, but they fall within contracts the Board has already approved.

As the executive session closed, those who had been excluded were welcomed back. David Steidley, as chairman for the day, read into the record a one-sentence summary of the 60-minute discussion, and the rest of the meeting agenda was then dispatched.

When I returned to my desk after the meeting, I found a fax of the signature page of the contract between the Siting Board and the local municipal agency. It had been signed that morning by the agency's chairman just before he left the country for a vacation. I called the entrepreneur and the local agency's executive director, who were no doubt in celebratory moods, to tell them they needed to sit down before hearing what I had to say.

That was Thursday; Friday was uneventful, and the week ended.

My hour-long TV show could have conveyed why working for government in such a week was "cool." It was cool because it was full of surprises that created situations none of us had faced before. Over four days, in this fairly staid, slow-moving field, New Jersey lawyers solved problems to agree on a draft contract; a waste management company concocted an innovative strategy that might solve a long-lingering national problem; and citizen volunteers on a state board who patiently attend monthly meetings and govern by consensus raised fundamental questions about a government contract literally hours before it was to be signed. Individual, well-intentioned people, each rising to the occasion to do what he or she thinks best, had collectively created a set of circumstances that puzzled us all.

Tune in for next month's episode.

November 11, 1997

Almost as startling as Governor Whitman's near-defeat in her reelection campaign last week is that Town Number Three is still in the running and looking perhaps more promising than ever. The two surprises are completely unrelated. Low-level radioactive waste disposal may well have to be addressed by the governor during the next four years, but it is one of the many issues on which any differences in how the candidates would handle them were not discernible during the campaign.

After last month's Siting Board meeting, when I had told Town Number Three's representatives of the new Barnwell proposal and the Board's discomfort with the draft contract we had agreed upon, they had concluded that the idea was dead. But after postponing it a few weeks, we went ahead with a previously arranged meeting and gathered once again in the office of the town's economic agency the day before Election Day. I explained that we still wanted to move forward if they did, and we began to develop a modified plan. During that meeting and several subsequent phone calls, the chairman of the Economic Development Commission stepped out from the shadows to assume local leadership for the project.

A memo the chairman had just received from the town's deputy mayor asking him to report on what his group was doing with this radioactive waste idea had seemingly annoyed him into action. It has made us all realize that this exploration is becoming a point of conversation and could become public at any time.

It also led to an amusing misunderstanding. During the meeting, the chairman made reference to worrying about the "green factor." I thought he meant anti-nuclear activists or environmental groups. The entrepreneur, however, thought he was referring to money and seized the opportunity to speak of his need to be paid soon. After looking puzzled for a moment, the chairman said, "No, not that kind of green. I meant Richard Green." Richard Green, he explained, was the deputy mayor and member of the Township Committee he felt was least likely to be supportive.

As a result of the phone call from Deputy Mayor Green, the chairman wants to seize the offensive and make the issue public himself soon. He would scrap our previous plan to try to award a contract quietly. Instead, he would issue a press release and some type of mailing to every house in town as soon as the Siting Board approves a revised contract. This change has the added benefit of helping to address the concern of some of our Board members regarding the extent of public discussion of the proposed process in this town. The Board can now have a full and open debate and vote on the draft contract. Long before the minutes of the meeting are approved and distributed, town officials will have widely publicized it.

The local agency will announce that it is beginning a study that will culminate in 15 or 16 months with a voter referendum. Thus, our agreement will go well beyond the four-month draft we almost signed in October. The chairman has stressed that he would have to be able to point to significant immediate benefits from the Board to make this fly. Reimbursing the town for expenses will not be sufficient. The Siting Board will have to agree to commit almost all of the discretionary money it is likely to have available in the planning process.

Up until this moment, the Siting Board has had no short-term incentives to offer towns. But now, as a surprise benefit of being part of the Northeast Compact Commission, we have identified a total of $750,000 the Board could award to this or other towns for public purposes of their choice over and above the funds we had always pledged as reimbursement for local expenses incurred during the siting process.

Since the early 1990s, the U.S. Department of Energy had been holding funds as collateral in case Connecticut and New Jersey failed to secure adequate in-state or out-of-state waste disposal access. When the states were able to continue using Barnwell between 1992 and 1994, the Department of Energy returned much of this money to the Compact Commission. The Commission, in turn, has been holding this rebate money in separate accounts for the two states. New Jersey's share, with interest, has grown to more than $750,000.

The Board began to look at this account more closely after the venture in Fairfield, when we realized how valuable it would be to be able to give unrestricted funds as an incentive or reward to towns in the early stages of the voluntary process. New Jersey's Siting Act gave the Board authority only to sign contracts for specific goods and services, not to issue grants. The Compact Commission, however, has much broader discretion.

Earlier this year, the Commission agreed to a request from the Board that it allocate New Jersey's share of these funds for municipal grants. This agreement specifies that the Commission will send funds from this account directly to a municipality if requested by the Board. Thus, the Board gained the ability to promise towns up to $750,000 for unrestricted grants even though it lacked the authority to issue such grants itself.

Our original thought had been to use this money to provide grants to several towns. Now, however, we are ready to put all our eggs in this one town's basket. If this plan works, it will be money well spent. If it doesn't, the plan will probably fail in the early months, so only a few of the agreed-upon milestones will be met, and therefore only a few payments will have been made. Thus, a significant portion of the $750,000 would be available to offer as an incentive to another town.

We are now on a very tight timeline. We have decided that the December 4th Siting Board meeting is the one at which this contract should be presented and voted up or down. The economic agency's press release and mailing will go out that day or the next. This feels rushed to us, but we want to avoid, if possible, the news story from the local paper that begins, "This reporter has learned ..." Over the next three-and-a-half weeks, the press release and mailing need to be prepared, post-announcement meetings need to be planned, more people supportive of studying this idea have to be identified, and we have to agree on the final terms of the contract.

We have agreed that $100,000 of the incentive money will be awarded to the town in February, two months after the public announcement is made. The town will then get monthly payments totaling another $400,000 during the remainder of 1998 and a final payment of $250,000 the day the referendum is held, regardless of its outcome. In addition, the local agency will prepare a new expense budget totaling up to $500,000 to cover a 16-month period that begins December 5.

I discussed this plan in executive session at yesterday's Board meeting. Reaction was generally positive with two members speaking so forcefully in support that I suspect anyone with doubts decided to keep them to themselves.

The date for this meeting had been changed to enable the Board to hear directly from George Antonucci of Chem-Nuclear about the proposal to keep Barnwell open for 20 years. While the company remains "cautiously optimistic," others seem to have much more doubt about its likely success than I had felt when first hearing about it. Both Rick McGoey and Jim Shissias, the two utility representatives on the Board, said they thought the Siting Board should continue its efforts while obviously monitoring what happens in South Carolina. McGoey gave one of the forceful speeches supporting moving ahead aggressively with Town Number Three. "December's board meeting is the moment the past eight years have been building to," he said. "We don't have to approve it, but if we vote it down, we should recognize that we are saying the siting process is over."

These comments made in closed session served as a pep talk for the rest of the Board and the staff. I was only sorry that the two members with the greatest concerns at last month's meeting weren't present for this one. We agreed to set up a small subcommittee to meet with them before the December meeting.

The town's economic agency chair and I are now frequently having what he calls "just Bob and John talking" talks. He told me today that he sees the two of us like Lewis and Clark heading out west and all but certain to run into many obstacles we cannot anticipate in advance. Since I often speak of any progress in this effort as moving into uncharted territory, we seem likely to be able to work well together, at least metaphorically.

He is making decisions that are removing most of my areas of previous discomfort. He now agrees that paying the entrepreneur a huge percentage of the money coming to the town would be perceived badly, and he is negotiating to lower this amount.

Similarly, he is concerned that sticking to the entrepreneur's plan of signing a contract that would cover a period extending back to September would fuel suspicions that we have all been plotting in a way that unfairly excludes the public. He is, therefore, agreeable to a contract covering only prospective work. He has made an arrangement to pay the entrepreneur out of other funds for the work that has to be done over the next few weeks before the Siting Board contract takes over. He is considering retaining him for only a few months at a time so that he can evaluate both his work and other opportunities that may come along once this exploration is public.

Although we have been preoccupied with this still unnamed town for months, the Board became a campaign issue in a different part of the state. In Hamilton Township, which is the city near Trenton that has been one of our secret hopes, the Democrats running against the incumbent Republicans learned that a municipal employee had attended several Siting Board meetings. They issued statements attacking the local administration for considering having a radioactive dump in their town.

This attack prompted Hamilton Mayor Jack Rafferty to write to Bob Shinn, Commissioner of the state Department of Environmental Protection, denying that the town had any interest. Several reporters who were covering the campaign and knew nothing about this issue called me for comments. When I pointed out to one that his paper had written an editorial supporting the siting process and encouraging towns to consider volunteering for the disposal facility, he showed absolutely no interest. I thought this could have contributed interesting perspective to his story.

Rafferty was reelected, but since he responded to the issue so definitively during the campaign, I cannot imagine any way he could now explore locating the disposal facility in Hamilton. Even though we never had direct contact with anyone in his administration, I assume someone had been sufficiently interested to assign the local employee the task of observing the Siting Board meetings.

It was simultaneously funny and sad to see any political party, but particularly the Democratic Party to which I belong, try to make an issue out of this, although perhaps I should be surprised it did not happen more often. Nevertheless, much to our surprise, the Hamilton employee was back for our Board meeting yesterday.

Somewhat more mysterious are news accounts that Mayor Joe Vas of Perth Amboy has announced that he has found "a new source of revenue" that will produce $2 million in additional funds for the municipal budget. When Jimmy Carter became President and Arlo Guthrie learned that a copy of his record, *Alice's Restaurant*, had been found in President Nixon's White House library, he amended his unusually long, political, and humorous story-song. He added a reference to the famous gap in the tapes that had recorded deliberations in Nixon's office, and asked, "How many things are there in the world that are 18 and a half minutes long?"

We are all wondering now how many sources of $2 million in annual revenue there are that a mayor might not want to mention the week before an election. No one from Perth Amboy has had any conversations with us, nor has anyone requested literature that we know of, although Perth Amboy does receive all the mailings we send to every town.

While continuing our various local courtships, we try to keep the U.S. Nuclear Regulatory Commission informed of what we are doing. Even though we are far from submitting a license application, we want to design and follow a path that makes the NRC's eventual review as short and smooth as possible.

Last month, I sent the NRC three of our recent more detailed publications. I received a call the other day from our contact there saying that it would cost the Siting Board $10,000 to $20,000 to have these documents reviewed. He explained, as he always has to, that Congress has reduced their appropriations so drastically that they are now a fee-for-service operation. As a result, they must charge for time and materials for any work they do, even for a state agency. I considered suggesting that we charge him $10,000 in production costs for each report we had sent and then cancel the debt if they review them for free. Instead, I told him we had just sent the reports for their library, and he told me he would find an hour or two to look them over and let me know if he saw anything problematic.

This is not how government should operate. Agencies at all levels, perhaps at the federal level most of all, should be available and helpful. They should have the resources to respond to other government agencies, as well as individuals and all kinds of profit-making and nonprofit enterprises. Their expertise and policies should be disseminated as widely as possible. If Congress is going to force government agencies to operate like this, maybe privatizing them wouldn't make much of a difference. That, I guess, is part of the reason why this particular Congress is pushing so hard in this direction.

One final thought for today. I am in this job because one member of the Board knew my work in the state Department of Environmental Protection and suggested that I apply. Towards the end of a phone conversation with him last week, discussing the situation with Town Number Three, he said, "Well, it isn't easy." Jokingly, I said, "I thought you told me this would be an easy job," and he said, "Well nobody thought you people were actually going to go out and find towns. We just thought you were going to have lots of meetings." I take this and similar comments as compliments. I know, however, that if Town Number Three makes it to a referendum and the referendum passes, the Board, the generators, the state Legislature, and others are going to want to at least take a long deep breath before signing a facility development agreement with the town and beginning the multi-million dollar site characterization work.

November 28, 1997

I heard a discussion on the *Derek McGinty Show* on *National Public Radio* this week about how presumably smart, successful people in Hollywood manage to make bad movies. The focus of this speculation was the movie *Flubber* starring Robin Williams. A caller to the show felt it was such an obviously bad idea that its failure at the box office should have been apparent long before it was made. One of the panelists responded that maybe a few people have too much power so that not enough people have to sign off before a movie is made.

People often have a similar reaction to problems in government, and I think it often leads to counter-productive responses. Most of my remaining hope for the voluntary siting process is based on the fact that decisions can be made by only a few people. They can, therefore, be made quickly and strategically. That ability has enabled us to reach an agreement with the mystery town that will become public next week. To my surprise, I am once again optimistic and excited.

Since the last Siting Board meeting, I have spoken with the two members who were not present, and both say they will support the plan. I have also met with Cynthia Covie, the assistant counsel in the Governor's Office assigned to monitor the board, and with Richard Sullivan, New Jersey's representative on the Northeast Compact Commission. Both are supportive. The day before Thanksgiving, I sent a memo describing the plan to the governor's communications and press directors, and to the Commissioners of the Department of Environmental Protection and the Department of Health, who are represented on the Siting Board.

On Monday, a requisite small legal notice will appear in a dozen newspapers around the state. Roger Haas, the Board's lawyer, has advised us that this must be done to avoid argument about whether we have complied with the Open Public Meetings Act. Such ads must appear in print at least 48 hours prior to the meeting, and they must describe all resolutions upon which votes are planned. Thus, this notice will name the town that is considering volunteering and the amount of money the Board will give them during the initial stages of the siting process.

We have no idea whether anyone will read the ad. Will it come to the attention of a reporter at one of the local papers who will decide to look into it? Will it be read by someone who will then decide to come to the Board meeting or perhaps post a notice on the web that will lead to lots of public attention by Thursday when the Siting Board meets? Or will no one notice it? And how do we plan for those various possibilities?

The chairman of the economic agency has decided that he will meet with the editorial board of his local newspaper on Monday, and inform them of the town's plan to explore the possibility of hosting a disposal facility for low-level radioactive waste. He will explain that he has negotiated an agreement with the state that will pay their town just to study something that they can later decide to take or leave.

The newspaper will then have the option of running the story the next day or of waiting until after the Siting Board votes on Thursday. The latter would be our preference, but this way, if someone picks up on the legal notice, there is less likelihood of the town being accused of being sneaky.

This approach emerged from meetings our staff and the local group have had in their office over the last two weeks. The meetings have been well-timed. The night before the first meeting, the local chairman had discussed the proposal with his fellow economic commission members, and the night before the meeting this week, he had discussed it, in closed session, with the entire Township Committee. He met with informal, but apparently unanimous, approval at both sessions, which has both amazed and excited us all.

Several of the Township Committee members reportedly became quite attached to the idea that the $750,000 the town will receive in benefits for the 16-month study process culminating in a referendum translates to $100 for each of the town's 7,500 residents. That is a happy coincidence and had occurred to none of us when we were struggling to arrive at an agreeable number. It is essentially the maximum amount available for this purpose.

Regardless of the number, the practice of a government agency paying a town to study something to which it can say no is, as far as I know, previously unknown in New Jersey. While I certainly hope we receive better reviews than *Flubber*, I know that we could not have reached this agreement if many more people had felt the need to play a role in the decision-making process. It is important that we continue to inform key staff in the Governor's Office and in the Departments of Environmental Protection and Health of the Board's plans so that they are not surprised and so they can object, or even help, if they wish. But if we had had to get approval from all of them for this plan, I don't believe we could have done it at all, much less in any remotely timely manner.

Just two weeks ago, I felt like we and the five people from this town were playing with a large set of dominoes. We have been hoping to get them all set up in an elaborate pattern, while fearing that they would all fall down in the first angry gust of public knowledge. That still may happen, but now it seems possible that the 16-month process we have outlined may actually occur.

The Siting Board will meet on Thursday and will consider a resolution authorizing me to sign a contract with the town's economic commission. Unless the Governor's Office suddenly decides to intervene, I am reasonably certain that the Board will approve the plan with no substantive changes. By the end of the week, if not before, this will be public information throughout the town. The first public meeting will take place next Saturday, two days after the Siting Board meeting. Soon we will no longer have to speculate about how local residents will react to this idea.

At the same time, events outside of this town could intervene in this process. Of greatest interest is that the draft being prepared for Governor Whitman to consider as her budget message in January currently includes a proposal to abolish the Siting Board, or perhaps to merge it into the Department of Environmental Protection. I have not seen the proposal, nor have I or the Board chairman been officially informed or consulted about it. I know about it only because a friend in the Governor's Office told me that it has been suggested by the Department of Environmental Protection.

Much as I feel annoyed, I think the people preparing the proposal have probably been right not to involve me because I would do what I am doing now, which is to work to have it removed. Still, it feels terrible to find that people in the Department I thought I was working with cooperatively are secretly

Carneys Point
Salem County – 1997

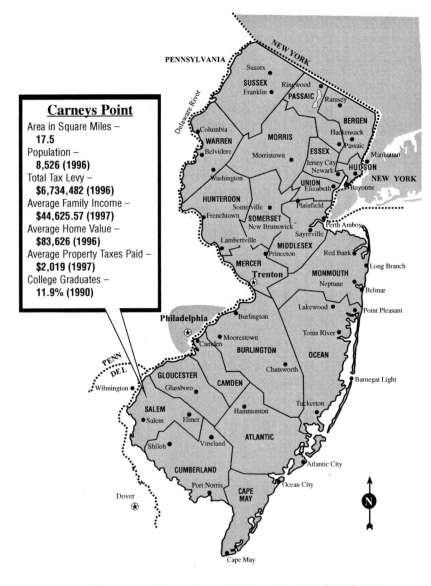

Carneys Point

Area in Square Miles –
17.5
Population –
8,526 (1996)
Total Tax Levy –
$6,734,482 (1996)
Average Family Income –
$44,625.57 (1997)
Average Home Value –
$83,626 (1996)
Average Property Taxes Paid –
$2,019 (1997)
College Graduates –
11.9% (1990)

Sources: N.J. Department of Treasury, New Jersey Almanac, NJ Legislative District Databook

plotting against the Board. Maybe when the residents of Town Number Three learn about the disposal facility proposal, this is the way they will feel about their local economic agency and the state Siting Board. I think, and certainly hope, their reaction will be different.

It appears that the Department of Environmental Protection proposal is based on the inaccurate assumption that abolishing the Siting Board could save the state money, because the Board has more than $8 million in its account. This money was raised from generators of low-level radioactive waste specifically to address the waste disposal problem. The money cannot legally be used to help fill other holes in the budget. I assume people working on the budget message know this, but some in the Department of Environmental Protection may think this is a surmountable legal obstacle. Part of the appeal of the proposal also may be to offer the governor an opportunity to reiterate her commitment to downsizing government by abolishing another agency, even a tiny one.

Another storm cloud on the horizon is a story a reporter at *The Record of Hackensack* has been researching for at least two months. She has requested all of the Board's publications and phoned several people who have nicely called to tell us of the conversations. Among the people she has called are representatives of nuclear utilities, the waste generators group, and the state Department of Personnel (to confirm the size of our staff). She also sent a survey to all 566 New Jersey mayors asking, among other questions, if their town "had been considered" for the disposal facility. Her wording makes us think she doesn't fully understand the process in which we are engaged. She has made clear that she doesn't want to talk with me, and our attempt to meet with her paper's editorial board was deemed "inappropriate at this time." Our guess is that she thinks she is writing an exposé of a government agency wasting a large amount of money on a hopeless, and perhaps dangerous or stupid, task.

It now looks like we will win the race with the state budget and with *The Record*. My strategy for derailing the budget proposal is simply to let many people inside state government know that we are actively working with a town, and that the process is extremely fragile. Thus, the memo I sent on Wednesday went to many more people than it might have otherwise. Once this town's interest becomes public, if it seems likely to last, members of the Board and staff will start talking directly with various people to try to get the proposed budget language changed. I don't think this will be a problem, though, of course, I could be wrong.

Our worry about the news story is that it could lead to statewide press and perhaps comments from legislators that would make the Siting Board look unstable. That would understandably be of concern to people in a town considering entering into a very long relationship. We plan to fax the article about Town Number Three's interest to the *Record* reporter as soon as it appears. Unless the *Record's* story runs in the next few days, she should then realize the need to rewrite it and perhaps change its focus. My guess is that the current version suggests that no New Jersey town will consider this project. Right now, it looks like we will have time to prove her wrong.

Finally, Chem-Nuclear's proposal to have its South Carolina facility solve all New Jersey's needs for the next 20 years is still alive, but according to the people with whom I speak, increasingly considered unlikely to become real. I have no idea, but I am happy to see it receive less attention as we move forward on a real possibility for a long-term home for low-level radioactive waste in New Jersey.

December 5, 1997

The schedule we had designed for Town Number Three worked perfectly, with one startling and totally unanticipated problem. Tuesday's edition of *Today's Sunbeam* announced that Carneys Point Township in Salem County was planning to enter into agreement with a state agency to explore hosting a disposal facility for low-level radioactive waste. The article was accurate and calm with quotes from local resident Bob Zonies, the chairman of the town's Economic Development Commission, and me. The front-page article appeared under a banner headline:

CP RADIOACTIVE WASTE SITE RULING COMES THIS WEEK

A casual reader of the paper would have had no way to know that there is ever any controversy associated with proposals for disposal facilities for low-level radioactive waste. For the first time in print anywhere, however, the reader would have been informed that some people think that Carneys Point should consider having one. Town Number Three now had a name and was moving forward.

On Wednesday, the *Sunbeam* provided additional accurate information under the headline:

DISPOSAL SITE STUDY FUNDS COME FROM N-WASTE PRODUCERS

We now felt well-prepared for the next day, when the Siting Board would meet and vote to authorize a contract with the Carneys Point Economic Development Commission. It seemed likely that the Board meeting would be attended by no one other than the usual small number of supportive interested observers. We had not heard from anyone who had seen the required legal notice we had nervously placed in 11 major newspapers across the state on Monday. We were in the unusual position of having spent hundreds of dollars for advertising that we hoped no one would see.

As we concluded our final pre-Board staff meeting at 4:15 on Wednesday afternoon, however, I received an e-mail from Mike Hogan, a top assistant and counsel to Commissioner Bob Shinn of the Department of Environmental Protection. Hogan's short message said Commissioner Shinn would like the Board to delay its vote for a month so that he would have time to review the details of the proposed contract.

I called Hogan immediately and explained that such delay was simply not possible. I told him of the expectations that had been raised in Carneys Point by our plans and the *Sunbeam* stories. I told him of the newsletter the Economic Development Commission had prepared for mailing the afternoon following the Siting Board vote, and of the public meeting at the local high school already announced and advertised for the following Saturday. I told him that the contract had been approved by the Attorney General's Office, and that Commissioner Shinn's representatives on the Siting Board had been consistently supportive of this plan in discussions over the last several months. Finally, I reminded him that he and I had discussed this plan in detail when we were at a conference together in October, and that I had again described it to him in detail in a memo in early November.

Hogan responded by telling me that Commissioner Shinn had just left for a meeting in Washington, and that he would pass this information along.

Many pages in most statutes creating boards and commissions are devoted to stipulating who the members will be and how they shall be selected.

Frequently, one or more slots are designated for the commissioners of major state departments, generally as a way of bringing in the expertise and perspective of that agency. The commissioners find themselves on multiple time-consuming boards and most often delegate the responsibility to a member of their staff, who is then recognized as a full voting member.

The Legislature, acknowledging that both health and environmental protection would be major issues in siting a disposal facility for low-level radioactive waste, gave two of the 11 seats on the Siting Board to the commissioners of the two departments responsible for those public policy areas. The Health Commissioner had designated Mike Lakat, a staff assistant in his department, to represent him, and the Commissioner of Environmental Protection had designated Gerry Nicholls, director of the division responsible for radiation programs. Because Gerry is often called away to regional and national meetings and conferences, the DEP had also designated as an alternate Jill Lipoti, the assistant division director who directly supervises the radiation programs. While the Health Department had changed its representative a few times, Nicholls and Lipoti had represented the DEP on the Board under all six DEP commissioners who had served since the Board's creation in 1989.

Early Thursday morning, Paul Wyszkowski, the Siting Board chairman, and I had our monthly pre-Board meeting. I described my previous day's communications with the Department of Environmental Protection.

As I was saying that I assumed my conversation with Mike Hogan had resolved the matter, Gerry Nicholls called. He reported that he had been instructed to ask the Board to table discussion of the Carneys Point agreement and, if that motion failed, to vote in opposition.

We were livid. I called Hogan, and he said repeatedly that he was only the messenger, but that we couldn't talk with Commissioner Shinn because he was in Washington. When I asked if he could at least allow Gerry to abstain should his motion to table fail, he said he would check and get back to me. By now, Board members were arriving for the meeting in an almost festive mood, believing that, as Rick McGoey had said at the last meeting, "Everything the Board has done for the last eight years has been leading to this moment."

Shortly after the meeting began, the Board went into executive session. The chairman asked Nicholls to explain the DEP position, and he repeated what Hogan had said to me on the telephone.

The governor's representative, Cynthia Covie, is included in executive sessions although she is not a member of the Board. She is part of a small Authorities Unit set up in the Governor's Office to monitor all the quasi-independent boards and commissions. Its role is to act as a conduit between the various generally small agencies and the governor, and to alert the governor to any issues of possible concern in the minutes she must review after each meeting.

After Nicholls spoke, Covie reported that Governor Whitman would like the Board to delay action until Commissioner Shinn's concerns could be addressed. She acknowledged that she had told me a week earlier that she was supportive of the agreement the Board and town representatives had negotiated, but now expressed the belief that a one-month delay was a reasonable request.

By this point, the tension in the room was intense. Two members responded that they thought the Board would lose all credibility with Carneys Point if it delayed action. They said that we might as well then vote to end the voluntary siting process because no other town would be likely to trust us either. I agreed that with all the publicity in Carneys Point focused on this Siting Board meeting, the process would not survive if the Board delayed its vote.

The Board returned to open session, and Mike Lakat made a motion to pass the resolution authorizing me to sign a contract with Carneys Point. Covie then whispered to him that he ought to check on the Health Department position on this matter, and he left to do so. It turned out that no one in his Commissioner's office knew anything about the Department of Environmental Protection concerns, but while on the phone he missed the key votes.

Nicholls then made a motion to table which, after a suspenseful minute of silence, received no second and died. The Board then discussed the Carneys Point plan and passed the resolution to enter into a contract with its Economic Development Commission by a vote of 7-1.

When I returned to the office, the reporter from the *Sunbeam* called. He asked whether the vote had been unanimous but, to my great relief, didn't ask who had voted against it or why.

Among the outcomes it had never occurred to me to consider was that the Siting Board's greatest step forward would occur over the opposition of the Department of Environmental Protection. Commissioner Shinn's stated concerns are just procedural, and we can quickly resolve them. But if this process gets off the ground in Carneys Point, we could spend much of the next six years answering questions about whether a location there could be safe if the Department of Environmental Protection had opposed a siting process in the town from the very beginning.

When the meeting ended, the Board, as it does each month, descended to our office to have lunch and then break into subcommittee meetings. The lunches are usually accompanied by lively and friendly conversation, but this day the mood was quiet and awkward with no one quite knowing what to say.

Our plan with Carneys Point had been to sign the contract the day after the Board vote, but as the governor's representative pointedly reminded us, officially the vote was not final until the governor approved the minutes of the meeting. This could not happen until the Board met again to approve them, after which the governor has 10 working days in which to veto any actions with which she disagrees. For most matters, such review is a formality, and we have often moved ahead on the basis of a vote before minutes were prepared and reviewed. Given Commissioner Shinn's opposition, however, and Governor Whitman's apparent openness to it, we had to delay the signing.

When the afternoon subcommittee meetings were over, I decided to go to the Statehouse and see if I could talk to people who might be helpful. We had a clear conflict between two state agencies. It seemed that a function of the Governor's Office should be to listen to both positions and then quickly make a decision.

"Going to the Statehouse" is an important skill that I have found difficult to master. It sounds impressive and even more so if you say, as I sometimes do, that I am "going over to the Governor's Office." For me, at least, it is usually like going to a party to which I'm not sure I have been invited. You don't make appointments with people and, for the most part, you don't drop in on them in their offices. Rather, you wander the halls trying to look busy while hoping to run into people you want to see. The best season for this sport is on the Mondays and Thursdays when the Legislature is in session because then the halls are full of others doing the same thing. You end up having many conversations with people who, like you, are simultaneously talking and scanning the hallways searching for the people with whom they really need to speak.

This particular Thursday, I was lucky and quickly found several of the people I thought might be able to help. One of them laughed when I described my saga and suggested I talk to another of the governor's top aides. After about an hour, I left feeling that whenever this conflict surfaces in the Governor's Office, there would at least be someone in the room who would have heard the situation described from the Board's perspective. And I felt confident that Governor Whitman would end up signing the minutes to ratify the Board's agreement to fund a 16-month study by the Carneys Point Economic Development Commission.

As governor, Whitman has forged her own path on environmental matters. She took office in 1994 promising to cut taxes and the size of state government, and the Department of Environmental Protection was one of the agencies hit hardest. Not only did it lose staff, but New Jersey's ridiculous Civil Service rules led to demoralizing reassignments for many of the survivors. On the other hand, she has reconstituted a moribund open space commission and installed as chair Maureen Ogden, a widely respected former member of the state Assembly. Whitman has charged it with developing a proposal to provide long-term stable funding for open space preservation, and she also has been an increasingly vocal supporter of New Jersey's controversial state land use plan.[58]

The governor's response to this situation, however, may be more determined by her attitudes about private industry and government. She has talked a great deal about supporting and enhancing industry in New Jersey. I imagine she will take notice of the private sector support for continuing the siting process. However, while Whitman clearly enjoys her job, she does not seem to have particular respect for the 60,000 state workers in her employ, nor for the procedures and rules under which they labor. She has adopted the Ronald Reagan technique of sometimes criticizing government as if she were not a part of it.

December 11, 1997

The events in Carneys Point have been occupying much less of my thought and time than the conflict with Commissioner Shinn, but they appear reasonably promising. The Economic Development Commission's public meeting on Saturday was attended by 100 people according to the *Sunbeam*, and by 50 people according to others. About 10 spoke against the study and two in favor, but the overall sentiment was reported by all to be strongly opposed.

Those of us from the Siting Board had stayed away at the request of the Economic Development Commission members who felt this first meeting would be better if it was just a conversation among neighbors. Bob Zonies, the commission's chairman, was leaving for a trip to the Galapagos Islands the next day. Before he left, he called to tell me about two issues raised at the meeting that he thought needed to be addressed quickly. Both related to trust in government.

The first concerned the local referendum. In the *Voluntary Siting Plan*, the Board had deliberately not required a local referendum. We thought some towns might want to proceed without one the way they do for virtually all other land-use decisions, and we felt that if the officials in one town chose to let their voters directly decide the issue, they ought to receive credit for that choice. It could give a mayor or other official more backbone for supporting this controversial study if he or she could say, "I know the state says this is safe, but you people feel strongly about it, and that is why I say you should have the final word."

Carneys Point was the first town to include a voter referendum as part of its plan from the beginning, but at the public meeting someone pointed out that such referenda are not binding. And this is true. Under the New Jersey Constitution, it is town governing bodies that have final decision-making authority. All the members of the Township Committee could vow that they would follow the dictates of a referendum, but when the vote came they would not be legally bound to do so.

So the first challenge Zonies gave me was to figure out how to reassure skeptics that if the residents voted against the disposal facility in a referendum, the Township Committee could not vote to approve it anyway.

The second concern was similar, though more difficult to address. What could stop the Siting Board from placing the facility in Carneys Point even if the referendum was defeated? Sure, the Board says it wouldn't do that, and that it has ended the process in other towns that said no, but look at all the money it is offering to spend here. Does anyone really think it would walk away from that? You don't get something for nothing in this world.

Monday morning, we met in our office with JoAnn Sumner and Al Telsey, who were respectively the Economic Development Commission's executive director and the heretofore-unnamed entrepreneur. Also with them was Michael Willman, a representative of the advertising agency the Commission had recently hired. The agency is called WMSH both after the initials of its partners and after its slogan, "We make stuff happen." When they had set up the meeting several weeks ago, I had at first been upset that people from a radio station were being invited to a planning session.

We went over everyone's impressions of Saturday's meeting and discussed the newsletter that had been mailed to all Carneys Point residents. Drafted by WMSH to be the first issue of a regular publication from the Economic Development Commission, it was a handsome, four-page, 9-by-7-inch production with a nice logo.

Bob Zonies — Chairman
Jerome Clement — Secretary
Adam Gagliardi — Treasurer
John Bibeau
Tony Booth
Dr. James Fields
Allan Wormack
M. JoAnn Sumner — Director

CARNEYS POINT TOWNSHIP
ECONOMIC DEVELOPMENT COMMISSION

Volume 1/Number 1/December 1997

This newsletter is published by the Carneys Point Township Economic Development Commission
460 Hollywood Avenue, Carneys Point, NJ 08069 • 609-351-9022/fax 351-0282

(FEATURE)

Carneys Point Wins 'No Obligation' Grant To Study Hosting Low-level Radioactive Waste Disposal Site

Municipality Will Receive $1.2 Million To 'Consider' Low-Rad Site;
Mayor, Township Committee, and EDC Promise Referendum

Carneys Point has been awarded a "no obligation" $1.2 million state grant to study hosting a low-level radioactive waste disposal facility.

The announcement of the award was made on December 4 thby the New Jersey Low-Level Radioactive Waste Disposal Facility Siting Board.

The grant, which does not obligate the Township in any way to accept the facility, will cover all of the costs associated with a year-long Township study of the feasibility of hosting the facility. Particular emphasis will be placed on structuring a guaranteed benefits package should the Township ultimately decide to host the site. In addition, merely undertaking the study will add an estimated $750,000 to Township revenues.

In announcing receipt of the grant, Township Mayor James Kain congratulated the Township's Economic Development Commission for negotiating the study grant.

"Commission Chairman Bob Zonies and his colleagues did a superb job for the Township in working with the Siting Board," said the mayor.

"They worked for many months to make this a complete 'win' for the Township. The bottom line is that we'll get $1.2 million just to take a good, long, no-holds-barred look at whether we want to even consider having a low-level radioactive waste facility here."

"And, if we do decide to host such a facility, it would mean at least $2 million per year in new revenues for the communityfor up to 50 years.

"But before we even get close to a decision, we're going to get paid just to 'think' about this for a year. We'll explore every detail and every nuance. We'll have exhaustive public hearings and debate. And then, when it's over, we'll have a Township referendum. The voters will ultimately decide this issue.

"Even though the referendum won't be legally binding, *the Township Committee and I have pledged to follow the referendum results.*

"As we see it, Carneys Point can't lose. We're getting the equivalent of $100 for every man, woman and child in the Township just to thoughtfully consider the issue," said Mayor Kain.

First Public Community Forum Scheduled:
Saturday, December 6th, 2PM
Penns Grove Regional High School Auditorium

(IN THIS ISSUE...)

ECONOMIC IMPACT: The Economic Development Commission's newsletter stressed the benefits the township would receive simply for considering the siting plan.

The newsletter's front-page feature story headline was:

CARNEYS POINT WINS 'NO OBLIGATION' GRANT TO STUDY HOSTING LOW-LEVEL RADIOACTIVE WASTE DISPOSAL SITE

MUNICIPALITY WILL RECEIVE $1.2 MILLION TO 'CONSIDER' LOW-RAD SITE; MAYOR, TOWNSHIP COMMITTEE AND EDC PROMISE REFERENDUM

Inside, the newsletter had a number of articles including a brief history of the Economic Development Commission and a "Tentative Schedule of Community Meetings" with nine events listed that all offered opportunities for public discussion about the disposal facility. On Thursday afternoon, it had been mailed to local postal patrons, and by Saturday, every household in Carneys Point should have received a copy.

At our meeting, we all congratulated Willman, the "W" from WMSH, for creating such an impressive publication. He then suggested a strategy to address the two issues that had been raised at Saturday's public meeting. Telsey, the entrepreneur, would draft language covering both issues that could be added to the unsigned contract between the Siting Board and the Economic Development Commission. The mayor and the Siting Board would then review the language informally and if both found it acceptable, they would publicly agree to amend and sign the contract. This sounded like a good plan to us all.

The mayor, Jim Kain, had already said he would not let the Economic Development Commission sign an agreement unless these dragons were slain. Over the next few days, he suggested that the whole issue be placed on a back burner for a few weeks while "everybody does their Christmas shopping."

December 21, 1997

Governor Whitman announced that she is reappointing most of the members of her cabinet, including Bob Shinn as Commissioner of Environmental Protection. She is also moving Judy Jengo, her trusted environmental policy adviser for the past four years, over to the Department of Environmental

Protection to serve as Deputy Commissioner. Before joining Whitman's staff, Jengo had been an aide to the Assembly Environment Committee.

Jengo's appointment is being widely viewed as acknowledgement that the governor doesn't fully trust Shinn as commissioner but has decided that it is easier to keep him than to find a replacement. Keeping him will also enable her to say that she has provided the Department of Environmental Protection with the continuity in leadership it has always lacked. He will be the first Commissioner to serve for more than five years in the agency's 27-year history.

This is an odd solution to a problem, though it is not uncommon in government. Rather than picking the best people to run agencies, giving them policy direction, and then asking what help they need to get the job done, New Jersey governors and other elected executives sometimes feel bound to appoint or keep someone about whom they have reservations. They seem to comfort themselves and feel more in control by placing someone they consider loyal to them in the number two spot. For one reason or another, such as age, experience, or a controversial reputation, this person is usually not considered politically viable to be given the top job.

This was not the first time that a Commissioner of the Department of Environmental Protection has been given no voice in the selection of his or her second-in-command. When Whitman's predecessor as governor, Jim Florio, was elected in 1989, he wanted to appoint a woman to head the department. He settled upon Judith Yaskin, a candidate considered by many to be poorly matched for that particular job. He then asked Michael Catania, a man with whom he had worked more closely, to serve as Deputy Commissioner. Four months later, Catania quit in frustration, and within a year Yaskin had been replaced as Commissioner and redeployed to the judgeship she had left to join the DEP.

Perhaps it says something positive about the success of the women's movement that this time the male Commissioner is the token and the competent person apparently being brought in to really run things behind the scenes is a woman. There must be a better way, however, than appointing a Commissioner to a very difficult job, and then depriving him of the opportunity to choose the person he thinks will do the best job in one of the very few positions that is exempt from Civil Service requirements. I imagine it will be hard for Commissioner Shinn not to feel sometimes that the adjoining office is occupied by someone sent by the governor to be a spy or parole officer.

What would happen if governors would just appoint someone they trusted, or talk directly to the offending official about what they could do better or differently? But, like people in most walks of life, even powerful executives apparently find it easier to maneuver around people causing problems than to confront them directly.

For the Siting Board, this means that Bob Shinn's concerns are not going away. On the other hand, now that the uncertainty hovering around his reappointment is gone, perhaps he will find time to meet with us.

I fear I have been making a pest of myself, sending memos and making phone calls to try to create a meeting between Shinn, members of the Board, and myself. I have been unsuccessful. Commissioner Shinn is clearly a busy person, and this is a busy holiday season, but it is nevertheless very frustrating. Taking him at his word that he wanted a month to go over the language of the contract, the weeks are ticking away. Although he didn't request a copy, I sent one to him, but he has offered no comments on it to date.

I do not, however, take him at his word. I have learned that he believes the Chem-Nuclear proposal to keep Barnwell open for 20 years is likely to be successful. He has reportedly concluded, therefore, that it doesn't make sense to commit a significant amount of money to finding a site for a disposal facility in New Jersey. Moreover, he apparently has an idea that the money in the Siting Board account could be used to buy space at Barnwell for New Jersey waste generators.

I say "apparently" because here, too, there is a lack of direct communication. The Commissioner and Mike Hogan talk with people in the Governor's Office. Some of them tell me what is being said. Then I call Hogan and he tells me that he is only the messenger and that we have to meet with the Commissioner, but that the Commissioner has no time available.

December 28, 1997

Last Monday night, the Carneys Point Township Committee voted 4-1 to end the siting process. They were holding a routine end-of-the-year meeting to pay bills, but apparently plans had been laid over the previous days. About 40 objectors to the agreement and no supporters filled the Committee Chamber. John Brandt, the Township Committeeman who had supported the process from the start, noted that the Economic Development Commission had followed the mayor's instruction not to push the issue until the new year. He said the town was giving in to demagoguery in listening to 40 or 60 objectors in a town with

8,500 residents, but he lost. And, as before in this voluntary process, months of planning were ended by one unexpected and somewhat spontaneous meeting. Bob Zonies called me first thing Tuesday morning to personally deliver the news, which I appreciated. He then called Wednesday morning, as Christmas Eve approached, to say that he was hearing from people who were angry at the Township Committee and were thinking of attending their next meeting and asking that the study be revisited. That would be great, but it certainly seems unlikely. Similar voices have popped up shortly after the end of the siting process in almost every town, but they have yet to lead to any local reconsideration.

The vote in Carneys Point would appear to resolve our specific impasse with the Department of Environmental Protection, but the larger question of the future of the Siting Board is now very much on the table. When the *Sunbeam* reporter asked my reaction to the Carneys Point decision, I said that the town had developed an innovative approach to siting and that another town now might want to adopt it. But I'm not sure the Board will want, or be allowed, to give another town a chance.

Then, on Sunday, out of the blue, *The Times of Trenton* ran a lead editorial entitled "A No-Lose Proposition," in which it described the money Carneys Point would have received for undertaking the study. The editorial began by saying, "One of the truly thankless jobs in New Jersey is that of the Low-Level Radioactive Waste Disposal Facility Siting Board," and concluded:

> *If Carneys Point decides not to take the board up on it ... surely there are a few towns around the state that can use some extra money and whose leaders aren't so fearful of their constituents that they would refuse to confront factual information and examine it objectively"*

While the editorial is an unexpected endorsement, it makes no sense for us to continue the siting process unless we can gain assurance that the Department of Environmental Protection will support the effort. That assurance could come if we could convince Shinn that he and Hogan seem to be the only people who don't work for Chem-Nuclear who still think their 20-year plan is going to work. Or it could come because Governor Whitman or one of her top aides instructs the Commissioner to be supportive. Either avenue would enable the Board to consider continuing the siting process, but neither seems likely to occur.

From what I can tell, and again I know this from the kindness of friends in high places rather than from direct conversation, people in the Governor's Office would rather not get involved in trying to build a disposal facility in New Jersey. Whether they agree with Shinn that the Chem-Nuclear proposal will work out or just see this as a problem they can ignore I don't know, but even before Carneys Point said no, apparently Governor Whitman had accepted Shinn's recommendation that she veto the Siting Board's minutes.

Moreover, I gather that I am now being defined as the problem. I believe this is being said to me as a kindly warning. I am told that Governor Whitman herself thinks I should have stopped the Siting Board from voting and "creating" this conflict. Or perhaps we just weren't supposed to be successful at all.

Apparently, once I knew Commissioner Shinn's views, I should have advocated them before the Board, even though I don't agree with them and even though I left the Department of Environmental Protection more than three years ago when I was hired by the Board. Commissioner Shinn even spoke at my farewell dinner. This is a notion of loyalty I don't understand.

I now feel angry and powerless. The process in Carneys Point might well have ended even if Shinn had been fully supportive, but maybe not. Within a few days after the December Board meeting, the entrepreneur had stopped work because he was concerned that he was never going to receive payment for all the time he had invested in this project. His decision was a direct result of his awareness of the dissension within state government. Perhaps his active involvement in Carneys Point in the two weeks following the Board's vote would have provided greater reassurance to local officials and kept the discussion alive. Perhaps he would have been present to rebut the charges made when the Township Committee ended the discussion.

My dismay derives much more from the process than the result. Whether the state should begin a siting process in Carneys Point was not a question with a clear-cut answer. A good discussion on the topic could have been conducted over the summer, in September or October, or even in November. The Board could have been convinced that a reasonable position would have been to put the siting process in New Jersey on hold as other states and regional compacts have done. We could watch to see if national developments might yield more secure long-term out-of-state disposal options. Even if the Board had not been convinced, it would not have gone ahead if the members had known that the Commissioner of Environmental Protection had objections that were supported by the governor.

Instead, one of the highest-ranking officials in state government chose to object for the first time just 17 hours before a vote that had been planned for months, a vote planned and endorsed by his staff. Moreover, he intervened disingenuously, acting as if his concerns would be addressed by minor word changes when really they were much more fundamental, and he raised them by dispatching an aide to send an e-mail and then making himself unavailable for discussion for weeks.

Rigidly following the steps of a process can lead to bureaucracy in the worst sense of the word and can be a justification for decisions that don't make sense. At the same time, there needs to be enough respect for process to enable government decisions to be predictable and trustworthy. Achieving a balance is one of the great challenges of government agencies. When I worked in the Department of Environmental Protection, we were often accused of maintaining too much emphasis on the process and ignoring its effects on a particular ecosystem or landowner. In this case, at least, I think the department faces a different charge.

January 8, 1998

As today's scheduled Siting Board meeting approached, no conversation between the Board and Shinn had yet occurred or even been scheduled. On Monday, I met with Cynthia Covie from the Governor's Office and agreed that we would cancel the meeting if she would create the face-to-face meeting we needed. By the end of the day, she had managed to schedule it for January 30.

I had told Covie that I thought the Board was at a crossroads and must choose between two directions. The first, which I was now certain the Board preferred, would be to use the model developed with Carneys Point and promote it statewide. Telling local officials that they could receive $750,000 for public projects of their choosing just for studying the possibility of hosting the disposal facility would make the Board's offer much more attractive than it had been. I thought there was a good chance that one or two of the towns that had previously said no after short public discussions might reconsider, and that there were other towns that might also step forward.

The only other choice I saw was to stop the siting process. Covie said she thought the Board should consider a third option of just continuing what we had been doing. When I asked if she meant continuing to try to site a facility but taking pains to be sure we didn't succeed, she said she wouldn't put it quite that way. I said I didn't see another way to take it.

The face-to-face meeting with the commissioner should have occurred immediately after the December Siting Board meeting. Having waited this long, however, it now makes sense to delay it until we can assess the success of Chem-Nuclear's Barnwell strategy. Chem-Nuclear has set January 16 as the deadline by which down payments for long-term space, or "burial plots" as most people are calling them, have to be received. By late January, we will know if this proposed 20-year "solution" has elicited sufficient support from waste generators to propel Chem-Nuclear forward to seek the concurrence it would then need from the South Carolina Legislature.

Gaining legislative support is going to be very difficult, particularly in an election year. Everyone I have talked to in recent weeks, however, feels that Chem-Nuclear is not even going to get to try because little generator commitment to its plan is forthcoming. The plan that had seemed so creative and sensible when I heard it described in Chicago in October is being seen by the nuclear utilities and industries that generate the most waste as a demand that they spend a large amount of money very quickly to bail out a private company.

By waiting a few more weeks for this overdue meeting with Commissioner Shinn, we should have much more definitive information to work with as we discuss the next steps for New Jersey. This seems particularly appropriate, because my impression is that Shinn and Hogan have represented to the Governor's Office that the Chem-Nuclear plan is going to resolve the problem.

January 20, 1998

The Chem-Nuclear plan didn't work. With the deadline less than a week away and few commitments in for disposal space, Chem-Nuclear announced a one-year extension. The goal will be to work more closely with waste generators to find ways to make the plan more attractive and to push back the need for legislative action until after this fall's gubernatorial campaign in South Carolina. The company still faces a significant shortfall for the current year and apparently plans to ask the nuclear utilities to help them muddle through.

Barnwell is now back to where it was several months ago. It is open for a duration that is unpredictable, subject to Chem-Nuclear's creativity, waste generators' various interests and perceptions, and politics in South Carolina. If Chem-Nuclear can't find enough money this year, the South Carolina Legislature could close the facility by July. If that doesn't happen, Barnwell could be shut down next January in the unlikely event that Jim Hodges, the likely Democratic candidate for governor, defeats David Beasley. On the other hand, Barnwell may remain open for decades, with each year presenting continuing uncertainties and speculation.

February 5, 1998 ... The End Game

On January 30, Siting Board Chair Paul Wyszkowski and I finally met with Commissioner Shinn and Mike Hogan. The meeting took place in the Statehouse and was chaired by Cynthia Covie. That morning, she had been promoted to direct the office in which she works, the Governor's Authorities Unit.[59]

Shinn began by saying that he thought Barnwell would continue to be available for New Jersey waste for some time. He felt that Texas and perhaps other states may soon open other disposal options. He concluded that he believed it was not possible to site a facility in New Jersey due to "the population density, geology, and things like that."[60]

Covie said her office wanted to avoid putting Governor Whitman in the position of having to defend a controversial project when the need for it was either not apparent or subject to dispute. She noted the continuing controversy surrounding plans to build a hazardous waste incinerator in Linden, not far from Newark Airport. The proposal has been debated for years, though many people doubt there is now a market for it and wonder why the private company involved, GAF, continues to pursue it.

Governor Whitman has been reluctantly pulled into this issue several times. Opponents petitioned her to veto minutes from the Hazardous Waste Commission which is supporting the facility. Then, when the New Jersey Turnpike Authority approved construction of a new exit ramp to accommodate objections raised by an administrative law judge, she was asked to veto those minutes as well. Twice, she has been placed in the position of having to endorse a project that may never be built. Her aide did not want her forced into a similar situation with a low-level radioactive waste disposal facility.

Wyszkowski and I then explained the risks of relying entirely on out-of-state disposal options, but it was clear that Covie was representing Governor Whitman and was going to back Shinn, and that he wanted the voluntary siting process to end. There was no point in arguing and little else to say. Shinn and Hogan could have apologized for not sharing their concerns about the Board's direction earlier, and for not doing so more directly and honestly, but they chose not to.

The meeting we had sought for close to two months was not very dramatic, but it provided the answer that had been needed since December when the differences between Shinn and the Siting Board became apparent. The question was about who was in charge on this issue. Now, we knew that it was no longer us. Although the New Jersey Legislature had established the Board to be "independent of any supervision or control by the department [DEP] or by the commissioner or any officer or employee thereof,"[61] Governor Whitman, through her counsel Cynthia Covie, had chosen to take the DEP Commissioner's advice over the Board's.

Part of the job of a governor or any executive is to review conflicting sets of advice and then make decisions. In government, where responsibility for an issue is frequently split among multiple agencies, the path to final decisions is often unclear. This becomes a serious problem, allowing problems to fester, opportunities for progress to be squandered, and ill will to poison professional relationships. One way to help reinvent government would be to devote more attention to this problem in the reorganization of existing programs and agencies, and in the design for new ones.

Now that we all knew Governor Whitman supported Commissioner Shinn's wish that the voluntary siting process end, we shifted the conversation from if to how. Covie finally acknowledged that the state budget the governor planned to unveil on February 10 would contain language concerning the Siting Board. I said that we would then suggest that the Board end the process at its February 6 meeting so that it would not be seen as simply reacting to the new proposed budget. Shinn agreed that this was a good idea.

Planning the end of the process, I believed, could contribute in a small way to improving government's reputation. For one thing, if the siting process had to be revived in the future, I thought the Board's job would be easier if it was seen to be more in control now. I also thought that maintaining that control, or perhaps the illusion of control, might also be helpful for other agencies facing difficult assignments now and in the future. If the end of the process was, in effect, announced by the governor, it would add to the skepticism of those who had argued that the promises of the Board could easily be negated by other political or governmental forces. In addition, the Board had asserted for three years that it didn't need to set an arbitrary date by which to end the voluntary process, contending that it would continually reevaluate the situation and cease the process if it stopped making sense. Now that we were at that point, the Board wanted to live up to that promise.

I returned to the office after the meeting and told the staff of the decision. They were more disappointed than surprised. Everyone believed in what we were doing and had wanted another chance to see if it could work, and everyone naturally began to contemplate their own futures.

By the time the Board met six days later, I had spoken with all the members, each of whom had received drafts of a resolution to "suspend" the siting process along with a proposed letter from the chairman to the governor. Some of them were very surprised and several were angry, but the meeting itself was largely uneventful. Paul Wyszkowski, as chair, summarized the meeting that had taken place in the Statehouse and the reasons why we had concluded that stopping was the best option. The resolution was read and approved with the support of all 11 Board members. There was one minor modification to stress that the Board would maintain a continuing role in providing education about the management of radioactive materials.

It was a bittersweet occasion, with everyone reluctantly agreeing that this was the right action under the circumstances. We were also aware that the ways in which we had all truly enjoyed working together were about to change. We took perhaps inordinate pleasure in the fact that the man who had started the small group called Citizens Helping to Oppose Radioactive Dumps (CHORD) chose this as his first Board meeting to attend, but left in a huff during a break when we could not immediately furnish the minutes he demanded of all past meetings. As a result, he missed the vote ending the siting process he had so vehemently opposed and only learned of it from the newspaper the next day.

That afternoon, we faxed a copy of Wyszkowski's letter with individual cover notes to each of the reporters who had covered this issue at some point over the past three years. Tom Johnson, the state's leading environmental reporter at the time, called immediately and said, "This is real news: 'Man bites dog.'" His large and prominent story in the Newark *Star-Ledger* the next day focused on the unusual fact of a government agency doing itself in. The headline read:

WITH NO TAKERS FOR A RADIOACTIVE DUMP, BOARD GIVES UP

Johnson wrote, "After nine years of fruitless effort costing nearly $7 million, a little known but powerful state agency has decided to do the honorable thing: Kill itself."

Writing that I was "philosophical about the agency's demise," he quoted me saying, "When we adopted our voluntary siting plan, a lot of people said no one would ever come forward. They turned out to be wrong. It was a case where there were many good leaders ... It just turned out to be there weren't many good followers."

The *Atlantic City Press* ran an editorial headlined:

A LAUDABLE RETREAT

The editorialist wrote:

> *Say a state agency was set up to perform a particular task that, it turns out, is impossible to accomplish and, as it happens, is no longer even necessary. What does the agency do? Why it does what all good bureaucracies do, right? It finds a nice corner to hide in and goes on paying a staff and shuffling papers and hoping no one notices that it isn't needed anymore.*
>
> *Well, guess what? That's exactly what the New Jersey Low-Level Radioactive Waste Disposal Facility Siting Board is NOT doing. This agency – to its eternal credit – has sent Governor Christie Whitman a letter saying ... 'You don't need us anymore.' The board ... deserves credit for admitting the obvious and doing what few bureaucracies ever do – close up shop.*

The newspapers that had covered the story for particular towns also ran lengthy and generally complimentary articles. The *Associated Press* wrote a story that appeared in many papers around the state, a small item was included on *New Jersey Network's* "New Jersey Nightly News," and it was reported on most major radio stations.

Two of the radio reports came from the same source, Gene Dillard, a skilled free-lance reporter who roams the Statehouse in Trenton with microphone in hand and then sells the resulting stories to as many stations as possible. In this case, *WHYY*, the major *National Public Radio* affiliate in Philadelphia, ran a lengthy story that included long excerpts of his conversation with me, while *New Jersey 101.5*, the talk radio station which has the largest audience of any station in the state, used some of the interview on its morning show to lament the fact that people don't trust government enough to consider $2 million a year for something that is really safe. They didn't seem to feel any need to link this distrust with their prominent talk shows that consistently bash governmental policies and officials.

The following Monday, *The Times of Trenton* ran its third major editorial about the Siting Board's work. I don't know if this writing had propelled any municipalities to give more thought to volunteering for the disposal facility. I do know it gave a tremendous boost to the Board and staff, and helped keep the siting process alive and vibrant. The editorial was entitled:

THEIR BEST SHOT

The paper concluded:

> It's not often that a government agency volunteers to reduce its budget, its staff and its role. But that's what the New Jersey Low-Level Radioactive Waste Disposal Facility Siting Board, to its credit, has done ... Its members and staff deserve the state's thanks for giving a frustrating task their very best shot.

February 12, 1998

The press coverage about the end of the siting process has been gratifying for several reasons. It does portray a small piece of government in a positive light, and it has also served notice to vast audiences that all of us need new jobs. Although the stories did exaggerate the Board's role in initiating its demise, that was a consequence of our discretion in not publicizing how the Department of Environmental Protection handled this issue.

There are, however, two significant issues that need to be addressed in the course of moving the Siting Board into hibernation, and the attention of the press may help gain better outcomes than would otherwise occur. The first is about money, and the second is about the Board's place in the Trenton bureaucracy.

The Board has successfully maintained sufficient funds in its account to be able to immediately finance a full-speed-ahead siting process if a town asked for one. Since that never occurred, it now looks like there will be approximately $7.6 million in the Board's account when the fiscal year ends on June

30. While the Board will continue to have expenses in its much more limited role, it is hard to imagine how this could total more than $300,000 a year. Decisions must be made about the remaining funds.

I was impressed by the strategy adopted by the Board under my predecessor, Sam Penza, to collect such large amounts of money in advance simply on the basis of a hope that the siting process would move far more quickly and successfully than most people expected. It was a credit to the utilities, corporations, hospitals, and colleges that generated the waste and had to pay for all of the Siting Board's expenses that they willingly accepted this strategy.

The one flaw was that the state Treasury Department, which held the money, assigned the interest it accrued to the general fund that supports other state government functions and services. The generators of waste thereby had not only been paying all the expenses of the Board and the siting process as the Legislature had stipulated in amendments to the Siting Act, but had also been seeing interest on millions of dollars diverted away into general operating funds for the State of New Jersey.

This is simply wrong and I have met no one who will argue otherwise. It is, however, the way New Jersey state government operates, and I have met almost no one who sees any prospect of this ever changing. It is one of those unfortunate realities that is somehow off the table for debate or discussion. If a proposal was made to handle the interest fairly, it would likely be in the context of the state's overall budget. There it would be smothered by all the other line items that people need to quickly comprehend and then debate.

It's different in Connecticut. There, the Hazardous Waste Management Service also collected large sums from generators in the early 1990s under an approach similar to that of New Jersey. But the agency keeps the interest and, in fact, has been able to run its program for several years now without reaching into the principal or assessing more money from generators. If New Jersey's practices were equally scrupulous, the Siting Board might consider keeping the $7.6 million in its account for the time being. That way, funds would be available to quickly restart a siting process if one were ever needed again, and in the interim a greatly slimmed-down operation could indefinitely live off the interest.

Because the interest goes elsewhere, however, the Board will want to retain about a million dollars to finance its reduced operations over the next few years and return the rest of the money to the generators. The mechanics for doing this are relatively straightforward. The Board collected a total of $16 million from generators as a result of fee assessments in 1992, 1993, and 1996.

The percentage of the total paid by each generator can be calculated and then applied to the amount the Board decides to return.

The Board can propose this distribution formula at one meeting, give each generator and anyone else who may be interested some time to comment on it, and then adopt it at its next meeting. This would be essentially the same process the Board used to collect the money in the first place.

The wisdom of this action may not be apparent to all. At our meeting in the Statehouse, Shinn seemed to bristle when Wyszkowski and I said the Board probably would move to return the money. He may think it should be used to reserve a limited amount of disposal space if Chem-Nuclear revives its long-term Barnwell proposal, or he may still think he could apply it to some of the many underfunded programs over which he presides.

I share the Board members' belief that the money should be returned and want to help them plan to have that happen. Stepping back a moment, it is odd to feel passion about returning money to nuclear power plant operators to prevent it from possibly being spent for valuable environmental programs. But this is an opportunity for government to keep its word and strengthen its reputation. Both as institutions and as individuals, those who have followed and participated in the Board's work are much more likely to be advocates for future needs of government if this money is not now diverted to some other purpose, no matter how worthy.

When the waste generators supported the 1991 legislative amendments that made them financially responsible for the siting program, they successfully lobbied for the addition of a qualifying sentence:

> In the event that the board ceases operation or that additional expenditures are not otherwise required, any moneys remaining in the fund shall be returned to generators in the same proportion in which the fees were assessed and paid.[62]

While this language seems to clearly obligate the Board to return the millions of dollars it does not now "require," I nevertheless sympathize with Commissioner Shinn. His ability to have DEP function effectively is dramatically limited by the strings attached to virtually every funding stream it receives. He can rarely move resources to respond to emerging crises or opportunities because each pot of money comes from different laws with different regulations and permitted uses.

That ought to be changed. If federal and state funds were allocated for less specific purposes, legislators could set annual, or better yet biennial, budgets that would better enable high-level administrators to fulfill the legislative missions assigned to them. The budgets could be based on reports they would receive from, for example, the Commissioner of Environmental Protection that would assess the environmental threats, problems, and opportunities facing the state and propose plans for addressing them. The budget would allocate a large lump sum to the Department of Environmental Protection, and the Legislature could ask or require the Commissioner to periodically discuss and debate appropriate changes in priorities. Then, at the end of the budget cycle, the Legislature could review the agency's success in fulfilling the agenda it had established.

Neither Congress nor most state legislatures, however, have been inclined in recent years to give up large amounts of authority to the administrative agencies they have created. In fact, they tend to veer in the opposite direction. But significant steps to increase administrative discretion seem essential if the agencies are to be capable of efficiently and effectively carrying out the laws and programs that legislative bodies create.

That requires reversing the current trend of thinking that people should only pay for the specific government services they are likely to use. Government would need to return instead to a more generous philosophy that charges taxes on the basis of ability to pay and then allocates the revenue on the basis of need, but in a manner sufficiently flexible to allow agencies to respond promptly to changing situations.

Still, the Commissioner of the Department of Environmental Protection should not be able to use the $7.6 million raised by the Siting Board to improve a park, hire more wetlands inspectors, or lower fees for air quality permits. This money was paid by one discrete set of individuals who were promised that it was for one prescribed purpose. To change the rules now would be unfair and dishonest.

The other issue we face in this transition period is less important, except that it is related to the money issue. The proposed state budget for the next fiscal year will include language requested by Shinn assigning responsibility for the Siting Board to the Department of Environmental Protection. This was not included in the budget document that was released to accompany the governor's budget address earlier this week. Apparently, it will be somewhere in the more detailed budget document to be unveiled by the state treasurer in a few weeks.

The Siting Board is likely to strongly resist this change, both because several of the members dislike the Department of Environmental Protection in general and because almost all of them worry that such an administrative change would increase the likelihood of more of the money raised by the Board finding its way to other environmental programs. They are also concerned that, should a siting process ever again be needed, it would have less public credibility if it was too closely tied to the Department of Environmental Protection. This is due in part to the fact that a disposal facility would need to receive a variety of state environmental permits. While there are real and often excessive walls between various Department of Environmental Protection divisions and bureaus, critics would be more skeptical if a disposal facility was seen as a DEP project.

I can see logic in some type of change and I have more respect than most Siting Board members for the DEP. Whatever staff the Board needs for its next phase is going to have a very limited set of tasks. Whether it is one, two, or three people, they are going to have fairly boring jobs and a lot of time on their hands. If a smaller Board staff was made part of a larger organization, the individuals could be assigned other tasks and still be available to give the Siting Board whatever support it needed.

I suspect that Shinn has his heart set on gaining closer control of the Board. It should be interesting to see how this issue is resolved.

CHAPTER 27

Postscript
November 1998

F inding a bureaucratic resting place for the Siting Board took six months. On August 10, Board Chairman Paul Wyszkowski and Commissioner Bob Shinn signed a two-page memorandum of understanding establishing that the Department of Environmental Protection would now provide staff support to the Board.

What seemed a minor issue took so long to settle because the Board and the leadership of the Department of Environmental Protection continued to disagree both about their future relationship and about the money. The Governor's Office decided to again back the DEP, but in this case lacked the clout or powers of persuasion to convince the Board to back down.

The battle was made more protracted and ugly by the way in which the Department of Environmental Protection attempted to win its position on the organizational issue. Continuing to avoid direct communication with Board representatives, in this case about how the Board should be staffed with its new reduced mission, DEP officials had a footnote assigning the Board to the DEP placed in Governor Whitman's proposed budget for fiscal year 1999.

The Board members were offended when they heard of this action, and then outraged when they read the actual words:

> *Such sums as may be necessary are appropriated from the ded-icated Low-Level Radioactive Waste Fund to the Department of Environmental Protection for administrative costs, which shall be solely responsible for providing administrative support to the .... Siting Board ...*

What the Board feared this meant was that the Department of Environmental Protection could spend money from this dedicated fund without their approval. When asked for an opinion, the Attorney General's Office provided a written statement that this analysis was correct.

Board members now felt their suspicions about the motives of Shinn and Hogan had been confirmed. From the day the two officials had seemingly cavalierly opposed the Carneys Point plan, most Board members had believed their intent was to get the Board's funds transferred into their Department's accounts to help pay for other programs. Now that the Department had written confusing but expansive budget language, no more benign explanation seemed available.

The budget language the Department of Environmental Protection had drafted now linked the organizational question to the fate of the unspent funds that the Board had collected. At its March meeting, the Board proposed that it keep $1.2 million to fund its reduced workload over at least the next two years and return the remaining $6.4 million to the generators. Shortly before this meeting, Shinn had named Hogan as his new representative on the Siting Board, replacing Gerry Nicholls, the official who had represented him and his predecessors since the Board was created. Everyone on the Board except Hogan voted in favor of the resolution returning most of the money to the waste generators.

Public notice was issued and, in May, the Board held a public hearing on the proposal. Representatives of the Chemical Industry Council as well as individual pharmaceutical companies and universities testified in favor of the proposal. Again, Hogan, on behalf of Commissioner Shinn, was the only voice in opposition.

On June 8, the Siting Board approved the proposal, and by telephone conference call later that week, it approved the minutes of that meeting. This procedure enabled the minutes to be quickly forwarded to the governor to begin her review period of ten working days.

Ironically, Board meetings since February had become much more acrimonious and tense than they ever were during the siting process. At each meeting, the gulf widened between Mike Hogan, as he tried to explain why Shinn opposed returning the money, and everyone else on the Board. At the same time, Cynthia Covie of the Governor's Authorities Unit, who could have served as a mediator, had lost any leverage she might have had by making it clear that she was going to always support Shinn's position, apparently no matter what it was.

Representatives of the waste generators attended each of these meetings and saw that the return of the funds they thought they were due was in jeopardy. They quietly took their case to several legislators, with the result that the budget the Legislature passed for fiscal year 1999 deleted the footnote Shinn had proposed to make the department "solely responsible for providing administrative support" to the Siting Board. In its place, the Legislature added a sentence reading:

> The balance of funds in the Low-Level Radioactive Waste Disposal Facility Fund ... shall be returned to all such generators in the same proportion in which the fees were assessed and paid.

This budget was passed on June 24, just six days before the new fiscal year would begin. The Legislature had upheld the Siting Board's position. While it had been a close call, the people involved in this process, from the waste generators to the League of Women Voters, might come away feeling a little more trusting of government.

But two days later, Governor Whitman vetoed that portion of the Board's minutes pertaining to the return of the money, and later in the week she vetoed the related new lines in the state budget. In a letter to Chairman Paul Wyszkowski, the governor wrote:

> While I am not philosophically opposed to the return of the dedicated funds to the generator community, I believe such an action is premature at this time. Specifically, I believe it is necessary to prepare a comprehensive plan and explore all options for the short- and long-term needs for the funds with sufficient input from affected parties before returning the funds. After the board has developed a comprehensive policy, I would be prepared to reconsider the issue.

When the dust had settled, an uneasy compromise had inadvertently been struck. Since the Legislature had deleted the language the Department of Environmental Protection wanted to add to the budget, and Governor Whitman had deleted the words the Legislature had substituted, the law had not been changed. Thus, the money would not be returned to the generators, but it could not be spent without the Siting Board's approval.

With this issue resolved for the time being, representatives of the two agencies were able to agree that the DEP's Radiation Protection Program would hire one of the Board's employees and then provide staff support to the Board. The Board would not employ its own staff, at least for the next fiscal year.

At the staff level, the Department of Environmental Protection had worked closely with the Board and could move into this role quickly and smoothly. With the siting process suspended, the possibility that the department would face a conflict of interest from both planning and regulating a disposal facility was remote.

New Jersey's low-level radioactive waste needs would be met as long as Envirocare in Utah remained open and the state retained access to the disposal facility in Barnwell, South Carolina. The Barnwell decision seemed to rest in the hands of South Carolina Governor David Beasley, who had reinstated access to the facility when he took office in 1995.

The surprising decisiveness with which Beasley reversed his state's policies towards Barnwell was among the strengths that were starting to propel him to the national stage. Even before he was up for reelection in November 1998, he had been chosen to chair the Republican Governors Conference and was being mentioned as a possible running mate for the Republican presidential nominee in the year 2000.

In September, the National Committee for an Effective Congress, concluding that he would likely be reelected, wrote:

> *Beasley, a Democrat turned Republican, won his first term with the help of the religious right. As governor, he has cut state spending on virtually everything except education and programs designed to promote business in the state. The key to this race may be how many of Beasley's fellow Republicans continue to support him after he changed his position concerning whether or not the Confederate flag should fly over the state capital...*

Whether or not the Confederate flag was the cause, Beasley lost to

Democrat Jim Hodges, the South Carolina House majority leader, in a major upset. Hodges, who campaigned on a pledge to raise funds through a state lottery, has a record of questioning the wisdom of keeping Barnwell open to the nation. The lottery might be an attractive alternative that would provide the state with equal or even greater revenue while evoking considerably less controversy.

If Hodges closes Barnwell to other states, New Jersey — and most of the country — will be back to the radioactive drawing board.

CHAPTER 28

The Problem Is Solved

J ust as David Beasley stunned South Carolina with his actions on low-level radioactive waste disposal shortly after he became governor in 1995, four years later his successor, Jim Hodges, also chose a path that could not have been predicted from the recent election. While his predecessor had reversed the state's 16-year crusade to close Barnwell, Hodges crafted a policy returning only partway to the position espoused by the two Democratic governors who had preceded Beasley, and by Hodges himself when he was a state legislator. There, according to Harriet Keyserling, who had served as one of his colleagues, Hodges had been "an effective spokesman in responding to the Republican leadership, speaking up against ... keeping the Barnwell nuclear waste dump open..."[63]

With surprising speed, Hodges formed a consensus behind a plan to keep Barnwell open for the use of radioactive waste generators in South Carolina, while gradually restricting access to waste from the other states. Under the Hodges approach, by the end of 2008 the only out-of-state waste accepted at the facility will originate in either Connecticut or New Jersey.

A NEW APPROACH: Elected in 1998, South Carolina Governor Jim Hodges quickly reached agreement with New Jersey and Connecticut to form a tri-state Atlantic Compact for low-level radioactive waste disposal. The move gave South Carolina the legal authority to eventually prevent all other states from using the Barnwell facility.

Photo courtesy of South Carolina Governor's Office

On June 10, 1999, six months into his first year, Hodges issued an executive order establishing a South Carolina Nuclear Waste task force and appointed former Congressman Butler Derrick as its chair. He gave the group two responsibilities: "To provide the people of South Carolina and the South Carolina General Assembly with a road map to discontinuance of South Carolina's role as the nation's nuclear dumping ground; and To recommend actions to ensure that future needs of South Carolina low-level radioactive waste generators are met."[64] The task force quickly issued a request to compacts, states, and other interested parties to submit proposals that would meet the Governor's two objectives.

In the course of framing their proposals, representatives from across the country traveled to Columbia to court the new South Carolina task force. Among those making the trip was a delegation from the Northeast Compact Commission and its two member states, New Jersey and Connecticut. This group had a unique set of attributes to offer.

Its most important asset derived from an accident of legislative drafting. The federal Low-Level Radioactive Waste Policy Act had encouraged groups of two or more states to form regional compacts to work toward regional waste disposal approaches. Each compact required separate ratification by Congress. Surprisingly, however, the language of the various subsequent compact statutes was not identical. The Northeast Interstate Low-Level Radioactive

Waste Management Compact Act, passed in 1986, was one of the few to authorize the representatives of the member states to accept new members without returning to Congress for additional federal legislation. The small size of the Northeast Compact Commission was also attractive. To accept a new member, it only needed the approval of its two commissioners — one each appointed by the governors of Connecticut and New Jersey. The few other compacts that also do not require Congressional action each have at least five member states whose representatives would have to agree. Moreover, some of the other states require that their legislatures approve any significant changes to their compact.

And, finally, of those few compacts South Carolina could join without setting off a Congressional debate, the Northeast generates by far the least waste — the same advantage the states of Maine and Vermont offered Texas in forming an equally unusual "region" more than five years earlier.

The Northeast Compact Commission's understanding that these were important factors to stress reflected one of the lessons learned from all the efforts around the country to resolve low-level radioactive waste disposal issues: That is, virtually every agency and legislative body that is involved, no matter how tangentially or at what level of government, has the potential to derail the process. If South Carolina sought to enter a compact that required Congressional action, Congress would inevitably delay and very possibly reject the agreement. Not only would it provide a stage for a wider discussion about nuclear safety, but concern would certainly be raised by representatives of the 35 states that would lose the access to Barnwell upon which they currently rely.

On December 15, 1999, six months after the task force was formed, Chair Butler Derrick submitted its final report to Governor Hodges. The unanimous recommendation was that the governor "immediately enter into negotiations with the [soon to be named] Atlantic Compact, which currently consists of Connecticut and New Jersey, to define the terms and conditions for South Carolina's membership in the Compact."[65]

It took just six months more for the South Carolina Legislature to place the task force recommendations into law, and for the state to petition and be accepted for membership by New Jersey and Connecticut in the formerly two-state Northeast Compact. On July 1, 2000, South Carolina entered the newly named Atlantic Compact. As the host state, it is entitled to two seats on what is now a four-member Compact Commission.

Despite the contentiousness the issue had caused in the past, this major change raised no immediate public discussion in New Jersey or Connecticut and was enacted in the South Carolina Legislature by voice vote with almost no discussion. After both houses had passed the bill, Governor Hodges said:

> Last year, I pledged to end South Carolina's status as the nation's nuclear dumping ground. This issue has plagued our state for decades. Today, we have moved a step closer to resolving this matter with legislation that puts the environmental and energy needs of South Carolina first.[66]

Joel Lourie, a member of the South Carolina House who had also served on the Derrick task force, called the new law "a real positive for South Carolina." He added, "This will allow us to shut our doors to the rest of the nation and will force other states to deal with their own nuclear waste issues."[67]

While the Atlantic Compact Commission immediately gains the power to close Barnwell to waste generated from outside the three member states, the legislation incorporated a recommendation by the task force that this not happen suddenly. Instead, in the first year, 160,000 cubic feet — the size of the projected national need — will be kept available for waste from outside of the new Atlantic Region. Each year, this figure will decrease, reaching 35,000 cubic feet in fiscal year 2008 and zero thereafter.

The bill passed by the South Carolina Legislature reflected agreements that the three states had negotiated once the task force chose the Atlantic Compact as its preferred option. Waste generators from Connecticut and New Jersey, for example, wanted not only the predictable, stable, long-term disposal site they had sought for 20 years, but also the freedom not to use it if they found a better alternative. Remarkably, South Carolina agreed, and the legislation gave the generators a flexibility that both Connecticut and New Jersey had expressly excluded when they were planning to build their own disposal facilities.

Also, the waste generators are guaranteed a price at Barnwell that is no higher than the price in effect in September 1999 and an assurance that they will not be charged a higher rate than generators from South Carolina.

In exchange for access to the Barnwell facility, South Carolina required the Northeast Compact to contribute $12 million. Connecticut and New Jersey are each providing $5 million, and the remainder is being split between Connecticut's two nuclear utilities, Connecticut Yankee and Northeast Utilities.

During the brief tenure of the South Carolina task force, its members had made clear to all its suitors that one of the factors they would weigh was the size of the dowry being offered. At the end of one session, Paul Wyszkowski, chair of the New Jersey Siting Board, asked how much money needed to be in a proposal to have it taken seriously. Butler, the South Carolina chair, responded, "Be as generous as you can."[68]

As it happened, New Jersey had its $5 million share readily available because Governor Whitman and Commissioner Shinn had prevented the Siting Board from returning the funds to waste generators in 1998. Otherwise, the Board would have had to issue a special fee assessment or the Legislature would have needed to include the money in the state budget for fiscal year 2001. Neither approach would have been difficult to achieve since Whitman and Shinn were supporters of the pending agreement with South Carolina, but neither would have been as easy as simply transferring the money from the Siting Board's account.

Were Shinn and his aide Hogan visionaries two years earlier when they convinced Whitman to stop the Board from returning these funds to the generators? Were the Board and I wrong? It's certainly a closer call than it seemed at the time, but I don't think so. I still believe it was a mistake for government to miss an opportunity to keep faith with some of the governed.

Whitman and Shinn can take some pride that their insistence that the state retain the funds made it easier to implement the Northeast Compact Commission's agreement with South Carolina. However, the Siting Board, the waste generators, and the Legislature can also take some credit for ensuring that the Department of Environmental Protection did not gain the power to spend the money without the Board's approval. Without their action, who knows if it would have still been available?

As the rest of the country slowly loses access to Barnwell, states in the Northwest will continue to be able to use the nation's other major disposal facility in Richland, Washington and, perhaps just in time, another out-of-state savior also may be rising on the horizon. Late in 1999, Envirocare, the private company originally licensed in Utah to accept only a limited class of low-level radioactive waste, proposed to expand its operations to dispose of the other classes of low-level waste as well. State approval, but no further local action, is needed for such a change. Envirocare must receive an expanded license from Utah's Department of Environmental Quality and endorsement by the state Legislature and governor.

. The more than 20-year history of this issue, filled with potential solutions and potent unforeseen obstacles, requires that any new "answer" be received with some skepticism. As the idea of an Atlantic Compact continued to evolve in the spring of 2000, it still seemed that some objection raised within or outside the Connecticut, New Jersey, or South Carolina state capitols or the Chem-Nuclear board room could quickly disrupt the emerging bucolic scene. It was not hard to imagine, for example, the growing national discomfort and outrage with the Confederate flag flying over South Carolina's capitol leading to pressure on Governor Whitman and Connecticut Governor John Rowland to shun any agreement until the flag was taken down. But opposition did not emerge.

On July 1, 2000, the South Carolina task force proposal became law, and on July 12th in Columbia, South Carolina, the Atlantic Compact Commission held its first meeting. Kevin McCarthy, Connecticut's Compact Commissioner, stepped down as chair so that one of the two new South Carolina members could take over, and the Commission voted to move its headquarters from Hartford to Columbia.

New Jersey, Connecticut, and South Carolina may never again have to worry about the disposal of low-level radioactive waste, and the rest of the country now has another eight years to find commercial solutions to this problem.

With this resolution, the lessons for public policy-making are murky. Little if any of the credit belongs to the federal law passed in 1980 and amended in 1985. The many compact commissions and state agencies it spawned don't have one new disposal facility to show for all their efforts.

Instead, the nation will rely upon disposal facilities in Barnwell, South Carolina and Richland, Washington that were already operating in 1980 and a facility in Clive, Utah, created by a private company operating entirely outside of the federally created compact system. Dramatic waste minimization by public and private users of radioactive materials, equally dramatic political upheavals in South Carolina, and unpredicted private initiative saved the day.

The swift satisfactory conclusion to this 20-year saga is astonishing to those who watched or participated in it. Yet, a solution that allows New Jersey, Connecticut, and most other states to rely on someone else to dispose of our waste also poses a warning. We did not learn how to publicly consider, discuss, and debate health and environmental hazards, nor did we find the words to acknowledge that not all risks which are measurable can or should be avoided. Instead, we only highlighted the pressing need to bridge the gap of distrust between our governments and the governed, and to seek new ways to talk rationally about risk and responsibility.

Next time, South Carolina may not come to the rescue.

PART 5

False Starts, Potential Solutions

CHAPTER 29

In-State Siting Options

T hroughout the years we traveled New Jersey making the case for an in-state low-level radioactive waste disposal facility, people would invari-ably offer advice. The ideas were usually thoughtful and appealing, but unfortunately barred by legislation or regulation or overwhelming lack of offi-cial support. Most involved building the facility in one of six types of loca-tions: contaminated land, an urban area, the Pinelands, land owned by the state or federal government, or land adjacent to one of the state's nuclear power plants.

Contaminated Sites — From a land use perspective, the disposal facility should have been placed on a contaminated site. Why lose another farm or piece of open space when New Jersey has many properties that, because of previous contamination, are providing no economic, recreational, or aesthetic benefits?

From a siting perspective, it should have been easier to gain community support for a site which would be cleaner and safer with the disposal facility than it was without it. When critics expressed "what if" fears based on exceedingly unlikely scenarios, supporters could point to far more likely risks that would arise from leaving the site in its currently degraded condition.

Technically, it would have been no more difficult than the two distinct actions of cleaning up a site and building a disposal facility. The Nuclear Regulatory Commission had informally advised the Board that such a facility would not be prohibited by its regulations. Their concern would be to ensure that all non-background radioactivity was removed from the site so there would be no risk that the monitoring of the disposal facility for possible leaks could become distorted. Coming from the NRC, an agency that seems to find it particularly difficult to speak affirmatively, this was close to outright encouragement.

It would have been expensive to remediate a contaminated site, even considering that acquisition costs would have been lower than for pristine land. But the Siting Board had authority to incorporate any clean-up costs into its overall budget for the facility and to then pass on those expenses to the waste generators, either through assessments in advance or higher disposal fees later.

Although waste generators worried about costs getting out of hand, they seemed supportive of exploring this possibility. If the added cost wasn't too great, it might be offset by reduced opposition that would result in fewer delays and an earlier opening date.

Late in 1995, this theoretical possibility became real for the Board. An official from Pompton Lakes in Passaic County approached the Siting Board's booth at the League of Municipalities Convention in Atlantic City. He asked whether the DuPont site in his town could be considered for the disposal facility.

The 570-acre Pompton Lakes Works, well known in northern New Jersey as "the DuPont site," was an explosives plant run by DuPont until its closure in 1994. The company then began to clean up the site as part of the settlement of a lawsuit filed by area residents accusing it of improperly dumping waste into Acid Brook, which runs through the town. By 1995, DuPont had spent more than $100 million on the clean-up, and local officials were starting to worry that the company would stop paying taxes on the land once it had satisfied its obligations under the settlement.

Soon after our conversation at the convention, I accepted an invitation from the Pompton Lakes Rotary Club to speak at one of its February luncheon meetings. The meeting was apparently unusually well-attended, and all of the questions and discussion focused on how, not whether, to locate the disposal facility on the DuPont site.

The mayor did not attend the meeting, but afterwards several people who had been there called or wrote to him. Each expressed the view that this could be a great thing for Pompton Lakes and recommended that the Town Council meet with the Siting Board. At least one of them sent the mayor a copy of the videotape he had picked up at the luncheon.

The Board never received any further formal response from anyone in Pompton Lakes. The enthusiasm of the Rotarians, however, encouraged the Siting Board to continue thinking about contaminated land.

With help from Foster Wheeler and the Department of Environmental Protection, the Board prepared a three-fold booklet on the subject. On the outside was a picture of haphazardly collected barrels and the words, "*If Your Town Has Contaminated Land ...*"

When a reader opened it up, it said, "*Here's How You Might Turn an Eyesore Into An Asset.*" Under the word "Eyesore" was a photo of a decaying factory, and above "Asset" was a computer sketch of the disposal facility.

Inside, the booklet cited New Jersey's "long history of industrialization that left thousands of contaminated sites and abandoned properties in its wake," and noted, "Your town may have one, or several, such sites."

The booklet included a brief description of the voluntary siting process and a summary of the Board's conversation about contaminated sites with the Nuclear Regulatory Commission. Under a heading of "*What Types of Sites Might Be Worth Exploring?*" the Board described various types of polluted sites that might be suitable and included five photographs of areas that almost anyone might admit would at least look better with a disposal facility.

The last section of the booklet asked, "*Okay, Where Do We Go From Here?*" The suggested answer was, "If you think your community has under-used, unattractive or otherwise burdensome property which could be restored to beneficial use, call the Siting Board." This was followed by the tag line, "*Consider the Possibilities.*"

I thought this was one of the Siting Board's best publications. It was dramatic, informative, clear, short, and visually attractive. We mailed copies to all of New Jersey's mayors, municipal clerks, planning boards, and environmental commissions. We handed it out at public meetings and from the Board's display.

We also sent the booklet to the New Jersey chapter of the National Association of Industrial and Office Parks and several other groups that might have members who own contaminated sites. The Department of Environmental Protection was helpful in displaying and distributing the booklet to people encountering their site remediation programs.

Good Ideas That Didn't Work

Not every proposal for addressing the nation's radioactive waste disposal problem came from government. A man named Thomas Stanley Huntington in Aztec, New Mexico sold "California Red Superworms" that, he said, would eat nuclear waste. His brochure offered four pounds of worms for $500. In 1998, he pleaded no contest to six counts of fraud and one count of issuing a worthless check.

An equally valuable, though better-financed, option was put forward by Dennis Lee from an organization called Better World Technology headquartered in Newfoundland, New Jersey. On September 23, 1997, Lee rented the huge Spectrum arena in Philadelphia for what his publicity called "The Most Important Show In America." Large newspaper ads and flyers, filled with exclamation marks, explained:

The most dangerous of all forms of pollution is nuclear waste! One pound of plutonium can wipe out every man, woman and child on this planet, and it's lethal for 50,000 years! An average nuclear power plant gets rid of 66,000 pounds per year and there is no foolproof way to store it safely! Come and witness a technology that can neutralize radioactive waste and make it totally harmless! We need lots of witnesses. There will be nuclear physicists present and we will conclusively prove this can be done.! Be a part of history. The demonstration is totally safe, but since big business has planned to get a $400 billion contract to store this waste, we need the people to come to protect us when we demonstrate this process![69]

Several web sites later reported that Lee's five-hour show mainly offered people the opportunity to buy franchises for his "free energy machine."

From this extensive outreach, we received no replies. Not one phone call or letter. I have never understood why. There would have been issues of liability and sequencing of clean-ups and commitments to work out, and some of them might have been insurmountable. But there must be some towns in New Jersey, besides Pompton Lakes, where it would have been worthwhile to consider the possibility.

Cities — Land use planning efforts in New Jersey and around the nation have adopted the mutually supportive goals of trying to revitalize cities while working to preserve open space and rural character. The strategy is to concentrate development in already developed areas. Locating a disposal facility for low-level radioactive waste in an urban area, rather than in a cornfield, could be a small step in that direction. An economically depressed municipality could turn unused, underused, or contaminated land into a ratable generating at least $2 million per year.

The first obstacle to pursuing this ideal was that few New Jersey cities have 100 or more acres of available land to support a disposal facility. The only possibilities were Camden and Newark, both of which have large areas long abandoned by industry. But informal contacts with leaders in and out of government in both cities revealed virtually no interest.

This certainly was not surprising. Cities would face all the siting difficulties experienced in relatively rural areas, plus concerns about real or perceived environmental justice. They would also have many more residents living in walking distance of any potential site. Furthermore, the $2 million annual payment that was such an incentive for small towns and suburbs seemed paltry to cities that were already receiving tens of millions of dollars in annual state aid.

The Pinelands — The protection of the New Jersey Pinelands is one of the great accomplishments of New Jersey state government in recent times. It is also a dramatic demonstration of how one individual — or two — can make a difference. It was after reading John McPhee's book, *The Pine Barrens*, that Brendan Byrne decided to invest much of his political capital as governor in a quest to radically restrict development in an area constituting one-fifth of the state's land.

Byrne's initiative at the state level was supported by then-Congressman Jim Florio, who introduced protective federal legislation to complement the more restrictive state law. Without enactment of these statutes, suburban sprawl would have already consumed much of this ecological treasure.

The two laws, however, defined the Pinelands differently. The federal delineation of the Pinelands National Reserve includes 173,000 acres not identified in the State Pinelands Area. To suggest putting any type of industry in the Pinelands sounds like heresy, but perhaps less so when specific sites in the additional federal acreage are considered.

Specific areas in the Pinelands were frequently suggested to the Siting Board. Locating the disposal facility on some of them would have violated the intent of the laws. But land on or near some other sites, such as Fort Dix, McGuire Air Force Base, and the Oyster Creek Nuclear Power Station, might have been able to accommodate the facility without threatening the integrity of the Pinelands.

The prohibition in New Jersey's Siting Act against considering the 1.1 million-acre Pinelands National Reserve, however, was unambiguous. While I imagined scenarios under which the Board and a willing municipality in the Pinelands might convince the Legislature to change the law, most people to whom I mentioned it thought I was delusional.

State Land – In casual conversations and occasionally in meetings, people would joke that the disposal facility should be placed in Drumthwacket, the governor's mansion in Princeton, or in "Trenton's governmental halls," as Tom Callinan had written in his song about the voluntary siting process. The idea was that this was government's problem and "they" should solve it themselves.

While the same attitude was sometimes behind the more frequent suggestions that the Board consider state or federally owned land, these comments were also offered – and taken — more seriously. The government must own lots of land it doesn't need, and couldn't the government just give itself permission to use it for the disposal facility?

Most of the land the state does own is permanently dedicated for open space and recreation, but some was bought for other purposes and might no longer be needed. In particular, we were aware that several of the large institutions built for people with developmental disabilities and mental illness were scheduled for closure. Each closure would add a tract of well over 100 acres to the state's surplus property inventory.

There may have been a time when a governor could have told the agency that ran the institutions, the Department of Human Services, simply to transfer the land to the Siting Board. The two agencies are part of the same government reporting to the same elected leader. Moreover, New Jersey, with no other officials elected statewide except United States senators, is widely considered to have the most powerful governor in the nation.

Today, however, such interagency give-and-take to solve public problems is rarely possible. In this case, the closing of several of the institutions had been planned for many years. Sam Penza, the Siting Board's first director, had explored the possibility of locating the disposal facility at one of these sites, and I subsequently did the same. He even had an inside track since he had previously been an assistant commissioner in the Department of Human Services.

We were both unable to stimulate any serious consideration of transferring the land from one agency to another. There is a provision governing the sale of state land that gives priority to proposals for parks and open space, but otherwise state policy is to seek the highest possible selling price for land no longer needed for its original purpose. Each of these parcels was considered likely to fetch a high price from private developers.

Once the land was sold, accounts for the Department of Human Services would be credited with the sale, giving the department more money for programs to help its clients. If Human Services gave the land to the Siting Board or agreed for it to be sold at less than fair-market value, it would hurt the people its staff is committed to help. Understandably, Human Services did not want to do that, and neither the Treasurer nor the Governor's Office was inclined to intervene. Even if they had, however, the voluntary process would still have led the Board to seek community acceptance.

Federal Land — Locating the disposal facility on a federal military base seemed like it would be potentially advantageous for the military and the host municipality as well as for the Board. Several of the large bases in New Jersey were considered vulnerable in the next round of base closures and had underused areas that might be suitable for the disposal facility. Perhaps adding this activity could alter a base's cost-benefit ratio enough for it to avoid closure.

Under the voluntary siting process, the Siting Board would still require an agreement with the community, but towns dominated by a military base might be more open than others to the safe disposal of low-level radioactive waste. The residents would be somewhat used to a variety of hazards, including tanks and guns, and might be well-situated to evaluate the relative risks posed by a disposal facility. In addition, since each base is a major contributor to the local economy, almost anything that might help keep it open probably would receive serious consideration.

New Jersey's two most prominent bases, Fort Dix and McGuire Air Force Base, could not be considered because both lie within the Pinelands National Reserve. The Board thought several others might be feasible, however. These included the Picatinny Arsenal in Morris County, the Earle Naval Weapons Station in Monmouth County, and Camp Evans in Ocean County.

Representatives from Picatinny had been following the siting process and had periodically attended meetings sponsored by the Board. The Arsenal had been aggressively seeking other tenants for the large tract of land the Army controlled, and some military staff were at least curious about the idea of adding the disposal facility.

More formally, however, there was no interest. The Siting Board sent repeated letters to the commanders at Picatinny, Earle, and Camp Evans. We sent them directly and also through the State Adjutant General, the title given to the head of New Jersey's Department of Military and Veterans Affairs.

All of the responses cited regulations and procedures which would not allow this suggestion to be entertained. The Navy Captain in charge of Earle, for example, wrote: "Based on the provisions of 10 U.S.C. 2692, the Secretary of Defense may not permit the use of an installation of the Department of Defense for the storage or disposal of any material that is toxic or hazardous and that is not owned by the Department of Defense."

In frustration, the Siting Board worked with the Northeast Compact Commission to write to U.S. Defense Secretary William Cohen on behalf of both New Jersey and Connecticut. If anyone in his position was ever going to grasp the potential to merge these public issues, it was Cohen. As a senator from Maine, he had been personally involved in his state's efforts to manage low-level radioactive waste.

"There is merit to opening a dialogue on this issue," wrote the Compact Commission's executive director, Jan Deshais. She continued:

> *The use of federal facilities as disposal sites may be beneficial not only for New Jersey, but throughout the country... We ask that you look at this issue from your perspective as someone who understands the difficult task faced by states in dealing with the management of low-level radioactive waste. The department should reconsider its policy and notify individual military bases that they can negotiate with a state or compact to accept waste from their state and/or region. This would enable bases to find new*

uses for their land as well as a solution for the management of the waste they generate...This issue is one that calls for new perspectives in the search for waste management options...

The Deputy Under-Secretary of Defense for Environmental Security, Sherri W. Goodman, responded. She did not explicitly close the door completely, but instead referred the states back to the base commanders who had already made it clear that they would pursue this unusual courtship only if ordered to do so. While not surprising, this response was disappointing.

Land Adjacent to Nuclear Power Plants – The one other frequently suggested solution was to build a disposal facility on land near New Jersey's existing nuclear power plants. At many public meetings, someone would note that most of the waste originates at the plants, and that the residents of the area have already gained some level of comfort with the issues associated with radioactive materials.

They might also add that any possible harm that could be caused by the waste is dwarfed by the risks already present from the power plants. Finally,

Photo courtesy of GPU Nuclear

NUCLEAR PLANT HOLDING TANKS: The Oyster Creek nuclear power plant in New Jersey already has a large holding area for its spent fuel rods and other high- and low-level radioactive waste.

some of them knew that the nuclear plants each had recently built interim concrete storage buildings with the capacity to store all the waste that they — and the entire state — would generate for many years.

The Siting Board representatives would respond by noting that, while nuclear power plants must be located near the water, the Nuclear Regulatory Commission requires that the waste be disposed outside the floodplain. We would note that this might seem illogical or ironic, but that it was based on the plants' operational requirements for water, and the subsequent need to keep any waste from seeping into waterways and flowing off site.

An added obstacle we usually mentioned was that the host municipalities for the power plants showed no interest in volunteering for the disposal facility. In the case of Lower Alloways Creek, home of the Salem and Hope Creek plants, this was absolutely true. The town received so much revenue from Public Service Electric & Gas Company that the prospect of another ratable did not seem to entice its leaders at all.

The situation in Lacey Township was more complex. This Ocean County municipality was home to Oyster Creek, the state's oldest nuclear power plant. The plant had been scheduled to close in 2009. But rumors grew that its operator, GPU Nuclear, planned to shut it down sooner, possibly as early as the year 2000.

The closing of Oyster Creek will have a major negative economic impact on the township. Even before the 2000 date was in the air, several business leaders contacted the Siting Board to learn more about the disposal facility. All of the available land adjacent to the plant, however, was in the Pinelands National Reserve.

This led to chicken-and-egg conversations. Board representatives would suggest that local officials approach their state senator, Leonard Connors, who was the original author of the Pinelands restriction. Perhaps he would champion a modification for land adjacent to a nuclear power plant. They would respond that it was the Siting Board that should get the law changed and then come back to talk to them.

What If New Jersey Was On Its Own?

It now seems likely that New Jersey may never need to build a disposal facility for low-level radioactive waste. As long as the Atlantic Compact holds, Barnwell continues to operate without significant problems, and the political winds in South Carolina don't shift too radically, this particular problem really is solved. But what if the situation does change and New Jersey again has to ask where low-level radioactive waste should be kept inside the state? Have we learned anything useful from going through the voluntary siting process? Maybe the answer is that there is no place in New Jersey for the long-term disposal of low-level radioactive waste. Perhaps radiation is such a tainted term that community acceptance for a new waste disposal facility can't be achieved in the foreseeable future. Perhaps scientific breakthroughs will be made that are sufficiently significant, or that at least seem sufficiently significant, to warrant completely new terminology that replaces the terms that now raise such fear. Perhaps the administrative and legal procedures that have been built into state and federal licensing and siting laws form a dragon that is impossible to slay. And yet, more low-level radioactive waste is being generated every day, and if no place outside New Jersey was accepting it, it would have to go someplace.

The first necessary ingredient for finding a site is the demonstration of immediate need. For the year that Barnwell was closed to waste from New Jersey, and the years leading up to that closure, the question was not whether the state should deal with this issue, but rather how. In New Jersey, politicians, reporters, Rotarians, and Sierra Club members all could agree something had to be done soon. The status quo of temporary waste storage at 100 locations was not acceptable. Had that situation continued, New Jersey's voluntary siting process might well have been successful.

When Barnwell reopened to New Jersey waste generators in the summer

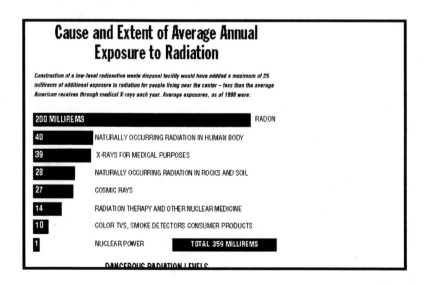

Cause and Extent of Average Annual Exposure to Radiation

Construction of a low-level radioactive waste disposal facility would have addded a maximum of 25 millirems of additional exposure to radiation for people living near the center – less than the average American receives through medical X-rays each year. Average exposures, as of 1996 were:

200 MILLIREMS	RADON
40	NATURALLY OCCURRING RADIATION IN HUMAN BODY
39	X-RAYS FOR MEDICAL PURPOSES
28	NATURALLY OCCURRING RADIATION IN ROCKS AND SOIL
27	COSMIC RAYS
14	RADIATION THERAPY AND OTHER NUCLEAR MEDICINE
10	COLOR TVS, SMOKE DETECTORS CONSUMER PRODUCTS
1	NUCLEAR POWER **TOTAL 359 MILLIREMS**

DANGEROUS RADIATION LEVELS

of 1995, the status quo became much more tolerable. Yes, future waste disposal was far from assured, but any sense of crisis was gone. There was no need for most people to become involved in a controversial issue and no need to research or challenge the arguments of opponents of a disposal facility.

If New Jersey ever again faces a disposal crisis and concludes that a facility must be built within the state, the siting experience of the late 20th Century would suggest that the first step should be a thorough examination of the possibility of placing it adjacent to an existing nuclear power plant or within a military base. Ideally, one of these options would be pursued by a group of states acting together so that one facility could be built for a significant region of the country.

Either initiative would confront three possibly immovable objects – the Nuclear Regulatory Commission's prohibition against building disposal facilities – though not nuclear power plants – at the water's edge, the Department of Defense's disinterest in accepting this type of multiple use, and the restriction in New Jersey's Siting Act against placing a disposal facility in the Pinelands National Reserve. On the positive side, community acceptance at such sites would probably be greater than at other locations in the state.

CHAPTER 30

Other States:
Utah, North Carolina, Nebraska, Texas and California

T hroughout the siting process, New Jersey's goal was for its utilities, universities, industries, hospitals, research labs, and other generators of low-level radioactive waste to be reasonably certain they would have long-term access to a disposal facility at an affordable cost.

The passage of the federal Low-Level Radioactive Waste Policy Act in 1980 gave the three states with operating disposal facilities the authority to form regional compacts and then deny access to waste from the states not included. Two of those states announced plans to do so, and the other decided to close its facility entirely. If no new facilities were built in the foreseeable future, most states would have no place to dispose of the low-level radioactive waste generated within their borders.

But two unforeseen events occurred. First, private entrepreneurs opened the Envirocare facility in Utah, and then South Carolina reversed its policy and allowed Barnwell to again accept waste from across the nation. By the middle of 1995, therefore, although no state or compact had opened a new facility, New Jersey generators of waste had access to two disposal facilities.

In retrospect, it might seem that at this point the Siting Board should have deemed the problem solved and shut down. But the Board, and its sister agencies in other states and compacts around the country, saw no signs that either or both of them could be relied upon over the long term.

Cartoon by Mark Wilson

The Board was always open to an out-of-state solution and continued to monitor developments at both Barnwell and Envirocare, while also keeping an eye on new facilities proposed in North Carolina, Nebraska, Texas, and California. Other faraway options were suggested at public meetings from time to time, but they seemed to have very little prospect of success. Sending the waste into space or depositing it on the ocean floor both seemed popular.

Occasionally, expressions of interest from other countries were reported to the Siting Board, but they were never taken too seriously. One problem was that they would face enormous bureaucratic obstacles. Another was the moral questions that would be raised if we considered exporting a problem that we deemed too dangerous or too difficult to discuss within our own borders.

Envirocare — The Envirocare facility in Clive, Utah is a private operation originally designed in the 1980s for the disposal of uranium mine tailings. The corporation owning it controls a huge amount of land that could accommodate all the radioactive waste the country might produce over the next 60 years. The company took a step in that direction in 1995 when it was given approval by the Northwest Compact Commission to begin accepting Class A low-level radioactive waste, which is relatively high in volume and low in radioactivity.

The immediate effect of Envirocare's expanded services was to provide low-level radioactive waste generators around the country with two disposal choices for much, though not all, of their waste. For the first time in many years, Chem-Nuclear's Barnwell facility faced price competition.

The 1995 change at Envirocare further eroded the sense of crisis over low-level radioactive waste disposal that already had begun to dissipate when Barnwell reopened to most of the nation, but it gave no comfort that a solution had been found for the long term. The State of Utah has not had serious financial problems and does not depend on tax revenue from the facility. The Siting Board felt that political opposition one day might lead the state to deny the license renewal periodically required for the facility.

In addition, unlike Envirocare, Barnwell accepted all classes of low-level radioactive waste. As a result, once Governor Beasley and the South Carolina Legislature reopened access to the Chem-Nuclear facility in July 1995, New Jersey gave more attention to South Carolina.

One final factor that seemingly lessened Envirocare's potential was a bizarre legal battle in 1996 that raised doubts about the company's long-term viability. Khosrow Semnani, then Envirocare's president, and Larry Anderson, the former director of Utah's Division of Radiation Control, filed lawsuits against each other. Amazingly, both acknowledged that Semnani had personally given Anderson about $600,000 during and shortly after the time in which Envirocare's license application had been under review by Anderson's agency. Anderson's suit alleged that Semnani owed him an additional $5 million, while Semnani charged that he was a victim of extortion.

While Governor Beasley's action in South Carolina further delayed the onset of a disposal crisis, at the time it did not promise a long-term solution. The policy the governor abandoned was one his state had doggedly pursued for more than 15 years — a policy that had helped make low-level radioactive waste disposal a national problem. The state's policy reversal in 1995 seemed unlikely to be permanent.

Other states seemed to offer greater potential for long-term disposal of

New Jersey's waste. When the Siting Board formally began its voluntary siting process in 1995, sites for new facilities had already been selected in North
Carolina, Nebraska, Texas, and California. All four states previously had been
given authority by the Nuclear Regulatory Commission to issue licenses for
low-level radioactive waste disposal facilities. New Jersey was in the minority of states that had not applied for this delegation.

Although each of the four was being designed for the exclusive use of the
members of a regional compact, the Siting Board hoped one of them would
eventually meet New Jersey's needs. Once one was operating, it seemed reasonable to imagine residents of that state becoming open to accepting waste
from other parts of the country. They and their leaders would find the facility
to be safe and relatively innocuous, and would be attracted by the additional
revenue they could receive by expanding its market.

This hope may have been unrealistic. The fear of waste from other states
has been a major issue in the debate over disposal facilities across the country.
Most supporters have felt forced to make specific pledges about the sources for
waste that would be accepted. It could be that the host communities in the
other states and their leaders might not be willing to revisit this subject for a
number of years and several election cycles, if ever.

We never had a chance to test this hope, however, because of a more fundamental problem: As New Jersey searched for a municipal volunteer within
the state, each of the proposals that had seemed so far advanced in other states
in 1995 either died or became moribund.

North Carolina — North Carolina, which had been chosen to be the host
state for the eight-member Southeast Compact, was supposed to build a facility that would replace Barnwell. Preliminary site screening began in 1988, and
by 1990, two sites had been selected to undergo site characterization. By 1998,
North Carolina had spent more than $100 million and was not yet close to
revising its license application to address the concerns of regulators. In 1999,
the North Carolina Low-Level Radioactive Waste Management Authority
adopted a plan to terminate its disposal facility development project.

Nebraska — The Central Compact Commission selected Nebraska as the
host state for a disposal facility for its five-state association, which also included Arkansas, Kansas, Louisiana, and Oklahoma. The firm of US Ecology was
hired in 1988, identified a site in Boyd County, and in 1990 filed a license
application.

The application was still pending for the four years I attended national meetings on behalf of the Siting Board. At those meetings, Nebraska was never treated as a serious contender. In fact, the Central Compact Commission director frequently had to deny reports of his program's death. I empathized, since representatives from most of the states without voluntary siting programs made clear their lack of confidence in our efforts.

It turned out that news of the Nebraska proposal's demise had only been premature. On December 21, 1998, Nebraska's Departments of Environmental Quality and Health, and Human Services Regulation and Licensure issued a joint denial of US Ecology's application for a license to construct and operate the disposal facility.

Texas — Texas invented the imaginative Maine/Vermont/Texas region of the United States in 1993 when the three states agreed to form a compact. Two years earlier, the Texas Legislature had identified a 400-square-mile area in Sierra Blanca, 80 miles southeast of El Paso, as the location for the disposal facility.

The Texas facility was generally viewed as the national frontrunner. Its major hurdle was thought to be achieving Congressional ratification of the three-state compact. This would enable the Texas Low-Level Radioactive Waste Disposal Authority to control the flow of waste in and out of the state. In bonding with Maine and Vermont, Texas deliberately chose partners who would send relatively small quantities.

Winning approval in Congress took several years, but the needed bill finally passed and was signed into law on September 20, 1998. Simultaneously, the Texas Waste Authority had been pursuing the license application approval required from the Texas Natural Resource Conservation Commission. While the acronym for this agency, TNRCC, lacked vowels, it was sometimes referred to as "Train Wreck."

A month and two days after the long-sought Congressional action approving the Texas Compact, the state Conservation Commission voted unanimously to deny the license application. Republican Governor George W. Bush responded: "I have said all along that this regulatory decision should be based on science and facts. The state's environmental officials have determined the site is not safe. Therefore, the dump will not be built at Sierra Blanca, period."[70]

California — The last of the four contenders left standing was California, and it appeared a good wind could blow it over. That breeze was provided by the election of a new governor, Democrat Gray Davis, in the 1998 election.

California's low-level radioactive waste saga is in some senses tragic. It was the designated host for the Southwestern Compact it formed with Arizona, North Dakota, and South Dakota. Like Nebraska, California chose US Ecology as its contractor and initiated a site screening process. By 1988, Ward Valley, which is flanked by desert on its eastern side, had been selected as the location for the disposal facility.

In 1989, US Ecology submitted a license application to California's Department of Health Services. In 1992, it was approved. One of the minor remaining steps was for California to gain full title to the site because part of it was on land owned by the U.S. Department of the Interior. By early 1993, Interior Secretary Manuel Lujan had issued a "Record of Decision" approving the land transfer, and US Ecology had paid the Department a down payment of $500,000.

Later that month, when Bill Clinton took office as president, the federal policy changed. The Interior Department returned US Ecology's money and asked the National Academy of Sciences to convene a panel to examine environmental concerns that had been raised. In May 1995, the Academy reported that the project presented no risk, but recommended that additional testing occur during the construction and operational phases.

Nevertheless, the Interior Department continued to refuse to transfer the land, and the rhetoric and legal challenges between Interior and California's Department of Health Services grew more and more heated. Ward Valley became a major subject of debate, demonstration, and speculation.

Prominent entertainers, including Robert Redford and Bonnie Raitt, joined first-term U.S. Senator Barbara Boxer, a California Democrat, in opposing the facility, while Pete Wilson, the state's Republican governor, was the leading supporter of the facility and the needed land transfer.

Even normally unjaded observers speculated that concern about Boxer's pending reelection campaign in 1998 was influencing White House policy. There were also rumors that the Interior Department's Deputy Secretary, John Garamendi, was opposing the land transfer with an eye towards returning to California to run for governor in 1998. Although eventually he did not run, the rumors had increased when an internal memo he wrote to Interior Secretary Bruce Babbitt on the subject became public.

Governor Wilson, wrote Garamendi, "is the venal toady of special interests (radiation business). I do not think Green Peace will picket you any longer. I will maintain a heavy PR campaign until the issue is firmly won."[71]

As this verbal battle continued, there were repeated rumors that US Ecology would not be able to afford to continue the costly fight. By the end of 1998, neither Secretary Babbitt nor President Clinton, two former governors who had been instrumental 20 years earlier in promoting the compact approach for addressing this issue, had taken any action to indicate that the land would be transferred while they were in office.

The New Jersey Siting Board continually monitored these events, as well as activities in the other states and compacts. The only other glimmers of hope came from other states pursuing voluntary siting programs. It was possible to picture Connecticut, or perhaps Massachusetts or Pennsylvania, one day accepting New Jersey's waste, but the opening of facilities in those states was at least as uncertain as one in New Jersey.

For New Jersey to stop its siting process in the hope that a disposal facility eventually would open in an unknown state would have been a risky gamble. The risk could have become even higher if the state or states that did open new facilities adopted the type of moralistic attitude that had guided South Carolina policy between 1985 and 1994. Waste from states that were judged to have made insufficient efforts to solve their own problems might be denied admittance.

And while South Carolina had reopened Barnwell to most of the country only months after New Jersey began the voluntary siting process, it had given no hint that four-and-a-half years later it might provide the reliable long-term disposal option the Siting Board was seeking.

CHAPTER 31

High-Level Waste: The Federal Experience

The Congressional decision to delegate responsibility for low-level radioactive waste disposal to the states and to retain the high-level waste problem for federal agencies to solve has inadvertently provided a controlled experiment testing the relative ability of state and federal governments to resolve a difficult issue. As the states and regional compacts raised controversy and occasional hopes but delivered no victories for low-level waste disposal, frustrated participants and observers regularly kept watch on the federal government to see if it would succeed in creating a home for high-level nuclear waste.

The Nuclear Waste Policy Act of 1982 directed the U.S. Department of Energy (DOE) to open one permanent underground nuclear waste repository by January 31, 1998. But the more than 15-year time frame proved insufficient, and the DOE currently projects an opening day in 2010.

The quest for a high-level waste facility had seemed to have a promising start. One year after the federal law was enacted, the DOE named nine candidate sites. By 1986, the list had been reduced to three and approved by President Ronald Reagan. Yucca Mountain, Nevada, located in a remote area 90 miles from Las Vegas, was highlighted as the first choice. One year later, Congress amended the Nuclear Waste Policy Act, removing a requirement that the DOE characterize three sites in detail and instead instructing the agency to focus its work on Yucca Mountain.

Nevada residents and officials, however, felt that Yucca Mountain had been unfairly selected over Texas and Washington, the other two finalists. They noted that the Speaker of the House of Representatives at the time, Jim Wright, was from Texas, and the second-ranking House member, Tom Foley, was from Washington. And they concluded that the small size and limited clout of their Congressional delegation was a more important factor than any scientific criteria.

The events since 1987 were succinctly summarized in a 1999 report prepared by the Department of Energy:

> *In the years since passage of the Nuclear Waste Policy Act and its amendments, the Radioactive Waste Management Program has faced changing legislative mandates, regulatory modifications, fluctuating funding levels, and the evolving and often conflicting needs and expectations of diverse interest groups. The real complexity of the scientific and regulatory challenge at the Yucca Mountain site began to be realized, and projected costs greatly exceeded initial expectations. It became increasingly clear that many of the expectations embodied in the Nuclear Waste Policy Act could not be met.* [72]

Now, as the original date by which a high-level facility was to be in operation recedes, more than 40,000 tons of used reactor fuel are being stored at nuclear power plants around the country. The Yucca Mountain site was vociferously opposed by Nevada's U.S. senators, Richard Bryan, a Republican, and Harry Reid, a Democrat.

Bryan is a former governor who was elected to the Senate in large part because of his longstanding opposition to having nuclear waste disposed of anywhere in Nevada. When he retired from the Senate in 2000, he was replaced by Republican John Ensign, who prominently touted his willingness to go "head-to-head with not only the powerful nuclear industry, but also the leaders of his own party" to keep radioactive waste out of Nevada. [73]

In recent years, proposals have been repeatedly advanced in Congress to establish a national interim storage site for high-level waste, also in Nevada, until a permanent facility is approved. Early in 2000, the Senate approved an interim storage bill, but Nevada's two senators convinced enough of their colleagues to oppose the bill that it did not receive enough votes to override an expected veto by President Clinton. The bill passed 64 to 34, but its sponsor, Senator Frank Murkowski, a Republican from Alaska, declared it dead at least until 2001. He said, "The [Clinton] administration is opposed to addressing this on their watch. It's perfectly logical considering Gore and Clinton and the environmental pressure."[74]

This federal track record on high-level waste gave pause to those frustrated by the poor showing of the states and regional compacts operating under the federal Low-Level Radioactive Waste Policy Act. Otherwise, throughout the 1990s, many would have been advocating that the responsibility for low-level radioactive waste be reassigned to the federal government.

PART 6

Fear and Distrust: The Modern Malady

There are no easy answers in the world we know.
Some do say the questions are improving though.

Lou and Peter Berryman from the song,

"Why Can't Johnny Bowl?"[75]

CHAPTER 32

Environmentalists and Risk:

Have We Met The Enemy?

I t made no sense for New Jersey, the most densely populated state in the nation and one that produces two percent of the nation's low-level radioactive waste, to plan to build its own disposal facility. For many people, this senselessness was only compounded by the Siting Board's decision to search for a willing host community with a suitable site.

The *Voluntary Siting Plan* the Board designed was like democracy as Winston Churchill had viewed it. It was only appealing when compared with the alternatives.

New Jersey's Low-Level Radioactive Waste Facility Siting Board knew how to build a disposal facility. It had access to sufficient technology, expertise, money, and legal authority. That it failed to do so suggests that some enemy defeated it, but who that might be is not immediately clear.

Credit is claimed, or "responsibility taken" as news reports say of terrorist attacks, by anti-nuclear organizations. In New Jersey, however, these groups were not particularly effective.

Press accounts of siting controversies often described the opponents as "environmentalists," but this was usually an incorrect categorization. Most established environmental groups, if they considered the issue at all, were cautiously supportive of the Siting Board's mission and approach.

Low-level radioactive waste disposal is a difficult issue for environmental activists. On one hand, it is an example of taking responsibility for the garbage we as a society produce. On the other, most of the waste comes from nuclear power plants, which most activists oppose. They would be troubled if resolving the disposal problem resulted in keeping some of the power plants open longer or increasing public acceptance for new ones. Some conclude that even if this type of radioactive waste can be safely managed, it could be a gateway drug opening a door to harder, more dangerous stuff.

The issue is also difficult for activists because few environmental groups can afford to have any paid staff. As a result, at least in New Jersey, the organizations tend to divide the wide array of issues among themselves, and then follow each others' leads. Thus, if the heads of 25 environmental groups all sign a letter expressing concern about a particular endangered habitat, it is likely that many of them have only a general familiarity with the issue. They endorse the letter because they share respect for the particular individual who prepared it.

In New Jersey, nuclear issues have been much less prominent than in some other states and no ongoing group has seen nuclear power as its primary focus. Greenpeace, for example, does not have an active New Jersey chapter. Hope Creek, the state's newest nuclear power plant, was approved in 1975. While the permit approvals were appealed at the time, no lasting group formed to maintain an exclusively nuclear focus.

Thus, when the Siting Board wanted to reach out to environmental groups during the formulation of the voluntary siting process, there were no one or two obvious organizations to contact. Instead, the Board invited representatives from more than 40 groups to various meetings and workshops.

Many of the organizations sent a representative to at least one of the Board's discussions, and a dozen of them remained active observers and occasional advisers throughout the siting process. Some of those who stayed were particularly intrigued by the possibility that the voluntary process might create unusually meaningful public participation with lessons that could be exported to other government programs.

The occasional participation and lack of opposition from organized environmental groups was surprising to many industry advocates, as well as

WISE
World Information Service on Energy /
Service Mondial d'Information sur l'Energie /
Weltweiter Energie Informationsdienst / Servicio Mundial de Información sobre la Energia /
Всемирная Информационная Служба по проблемам Энергии

Welcome to Nuclear Information and Resource Service & World Information Service on Energy

NIRS/WISE is the information and networking center for citizens and environmental organizations concerned about nuclear power, radioactive waste, radiation, and sustainable energy issues. We're located at 1424 16th Street NW, #404, Washington, DC 20036; 202-328-0002; fax 202-462-2183; e-mail nirsnet@nirs.org Our Amsterdam office is at P.O. Box 59636, 1040 LC Amsterdam. The Netherlands;31-20-6126368; fax: 31-20-6892179; wiseamster@antenna.nl Web: www.antenna.nl/wise. Our Southeast U.S. office is NIRS Southeast, P.O. Box 7586, Asheville, NC 28802;828-251-2060, nirs.se@mindspring.com.Worldwide NIRS/WISE relay offices.

| Get Involved! | NIRS Campaigns | Int'l News & Info | About NIRS | Fact Sheets |
| Links | U.S. News & Info | Nuclear Monitor online | Actions | Press Releases |

U.S. ENVIRONMENTAL GROUPS URGE DOE TO DISQUALIFY PROPOSED YUCCA MOUNTAIN WASTE SITE
150+ GROUPS SIGN LETTER TO RICHARDSON, ALBRIGHT, URGING U.S. TO PREVENT ATOMIC WASTE SHIPMENTS TO RUSSIA
EBRD Approves Loan for K2/R4 reactors in Ukraine, December 7, 2000
NEW! Available for the first time on MP3, two great songs from singer/songwriter Tom Neilson.
U.S. News and Information
International News and Information
NIRS seeks fundraising consultant

WEB FOE: The Internet played little role in siting controversies in the mid-1990s, but by the end of the decade, this Nuclear Information Resource Service web page provided opposition research and immediately available ammunition for residents just learning a low-level radioactive waste disposal facility might be considered in their area.

reporters, who had thought this issue would fit into a business vs. environmentalist mold. It didn't.

Nevertheless, the media often continued to refer to the opponents as "environmentalists." In reality, the opponents fell into three categories. The largest category was the local opposition, which grew as individuals almost spontaneously coalesced around the first people in town to express strong opposition.

The second group was the Nuclear Information Resource Service (NIRS). This Washington-based organization is much less well-known than Greenpeace, but is the leading national voice arguing that disposal facilities for low-level radioactive waste should not be built, at least not in the manner any state or compact has yet proposed. NIRS describes itself as "the information and networking center for citizens and environmental organizations concerned about nuclear power, radioactive waste, radiation and sustainable energy issues."[76]

The influence of NIRS grew as the Internet spread. By 1997, within days of learning that someone was suggesting that their town consider volunteering, local opponents who had not heard of the group a week earlier were distributing literature found on its web page. As far as I know, a representative of NIRS only came into New Jersey twice during the three years of the voluntary process.

The third form of opposition came from Citizens Helping Oppose Radioactive Dumps (CHORD), a group that had been formed by several residents of Roosevelt when their borough briefly considered hosting the disposal facility. The group's chosen role took the form of appearances at a few of the meetings other towns hosted with the Siting Board. They also wrote occasional open letters to newspapers.

Those of us on the Board staff never quite understood CHORD's position or motivation. We repeatedly asked to meet with them, perhaps at one of their meetings. Finally, in July 1997, they replied:

> *We acknowledge your offer to meet with CHORD, however, because we do not feel that such a meeting would further the objective of open public debate, we politely decline. CHORD has adopted a precept opposing private meetings between parties to the siting of the LLRW storage facility. As we have done in the past, CHORD will continue to address pertinent issues in a public forum.*

To the Siting Board, opposition from groups like these was to be expected. The real story was that they were not joined by the state's more well-known and established environmental groups. NIRS and CHORD formed the iceberg of organized opposition, not just the tip.

How could the Board demonstrate that most environmental organizations were not, in fact, opposed to the construction of a disposal facility? We faced the same problem that had made it difficult to show that outspoken scientific critics did not represent the general consensus in their field. We tried to replicate the same approach.

In the spring of 1996, we asked environmental organizations to sign a statement of support for the voluntary siting process. The draft wording was different from the one that scientists, doctors, and educators were endorsing. The 275 signers of that statement had agreed that "...a disposal facility for this material would cause no health or safety hazards if care is taken in the selection of a site and in the design, operation, monitoring and eventual closure of the facility."

The statement offered to the environmental groups attested more to the siting process than to the ultimate safety of a disposal facility:

> We don't know if New Jersey's low-level radioactive waste disposal facility is right for your town, but we believe it may be worth exploring. As environmental leaders in New Jersey, we have formed a number of impressions and conclusions that may help others consider this important issue: Radioactive materials are in use in New Jersey today at nuclear power plants, hospitals, industries and colleges and universities, and we don't yet know where to put the waste...
>
> You don't have to like or approve of nuclear power to consider accepting the disposal facility. Most of us believe that New Jersey and the country as a whole should invest in alternate sources of energy, that we should build no further nuclear power plants and that we should shut down the existing ones when their NRC licenses expire. But our use of electricity from those plants until then is generating low-level radioactive waste, as is our use of nuclear medical procedures, pharmaceuticals and other industrial products and processes, and academic research.
>
> As environmentalists, we believe that we have an obligation to ourselves and our children to participate in finding a location for the safe long-term disposal of this waste.

One immediate benefit of this initiative was to give the Siting Board an opening to discuss the voluntary siting process with the trustees of several environmental groups. Even if they didn't sign the statement, it couldn't hurt for them to know more about the issue.

Eventually, the boards of two large groups, the Association of New Jersey Environmental Commissions and the New Jersey Environmental Lobby, voted to sign the letter, as did the smaller Ocean Nature and Conservation Society. We had promised all the groups we contacted, however, that the letter would not be released unless at least 10 groups signed on. Despite many conversations and much trading of possible modifications to the text, the Board never received a fourth signature, and the statement was never released.

This pledge to wait for 10 signatures had been offered to overcome the fears that group leaders might have had of being the only name on the page and in subsequent news articles. I don't know whether the three groups would have signed without it, but we saw no way to take back the promise.

With more endorsements, the statement could have been very helpful. Even without it, however, one of the surprises of New Jersey's experience was the very limited role played by organized opposition groups. But if they weren't significant, where did the opposition come from?

CHAPTER 33

Fear of Risk —
It Could Happen

Many of the people who actively opposed any consideration of locating a low-level radioactive waste disposal facility in their town had rarely, if ever, been active in community or political debate before. Most of them seemed motivated by gut-level fear at the prospect of radioactive waste coming near them. They found credible the articles, reports, and anecdotes that supported their initial fear, and were suspicious of most opposing information.

"The public's antipathy toward radioactive waste disposal facilities is difficult to overstate," notes Bill Newberry, who has worked on radioactive waste disposal for the states of Vermont and South Carolina and for the federal government. "Responses to a 1991 word-association survey of attitudes toward a nuclear waste repository revealed pervasive qualities of dread, revulsion, and anger — the raw materials of stigmatization and political opposition."[77]

Studying how people respond to risk has emerged as a minor academic discipline. Terry Davies, director of the Center for Risk Management, has written about attending a meeting of "about a dozen organizations, of which four had 'risk management' as part of their name."[78]

The general tenet of this field is that most people worry about the wrong things. Studies and reports repeatedly agree that the overwhelming cause of premature death is behavior each of us can control: smoking, overeating, alcohol abuse, lack of exercise, not wearing seatbelts, and failing to control high blood pressure.[79] Yet, it is the risks that individuals can't control, even though they are much smaller, that raise much higher levels of concern.

"A great deal of fear is to do with uncertainty," writes British journalist Jeremy Hall in his book *Real Lives, Half Lives: Tales From The Atomic Wasteland*. "If there is a slither of uncertainty the fearful mind is set off, because the only acceptable risk to a fearful mind is zero risk, and science does not talk about zero risk." Hall adds that if a risk is involuntary and beyond our control, it "in turn increases fears that the government and the industry are lying to us in a grand conspiracy; the result being, whether consciously or unconsciously, that pro-nuclear statements are 'big business' statements and anti-nuclear statements are 'public interest' statements."[80]

The issue of control arises even when comparing risks that all involve radiation. Radiation from X-rays and from high radon levels stir different feelings than radiation from a disposal facility. People may have several X-rays a year or take no action to ventilate their house to reduce a high radon level, but are then terrified by the prospect of living near low-level radioactive waste. Apparently, the perception that they could someday choose to reduce these exposures makes them more acceptable than the prospect of receiving a much lesser dose, if any, from a disposal facility.

Most opponents of the disposal facility seemed to accept that the chance of it causing a health or environmental problem was small. But when they asked "Why take the risk, no matter how small?" the Siting Board had no answer that proved convincing.

To begin with, many people apparently believe any problem involving radioactive materials would probably result in a catastrophe. Images of Hiroshima, Nagasaki, and Chernobyl, remembrances of the fears raised by Three Mile Island, and perhaps decades of movies with radioactive monsters have left an indelible impression.

While the critics may admit that the odds are low, they feel the price of betting wrong will be huge. Just as normally sensible people flock to buy lottery tickets when the jackpot is enormous even though their chance of winning is infinitesimal, the prospect of a disposal facility sends them running in the opposite direction: The perceived magnitude of the penalty for something going wrong is not worth risking the minute chance of such an event occurring.

HOME-GROWN OPPOSITION: This flyer from South Harrison is typical of opposition leaflets that would spring up in town after town.

Few people are aware how routinely radioactive materials are handled, or that accidents do occur without causing serious or long-term damage. For example, as the Siting Board wrote in its pamphlet, *Transporting Radioactive Materials in New Jersey,*

> *Small quantities of radioisotopes are shipped, sometimes twice a day, to New Jersey's many pharmaceutical firms just about every day of the year. An express mail carrier will pick up this material in a van or truck at Newark Airport, take the Turnpike to the appropriate exit, then follow state, county and local roads to its destination.*

Not In My BackYard

The British musical group *ARTISAN* was inspired to write this song by protests alleging that a wind-power project proposed in England would create too much noise and be harmful to birds. In their album notes, they write, "Whether it's a bypass, a home for less-able adults, or a clump of windmills, most of us think the ideas are good but want to avoid living with the realities."

NIMBY

Words & Music by Brian Bedford[81]

Wanna have a smoke free, choke free
Vege-burger, cup o' tea
Hey, man. Where's it at?
Where you gonna put that?
Over there, I don't care
Anywhere but my backyard

You can have a windmill, landfill
Jerry rig a big drill
No hope, isotope
Hanging from a big rope
Fall out no doubt
Keep it out of my backyard

CHORUS: *Over there don't care*
You can put it anywhere
On a beach out of reach
I can preach a good speech
Put it down out of town
Find a pool and let it drown
Up a tree in the sea
Lock the door and lose the key

You can have a highway, byway
Keep it out of my way
No grass by-pass
Burn your gas and spend brass
Don't brake overtake
Anywhere but my backyard
Dig a pit bury it

You can have a bad house, mad house
Prison full of posh grouse
Jolly good knew you would
Stick it full of blue blood
Old folk can't cope
Keep em out of my backyard

Push it through an exit
Overflow don't know
Give a push and let it go
No fear too near
You can make it disappear
Find a screen paint it green
Stick it in a magazine

CHORUS

You can keep an old aunt, pot plant
Find a heart and transplant
Gene clean mean green
Keep it like it's always been
Unrest protest
Keep 'em out of my backyard

You can have a long chat, fancy that
Standing on a door mat
Old slum welcome
Haven't got a crumb chum
Move on have they gone
Anywhere but my backyard

CHORUS

A prejudice, although usually considered in terms of a category of people, can be applied to any class about which someone forms a prejudgment based on limited and often inaccurate information. Few people ever have reason to think about whether a disposal facility for low-level radioactive waste could be safe. Nevertheless, the first time the question is raised, most have an immediate opinion.

Public perceptions about radiation are not irrational. They certainly are not as ill-informed as prejudices against racial and ethnic groups. Radioactive materials are dangerous, as proven by several universally known events. Atomic bombs killed people in Hiroshima and Nagasaki at the end of World War II, as did the explosion at Chernobyl in 1986. The accident at Three Mile Island in 1979 injured no one, but the harm that was narrowly averted induced a lasting terror for much of the nation.

New Jersey's Siting Board was contending that there are major differences between atomic weapons and horrendously designed and maintained nuclear power plants on the one hand, and a well-engineered and publicly monitored disposal facility for low-level radioactive waste on the other. For one thing, we were saying that scale and magnitude matter.

Many substances can be extremely dangerous in large doses, but are safe and even beneficial in small controlled amounts. Ingesting a bottle of aspirin, for example, would be fatal, but with a childproof cap it can safely be kept in the home. A health physicist named Bernard Coen has a longstanding offer to eat pure plutonium if a nuclear-power critic will eat an equal amount of pure caffeine. He calculates the risk from the two substances to be about equal.[82]

No matter how cleverly or cutely these analogies are framed, however, they seem to have little or no impact on public perceptions of the danger of radioactive materials. The words "radiation," "radioactive," and "nuclear" have too much power. It is noteworthy that the name of the powerful medical scanning technique we now know as "MRI" was changed from "nuclear magnetic resonance" to "magnetic resonance imaging" before it gained widespread usage.[83]

The Siting Board and other supporters of disposal facilities also were trying to communicate that government has acknowledged and responded to past problems and fears concerning radiation. The regulations and procedures applicable to low-level radioactive waste disposal were changed and dramatically tightened during the 1970s and 1980s. The Barnwell facility had leaked small amounts of radiation into a stream, but the Nuclear Regulatory Commission's revised regulations would not allow anything like that original facility to ever be built again. Changes like these that were intended to be responsive to public concerns, however, seemed to provide absolutely no comfort to people who were fearful to begin with.

At the Siting Board's open houses, people would invent scenarios and then dare Board members and staff to say they were impossible. A person would ask, "What would happen if a plane crashed into a concrete bunker filled with radioactive waste and exploded?" We would explain that while the plane and its contents might explode, nothing in the disposal facility could. And they would say, "But what if explosives had mistakenly been disposed of, and the monitoring devices at the facility had malfunctioned so they weren't noticed?" We would head down the road of saying that this was an extremely unlikely set of events. And they would say, "Well, it could happen, couldn't it?"

That points to a fundamental challenge in communication about risk. If someone is predisposed to be worried, degrees of unlikeliness seem to provide no comfort, unless one can prove that harm is absolutely impossible, which itself is not possible. Honest people must admit that almost any scenario, no matter how farfetched, could happen. As Albert Einstein is reported to have said, "No amount of experimentation can ever prove me right; a single experiment can prove me wrong."[84]

Throughout the voluntary siting process, people would tell us that their fears were fueled by their conviction that the government had lied to them, or misled them, in the past. Things they had thought would be harmless had turned out to cause harm. Most often mentioned were pollution and contamination caused by industry, both in general and at specific locations. Some of the incidents cited were far from New Jersey, and many were known only anecdotally but, like the monster movies, they had made a lasting impression.

On the other hand, the fact that some widespread fears have proved to be unfounded seems to have had no impact on the evaluation of present and future risks. The American Council on Science and Health documents "the greatest unfounded health scares of the last five decades." Its list includes the chemical alar used to ripen apples, electric blankets, and asbestos in schools, all reported widely — and in error — to be likely to cause cancer.[85]

The most common response to such misconceptions from people concerned about public understanding of risk is to target education to correct them and equip consumers to better evaluate information. The Siting Board joined in this crusade, deciding to maintain education as one of its central functions even when it suspended the siting process.

"All in all," however, as Michael Gerrard has noted, "public information campaigns have failed in their purpose." As a lawyer, Gerrard has represented a number of community groups opposing waste facilities. In his book, *Whose Backyard, Whose Risk*, he writes, "Although such campaigns can increase perception of risk, there is little or no evidence that they can reduce this perception. Media blitzes have increased public concern over AIDS, radon gas, drunken driving, and other threats, but there are few examples of successful campaigns to allay preexisting fears."[86]

The mantra that often accompanies needless worrying is "better safe than sorry." Yet, the story of the national effort to remove asbestos from schools shows there can be costs to such thinking. The risk of a child getting cancer from the asbestos insulation that schools installed to retard fires was eventually estimated to be one-third the risk of being stuck by lightning. Yet, the nation has spent more than $6 billion to remove asbestos from schools.[87] Maybe schoolchildren would have been safer if the asbestos had not been touched and the money had been spent to create smaller classes.

The problem underlying most discussions of risk was nicely captured by a headline in the *Asbury Park Press*. The paper reported on a study by the National Academy of Sciences analyzing more than 500 studies on the subject of health risks caused by power lines. "There is no convincing evidence that residential exposures to these fields are a threat to health," the panel concluded. "While research should continue in certain areas, the scientific evidence gives no cause to worry about exposures to electric appliances or power lines."

The headline was:

POWER-LINE PARANOIA
EVIDENCE SUGGEST NO LINK TO CANCER, BUT DON'T EXPECT PROOF

CHAPTER 34

Fear of Government

I t is a role of government to decide what is safe. No other sector of society can make practices and products illegal or impose limits on their size, extent, or effect. Scientific studies, organizational web pages, and journalistic exposés can affect individual behavior, but their impact is usually far more significant if they cause laws to be passed, regulations to be written, or administrative practices to change.

Legislators and government administrators hold public hearings to weigh often-conflicting evidence, after which they use their best judgment to make decisions. But increasingly their judgment is trusted only by people whose views they endorse, and only until the next issue arises.

The laws passed since the Vietnam War and Watergate, including all of the modern environmental and land-use statutes, have tried to rebuild trust in government by including requirements for public participation. As the distrust has only continued to grow, legislatures and agencies have responded by making the rules mandating participation ever more extensive and elaborate.

New Jersey's Low-Level Radioactive Waste Siting Act, and the Hazardous Waste Facilities Siting Act on which it was modeled, represented major steps in this progression. It wasn't enough for the Legislature to establish citizen siting commissions with citizen advisory committees to look over them. The two laws also had to define specific ways in which public involvement was to permeate the process. In both acts, this concluded with a hearing before an administrative law judge who would take several years to revisit every decision already made by the commission and advisory committee.

New Jersey was not the only state following this path. As Bill Newberry writes, "In an excess of good faith, most states established quasi-state agencies to develop the new sites and new departments within existing agencies to watch over the new agencies as they developed the sites... Most states compounded the burden of state agency involvement by adding an amazing array of new caveats and procedural hurdles."[88]

In New Jersey, the 11 members of the Low-Level Radioactive Waste Disposal Facility Siting Board and the 13 members of the Radioactive Waste Advisory Council concluded that even the procedures in the Siting Act would not be sufficient to give credibility to a government agency addressing such a controversial issue. The voluntary siting process was their response, a mechanism designed to invite the public into every aspect of facility siting and decision-making.

Richard Sullivan, as New Jersey's representative to the Northeast Compact Commission, surveyed the low-level radioactive waste disposal efforts across the country. He noted that "the whole system is treating the cure for the problem like it is the problem."[89]

It was something of a mixed message for the Legislature and Siting Board to construct this complex civic infrastructure and then say it was needed to support a safe, low-impact, light-industrial development. The effect was like erecting a huge bandstand to watch commuters go to work. People attracted by the construction would assume that the line of cars with drivers sipping coffee couldn't possibly be the whole show.

Peter Sandman of the Harvard-MIT Public Disputes Program makes a similar point:

A facility that is dangerous enough that you have to find the safest spot in the area for it is dangerous enough that no reasonable person would want that spot next door. You can't set up scores of siting criteria that prohibit the facility near waterfowl and hospitals, spend years and millions of dollars finding just the right site, and then tell me I'm being silly to object to it in my backyard.[90]

Nevertheless, in New Jersey, the voluntary siting process did positively impress some people. It led them to greater openness to the possibility that the disposal facility might be safe and that the Siting Board might be worthy of trust. But it was simply an attractive island in an otherwise threatening sea.

Much of the public seems to confront government with demands and expectations that are fundamentally at odds with each other. Promises of perfection and no risk are demanded, while incompetence and corruption are expected. When the reality is almost always in between, alienation from government only grows.

Increasingly, it seems as if government has been set up to fail. Society has usually assigned tasks to government only when the private sector has already proven incapable of handling them. Running railroads and protecting people from abject poverty are two examples. Then, when the government's programs lack the efficient elegance romantically attributed to business, they are cited as examples of the inevitable incompetence of the public sector.

This is a serious problem without an easy solution. How can trust in government be restored, created, or reinvented? Certainly, cutting funds and staff until "they" get it right is not the answer, although that is an approach sometimes pursued by frustrated and inexperienced government critics.

Another counter-productive response would be to continue to encumber decision-making processes with ever-increasing steps and levels of review. Significant decisions by government agencies should be preceded by public education and discussion, and officials should modify or drop proposals in response to good information or arguments presented to them. It is a mistake, however, to believe that if enough additional steps are mandated, a process can be perfected that will receive applause from advocates whose position is not adopted.

Moving in the other direction, however, could have a positive impact. If the overall public decision-making system was less lumbering and provided a more predictable schedule and endpoint, respect for government could grow, and almost everyone involved would benefit. Only those who believe the status quo is always preferable to any change government might devise would be disappointed.

The current system for approving the construction of most controversial land uses requires that the final decision be made on a piecemeal basis before multiple agencies over many years. Most of the evolving thoughts and techniques of risk communication are not applicable. Even if the Siting Board had been more successful in a New Jersey town, the Nuclear Regulatory Commission hearings on New Jersey's license application would not have occurred until four or five years after local discussions had begun. By then, the townspeople who had been actively involved in all the discussions might have stepped down as local leaders and perhaps moved.

Richard F. Paton, who has been involved from several vantage points with siting disposal facilities, has written:

> As with any public policy decision, new stakeholders and new perceptions of the problem develop when there is a long time delay from the setting of the policy, and its ultimate problem is even more pronounced if the ramifications of those actions are felt by individuals or groups that have had little or no direct involvement or stake in the establishment of the policy in the first place. [91]

To rebuild trust, government agencies need to have more authority and flexibility to respond to community concerns. Public participation in any form will actually reduce trust and respect if government representatives can only respond to good ideas suggested by the public by explaining why they lack the power to implement them.

Many people did praise the voluntary process and say that the members and staff of the Siting Board all seemed decent enough. Then they would ask how they could be sure that our successors would be trustworthy or that the governor, Legislature, or courts wouldn't someday change the rules.

The only response was to explain the protections that could be built into an agreement between the Siting Board and a municipality that would make such changes difficult. But, as with the worrisome scenarios associated with the waste itself, potential problems could not be absolutely ruled out.

The fear that government policy might change was, in fact, harder to refute. It was a far more likely possibility than any likelihood that the waste would cause harm.

Skepticism is often a wise and useful stance to maintain, whether about government, science, or professed expertise in general. Society's ability to make complex and controversial decisions can be dangerously jeopardized, however, when virtually any proposed solution is met by a cynical attitude that is not open to new or challenging information.

A whimsical step toward rebuilding trust may be found in a proposal by the musical group, *LEFT FIELD*. They have suggested establishing an Institute for a Healthier Skepticism. The Institute would be "home to a new belief system that provides a safety net and a soft landing for those individuals who find themselves on a breath-taking ride from deep faith to utter cynicism, and the resultant serious implications for society at large." [92]

PART 7

'Politics Is More Difficult Than Physics'

- Albert Einstein [93]

CHAPTER 35

Making Government Work in a NIMBY Age

S ociety gives government the tasks that are important and that otherwise cannot or will not get done. These assignments invariably involve some combination of entrenched problems, strong competing interests, and intense controversy. They are difficult and generally not amenable to cost-effective solution.

When a job appears likely to be both doable and profitable, it goes to the private sector. Most construction of housing, stores, office buildings, and factories meets this test. There is a general recognition, however, that private developers will not do the work in ways that will assure the protection of environmental resources, and workable and livable land use patterns. As a result, from the early days of local zoning through the environmental statutes of the past 30 years, government has been assigned the review and approval of some private land use proposals. Government also is given a major role in the financing, design, and construction of facilities whose profit incentive is insufficient to attract private builders. This includes schools, prisons, ports, railroads, and waste disposal facilities, as well as housing and other accommodations for people who are significantly less healthy or wealthy than their prospective neighbors.

One might expect that this workload, made up of responsibilities too challenging for the private sector, would be accompanied by an organizational structure and procedures designed to help government address it. But laws and regulations governing most public programs generally go in the opposite direction, placing increasingly high hurdles between the starting line and the goal.

New Jersey's Low-Level Radioactive Waste Disposal Facility Siting Board was able to operate with more authority and flexibility than most government agencies. The ways in which it benefited from that design, as well as some of the difficulties it experienced, point to changes that could increase government's ability to operate effectively.

Some of the lessons the Board's experience can offer are specific to siting controversial land uses, like "It's better to hold a public meeting in a room that proves too big than one that proves too small." Others fall within the more general themes of time, openness, leadership, and the need to improve government so it becomes better equipped to actually accomplish the tasks assigned to it.

Siting Controversial Land Uses

New Jersey's Low-Level Radioactive Waste Disposal Facility Siting Board worked to address a public problem with professionalism, creativity, and persistence. It included a wide variety of individuals and groups in its work so that public awareness and understanding of the problem and possible solutions increased. It listened to the reactions and suggestions it received and in response made major revisions to its plans. It operated without a hint of financial impropriety, and it won universal respect for placing itself on inactive status when its mission and mandate evaporated.

The Siting Board's success came from the authority and structure it was given by the New Jersey Legislature, and the approach to the specific issue and to governing in general chosen by its members and Advisory Committee. It benefited from having almost exclusive responsibility within the state government for the issue it was to address, by having a large amount of control over the funds assigned to it, and by being able to hire staff members on the basis of their ability to carry out the mission of the agency. The Board also had adequate financial resources. The Board's record, however, must be placed in context, for success is not necessarily part of the obvious eulogy for the agency.

It certainly would be more self-evident that the Siting Board deserved notice and analysis if a low-level radioactive waste disposal facility was operating or at least under construction in New Jersey. Then, there would be no question that studying the experience of this small agency would be likely to offer valuable lessons for siting other publicly needed but locally unwanted land uses, and perhaps for resolving other contentious issues as well.

In that sense, the Board's results were meager. Eventually, it played a major role in helping design and secure an agreement creating an Atlantic Compact with South Carolina, thereby meeting its initial objective of providing New Jersey with a reasonable guarantee of long-term low-level radioactive waste disposal. Also, without its efforts, New Jersey might not have retained the ability to use the Barnwell facility for all but one year. Overall, however, the final result is not exciting or earthshaking: New Jersey's users of radioactive materials in 2001, just as in 1981, can send the low-level radioactive waste they generate to South Carolina.

The Siting Board's more instructive, though less tangible, accomplishment was to operate in a manner that delivered more than was expected of it. After acknowledging that some of its initial premises were wrong, the Board creatively and gracefully changed direction. It then succeeded in enticing groups of citizens in 12 municipalities to seriously and publicly consider bringing a disposal facility for low-level radioactive waste into their community.

The failure to eventually reel in an agreement with one of them was determined largely by external events beyond the Board's control that gradually downgraded a looming crisis into merely a potential problem. The sense of urgency that might have helped build a movement strong enough to overcome the fear of anything radioactive dissipated. In the end, New Jersey did not need to have its own disposal facility.

The Board was addressing an emotionally charged issue that elicits unusually strong opinions. "I haven't felt this way about an issue since Vietnam" was the way one objector to a disposal facility proposed in New York state expressed her feelings to a *New York Times* reporter in 1990.[94] Others have observed that the only issue to generate positions of comparable intensity is abortion.

Feelings about radiation, waste disposal, and land use controversies in general, however, are distinctive because they ignite passion only on one side. Both abortion rights supporters and anti-abortion activists write letters, speak out, demonstrate, vote, and even disrupt or change their lives to help further their cause, as supporters and opponents of the Vietnam War did before them. Not everyone feels equally passionately, but each side has some proponents who almost literally will not sleep until their viewpoint is victorious.

Advocates for a particular land use, however, are usually much more tentative. In each town the Siting Board visited, being "in favor" of the low-level radioactive waste disposal facility meant thinking that it was worth looking into and being open to the possibility of someday endorsing it if everybody's questions about health and safety were satisfactorily answered and if significant compensation to the community was guaranteed. Local supporters were curious or interested, maybe even enthusiastic and a little excited, but rarely so much so that their slumber was disrupted.

This dynamic is not unique to consideration of waste disposal facilities. It is found in deliberations over many of the land uses that are needed by society at large, but not by any individual neighborhood or municipality. But injecting radiation into the discussion moves it into a different dimension.

When other proposed land uses raise controversy, with vocal and passionate opponents almost always far outnumbering outspoken supporters, the backers often have counter-arguments that can help them maintain their position and seek converts to it. Advocates for prisons probably have been the most successful recent purveyors of locally unwanted land uses because they have been able to promise the host community large numbers of permanent jobs, many requiring little or no advance training.

Transportation projects also offer many jobs, though usually most are only during the construction phase. However, area residents, even if they are concerned about accompanying environmental or growth impacts, can also envision how they personally would use the finished product. They can become open to considering changes in a proposed road design, for example, that would respond to their concerns while still enabling the road to be built.

Another positive factor that can be added to the debate over most land use controversies, but less easily for low-level radioactive waste facilities, is the example of other success stories. Community representatives can visit group homes for people with developmental disabilities and talk to residents, neighbors, real estate agents, and police officials. They can learn firsthand that the impact of these homes on a community is little different than the impact of their own home. They can travel to prisons and see that no one has escaped and that the jobs promised to the community were delivered.

Concerned citizens have a much harder time observing comparable low-level radioactive waste disposal facilities. The only one operating in the eastern half of the United States is in Barnwell County, South Carolina. It is not conveniently located for mildly curious residents of another state. While most people who do make the trip come away impressed that the surrounding community has experienced no negative health effects, only very minor environmental impacts, and no diminution of real estate values, their positive feelings are hard to convey to their neighbors who did not accompany them.

Skeptics may agree that the Barnwell facility has operated safely throughout its 30-year history, but they will quickly note that three decades is just a small percentage of the time it will take for all the radioactive material brought onto the site to decay to background levels. They will say that if the facility had opened when Columbus first landed, perhaps an examination of it today would give them comfort that it could operate safely in their community.

Traditional land use decision-making allows advocates, generally with a financial stake in the outcome, to present proposals to juries of local officials and residents. These panels, constituted as planning and zoning boards, evaluate how the proposal complies with the town's plans and regulations, and how it relates to their vision of their community. They hold public hearings and make decisions. The assumption is that this system, replicated in hundreds and thousands of municipalities and counties across the country, will yield the various types of development society needs.

As the nation's growing population has increased density, and as society has spawned more diverse and conflicting land use needs, this system has become increasingly less reliable. Federal and state laws and judicial mandates have been necessary to force local governments to consider the protection of natural resources and the prevention or reduction of congestion, and to accommodate a range of needed but locally unwanted land uses.

When it passed the Low-Level Radioactive Waste Policy Act of 1980, Congress acknowledged that, without a federal requirement, neither local nor state governments were likely to step in and provide what the nation needed. The federal law and the Amendments Act of 1985 that followed, however, did not provide much direction. After implicitly acknowledging that the combined efforts of the private sector and local boards and councils would not be sufficient, all it did was to provide incentives for states to band together in compacts to solve the problem.

The states were left to define their own regions, and the resulting compacts, as well as the remaining unaligned states, then had to create their own site-selection processes. Most began with efforts to establish environmental and social criteria to support a search for the "best" site or sites in the state or region. After selecting a site, they would try to persuade local residents that their area was chosen fairly and that the disposal facility would be good for them. At the same time, they would prepare to employ the powers of their state legislatures to preempt local regulations, take ownership of the land, and build regardless of local attitudes.

While New Jersey's Siting Board started down this path, it got underway more slowly than the agencies in some other parts of the country. Inadvertently, Board members were given the opportunity to learn from the experience of others. Also, the Board was able to observe the New Jersey Hazardous Waste Facilities Siting Commission's lack of success with its "decide, announce, defend" formulation.

The move to a voluntary approach was visionary and innovative. By the end of the process, the Siting Board had received inquiries from residents of more than 45 towns, almost 8 percent of the state's 566 municipalities. In a dozen of those communities, the inquiries led diverse groups of people to publicly support consideration of inviting a disposal facility for low-level radioactive waste to be their new neighbor. The process in each town was different from the previous ones, both because the towns were not the same and because the Board tried to incorporate lessons it had learned as the process continued.

In Carneys Point, which turned out to be the final stop on New Jersey's voluntary siting tour, local leaders and the Siting Board arrived at an outline for a program that, under the right circumstances, might prove successful for forging siting agreements for controversial developments with willing municipalities. The fact that the facility is not now under construction in Carneys Point is due to a unique set of circumstances. These range from the Siting Board losing the support of the DEP Commissioner and the governor to the chairman of the local Economic Development Commission beginning a previously arranged vacation just as the siting process was unveiled. Those particular circumstances will not occur again.

While voluntary siting did not produce a disposal facility for New Jersey, and has had few, if any, successes nationwide, it would be premature to forsake this approach as a possible technique for seeking locations for unpopular projects. As former U.S. Environmental Protection Agency Administrator William Ruckelshaus said in 1997, "It is no exaggeration to say that movement towards a more collaborative, inclusive way of addressing environmental and natural resource problems may hold the only real hope of releasing us from a self-destructive gridlock." [95]

Time Can Change Everything

One of the biggest challenges in the siting and construction of controversial land-use projects, as with most major government undertakings, is time. The length of time between the identification of the need for government action and the implementation of a specific response is usually so long that the initial actors, premises, compromises, and even realities can change dramatically.

In a voluntary siting process, one of the most critical questions is how much time community residents are willing to devote to being involved. In New Jersey, very few people were willing to suddenly enroll in freewheeling and possibly multi-year seminars on radioactive waste, disposal facilities, and the philosophy and mechanics of voluntary siting.

New Jersey's disposal facility siting acts for hazardous and low-level radioactive waste were each drafted with great input and buy-in from leading environmental and industry group representatives. Five and ten years later when the laws were being implemented, however, those representatives, particularly in the environmental area, had moved on. Their successors had not been party to the compromises and agreements embodied in the law, and they felt detached and sometimes hostile to them.

True, representatives of some environmental groups were interested in the voluntary siting process and participated in a few of the Siting Board's workshops. But when small groups of people suggested the disposal facility might fit into a particular community, the Sierra Club and the others who had endorsed passage of the laws were nowhere to be seen. They did not rush in to support the vision or courage of the community members considering meeting the statewide need their group had helped define nor to challenge the misinformation being spread by each fledgling local group of opponents. Radioactive waste disposal was no longer on their front burner, and the current environmental activists, with only a few exceptions, felt no responsibility to help achieve the solution their predecessors had endorsed.

Similar turnover takes place in legislatures. Early in 1998, when the Siting Board was considering stopping its search for a site, only 33 members of the New Jersey Legislature remained from the 120 who had been in office in 1987 when the Siting Act was negotiated and approved. There was institutional memory among key legislative staff members, but only among a handful of the legislators who had participated in the initial debate and continued to follow the issue. This problem would only be worse in states, unlike New Jersey, that have enacted term limits.

The most significant effect of time is that situations and options evolve and change. Both the Hazardous Waste Facilities Siting Commission and the Low-Level Radioactive Waste Disposal Facility Siting Board were created to address exigent crises. Virtually everyone paying attention to waste disposal in the 1980s, from the Legislature to the press to the activists on all sides, and those who served on the siting agencies they spawned, agreed that something had to be done or the existing capacity would disappear and a vast array of New Jersey enterprises would suffer. But in the years that followed, unexpected events intervened.

The impact of the passage of time could be positive if it prevents the construction of something potentially dangerous that isn't needed. On the other hand, societal problems can fester while a quest for a perfect solution engenders an endless search for more complete and up-to-date data.

For low-level radioactive waste, three unanticipated events changed the situation facing New Jersey and other states. First, the dramatic increases in the cost of disposal led waste generators to institute extensive waste minimization programs. Their initiatives were so effective that the annual volume of low-level radioactive waste generated in New Jersey fell from 200,000 cubic feet in 1980 to about 20,000 by the late 1990s.

The second factor was the opening of the Envirocare facility in Clive, Utah. By 1998, this private sector venture, totally independent of the regional and state agencies created under the federal Low-Level Radioactive Waste Policy Act, was accepting 76 percent of the volume of the nation's low-level radioactive waste. Although this huge quantity contained only a miniscule 127 of the 334,563 curies of radioactivity that needed disposal, Envirocare's operation had the effect of reducing the amount of waste going to Barnwell, thereby effectively extending Barnwell's long-term capacity.

The third change, the reopening of Barnwell in 1995, was the most unexpected. Much of the impetus for the federal law, and the entire rationale for the New Jersey Siting Act, had been South Carolina's repeatedly expressed desire and plan to close the East Coast's only disposal facility.

These three changes meant that waste generators in New Jersey still had a problem, but the magnitude was greatly reduced. For the first time, they had two disposal options for the bulkier, less radioactive waste, and they could comparison-shop between Barnwell and Envirocare. The generators also knew that if one or both of these facilities closed, their available storage space on-site could serve their needs for a longer time due to the decreased volume they were generating each year.

The boards and commissions involved in trying to figure out what do with low-level radioactive waste continued their efforts. For the most part, they saw no reason to stop because they believed that new disposal facilities would be safe, economically beneficial to a local community and the state, and environmentally and ethically responsible. It might no longer be absolutely essential that the facilities be operating within five or ten years, but they still offered the only reliable long-term disposal option.

The impact of the unanticipated changes was that the siting agencies lost much of their ability to enlarge their base of support. When access to Barnwell had been cut off, the New Jersey Siting Board had carried to every meeting and event a map showing the 100 locations in the state where low-level radioactive waste was generated. We were able to grab people's attention by pointing out that each dot on the map also represented a place where the waste was being stored at that moment. The question was not whether New Jersey should have a disposal facility, we would say, but whether one carefully located, designed, and operated facility would be better than the 100 interim — and less well-supervised — storage rooms and facilities now located around the state.

This was not an argument that would necessarily lead residents of any particular town to volunteer to host the disposal facility, but it was a reality that could appeal to a variety of influential civic actors. Journalists, state legislators, governor's and commissioners' staffs, corporate executives, and leaders of statewide and regional organizations would be more supportive of efforts to place a disposal facility in one New Jersey town knowing that it would remove the waste from many others. This meant that a town that did volunteer would receive more outside support, perhaps including added benefits for hosting the facility.

Once South Carolina enabled Barnwell to reopen for the nation's waste in July 1995 and then maintained that policy for a few years, New Jersey's voluntary siting process was doomed. There was no longer an imperative that would attract and retain the "good government" support necessary to crack the wall of public prejudice against anything radioactive.

Another way in which time worked against the voluntary siting process was much more local and personal. As New Jersey's Siting Board envisioned the process, area residents, including many previously uninvolved in local governance, would read articles and reports, attend workshops and discussion groups, listen to talks by their neighbors and outside experts, and even take a few days to fly to South Carolina to visit the Barnwell disposal facility. They would not only be seeking reassurance, but they would be formulating the terms and conditions under which the facility would be acceptable to them. They would be learning enough so that they could specify — or at least choose among several options — the design features, operating restrictions, and compensation package they considered essential.

In short, we thought that many area residents would invest time in this process comparable to what they might devote to their children's college application process. We also thought that public confidence in an eventual decision to host the disposal facility would increase because of the years of review that preceded the decision. And, finally, we hoped that people would enjoy meeting and interacting with their neighbors in this way, and would gain some of the social excitement and benefit that participants in opposition groups often experience.

These expectations were wildly unrealistic. Not only are very few people intrigued by an invitation to spend many evenings over a span of months and even years in meetings, but the prospect of such a time frame proved unnerving to even the most ardent local supporter of a disposal facility. The amount of civic disruption caused by consideration of radioactive waste can be enormous.

Shortening the decision-making period would help. Although it might reduce the number of opportunities for public participation, it need not lessen the public's input and involvement. There is no reason to believe that the quantity of meetings and hearings, and the number of months over which they occur, has any direct bearing on the quality of deliberation or public confidence in the eventual decision.

A voluntary siting process also probably would be better received if the sponsoring agency offered an initial proposal more complete than the one the Siting Board put forward. This would include a detailed schedule for the local consideration process with a much shorter projected time frame, as well as a more precise image of what the disposal facility would look like and how it would operate. Most people are more willing to critique specific proposals than to invest time in writing from a blank slate.

Some of the residents we met simply scoffed at the idea that they or their town council would know what was necessary to make a disposal facility safe. The notion that this was up for local debate among their neighbors, while meant to be reassuring, was actually frightening. In addition, with virtually any possibility seemingly on the table, people who were fearful to begin with were almost encouraged to focus on those that seemed worst to them.

Similarly, the Siting Board was proud of the discretion it left to the host community for designing and negotiating the benefit package it would receive.

We did transform the formula included in New Jersey's Siting Act into a promise of at least $2 million a year. This was much easier to communicate and visualize than the words in the law which read in part:

> ... the facility shall ... pay to the affected municipality a sum equal to the amount which would annually be due if the land on which the facility is located and any improvements thereto were assessed and taxed as real property subject to local property taxation. ... The owner or operator of the facility shall ... file ... a statement ... showing the gross receipts from all charges imposed during the preceding calendar year upon any person for the disposal of low-level radioactive waste at the facility, and shall at the same time pay to the chief fiscal officer a sum equal to 5% of those receipts. ... The municipality in which the facility is located may petition the board for approval to collect an amount in excess of [this] amount. . .[96]

The Board chose not to be more specific about the benefits because it did not want to presume to guess how different municipalities would choose to spend $2 million a year. But in this era of mistrust of government, vocal residents in most towns visited by the Siting Board elicited knowing nods from their neighbors when they expressed doubt that they would ever see any benefits from the money.

This doubt took two forms. First, there was distrust that the Siting Board would fulfill its promise to deliver the funds. A potentially satisfactory response to this concern was in the plan the Board developed with Carneys Point Township. Early in the consideration process, the Township would have received grants that were sufficiently large that residents could experience some benefits before they had to decide whether to accept the disposal facility.

This pledge could be even more effective if the state agency encouraged local leaders supporting consideration of the disposal facility to articulate specific uses for these funds as part of their public introduction of the whole subject. This would strengthen the leaders' position when defending why they were subjecting their town to such a divisive debate.

Rather than saying that the community will receive a specified sum within a specified number of months just for studying the issue, local leaders could announce, for example, that the town would be able to reduce or eliminate a local tax for a year, acquire parkland, or improve a library or intersection. If residents witnessed the promised benefit being delivered as the siting discussions got underway, skepticism about the integrity of government in the siting process might be reduced.

The second doubt raised by the promise of local rewards rose from the low esteem in which some people held their local officials and their distrust that they would spend the money wisely. In the same way that early benefits could inspire confidence, the proposal probably would be better-received if the local officials putting it forward were as specific as possible about how they would like to spend the more substantial funding the town would receive if it accepted the facility.

Perhaps if a town's residents were presented with two clear options, some of the public debate could focus on choosing between them instead of just saying yes or no to the facility. These options also would be subject to revision through the local deliberation process.

A voluntary siting process that acknowledged how busy most people are and how little time they want to devote to any aspect of civics would present area residents with a different question than the Siting Board had suggested. The opening salvo we had suggested local proponents use when publicly raising the possibility of their town hosting a disposal facility had been, in effect:

> What do you think about placing a disposal facility for low-level radioactive waste in town? We don't know what it will look like. In fact, you can help decide whether it is placed above or below ground. We don't know where it will be located: Where do you think would be a good spot? And the town will receive more than $2 million a year for 50 years: Do you think we should use it for tax relief or other needed services? What do you think we most need?

A preferable alternative could be:

> We want to know what you think of the following proposal: Our town agrees to let this disposal facility be built out on that vacant land on Elm Street where the factory used to be. It will look like three long warehouses and a small office building. Trucks going to and from the facility would probably get off the interstate and take Route XXX directly to an entrance road onto the property.
>
> We'll have two townwide referenda on this issue. In the first one, we'll choose the benefit package we want to attach to the proposal, and in the second, we'll decide if we want to take the facility.

One way to use the revenue from the facility would be to allocate 40 percent to lower property taxes, 40 percent to set up a land acquisition fund, and 20 percent for local projects. That would mean that in the first five years after the facility was in operation, the average homeowner would save $XXX in taxes, we would preserve YYY acres, and we would able to add one extra police officer to the force and finance a community gym and the addition to the school.

The other choice would be to set up a fund that would guarantee up to $15,000 towards college tuition for every township resident to graduate high school: Do you realize what that would do for our dropout rates and for our property values?

The siting process might also be more effective if the idea of articulating multiple choices was extended to other major options concerning the facility. The residents of the 12 New Jersey towns in which volunteering to host a disposal facility was publicly explored were confronted with a yes or no question: Should everything stay the same or should we take this facility? The proposal might get a more complete hearing if a way is found to frame the discussion so the status quo is not such an imposing candidate.

Labor-management deliberations, frequently cited as a model for resolving contentious disputes, can't really be replicated for land use situations. The accepted recognition in such discussions, missing from most land use controversies, is that the final resolution will involve compromise on all sides and will cause some changes. People trying to stop a land use proposal usually come together on an ad hoc basis with the sole mission of not allowing something new and thereby keeping their area the same as it has been. They are held together as a group by their opposition, and only very rarely are any of their members in a position to negotiate, accept a compromise, and commit to a binding resolution.

The Siting Board did encounter two municipalities in which the public debate did include some recognition that, with or without the disposal facility, change was in the air. In Bethlehem Township, a substantial number of residents concluded that accepting low-level radioactive waste might be the only way to stop a large new housing project. As a result, the Siting Board's meetings and open houses were different there than in most towns. Many conversations included discussions comparing the added traffic and other secondary impacts that would result from the new housing as opposed to the disposal facility.

In Lower Township, a small number of officials and residents tried to apply this concept more formally when they planned to raise the disposal facility as one among several proposals a new citizen committee would consider as ways to avoid an impending huge tax increase. The group would evaluate the town's finances to determine the magnitude of the problem. At the same time, they would form task forces to examine each of the options anyone could suggest for addressing the situation. The disposal facility would be explored, but so would additional housing developments, shopping centers, retirement communities, and tourist attractions, as well as the impact on current residents of simply paying the projected tax increases.

When this committee presented its report to the Township Committee and the town's voters, it would be saying:

> *This is the way the town is going to change. If we do nothing, the average homeowner will see taxes rise by $XXX per year. Alternatively, here is a list of new developments we could consider with our understanding of each of their pros and cons in terms of health and safety, impact on our schools, traffic, and other benefits and detriments.*

The plan never got tested because the first meeting of the Lower Township citizens' committee focused almost entirely on the disposal facility. Before the tax situation was documented and publicized and before any other options for reducing taxes were named, low-level radioactive waste was dominating the headlines.

Lower Township's initial idea, however, could be an element of a winning strategy. The public discussion can be started by comparing the consequences of following a number of paths, including taking no action, for the town's future. Such a strategy would lay out specific descriptions for each option, and then allow extensive public discussion and debate with multiple opportunities to ask questions and propose revisions before setting a time for the local governing body to approve a referendum asking voters to accept or reject each alternative.

This approach would acknowledge that consideration of a low-level radioactive waste disposal facility captivates most residents and effectively turns them into decision-makers. It would empower them to shape the town's future rather than to just say no to one proposal. Public discussion would be focused on choosing among a series of imperfect choices, just as we do in considering which candidates to vote for and in making decisions in many other aspects of our lives.

Openness in Government

There is one major problem with presenting residents of a town with a proposal tailored specifically to their situation: It requires that considerable discussion and research at the local level occur quietly before the concept is presented to the general public. An individual or small group has to identify one or several targeted sites and itemize a wish list of benefits. If the Lower Township approach is used, they also have to do the same for the status quo and for a number of other options, all of which will be compared with the disposal facility proposal.

For traditional developments, detailed planning occurs within the private development team and then during local planning board meetings. If a reporter happens to be present at one of these sessions, the potential development may or may not be mentioned in a small article summarizing the meeting. Weeks or months later, when the proposals are formally considered, a few may sail through the approval process, while others become the focus of many public meetings and perhaps disputes. Almost all of them, however, will have been initially prepared and discussed with local officials without significant public attention or publicity.

This path is not available for consideration of a radioactive waste disposal facility. There is no way to raise this topic gradually. Any open discussion, or even mention of the concept, is likely to result in front-page headlines usually closely followed by counter-organizing and angry meetings. Before a proposal specific to the community can be developed, the community debate has taken place and the idea has been rejected.

To the extent that local officials do find a way to have quiet preliminary discussions on the issue, they then become subject to almost inevitable criticism when the proposal is unveiled. How can they say this is an open process when they plotted in private? If this is so safe, why were they afraid to talk about it openly?

The New Jersey Siting Board was accused of secret collaborations without reaping the benefit such discussions could have produced. In every town, the Board and local leaders were condemned by some for having had any conversations before the matter was completely public. Each time, some people felt that no matter when they had learned of the proposal, they should have been told earlier. It seemed as if they were saying they did not want their introduction to the proposal to coincide with the first time they heard about it.

In Roosevelt, where the only contacts before a large public meeting were two phone calls between the town clerk and Board staff to arrange a date and location, residents received generous applause when they complained that this idea had been sprung upon them, and that they had been excluded from whatever had occurred previously. The phone calls, apparently, should not have been made until everyone in town had a chance to consider the idea.

The reaction was similar but interestingly no more vehement in South Harrison, when residents there learned that several town leaders had been meeting with Board representatives for more than a year, and that they had created a 30-member advisory committee before making their interest public. Had the process in Springfield lasted longer, it would have been instructive to see the reaction to the Planning Board taking a trip to Barnwell before deciding whether the idea had sufficient merit to discuss publicly.

In the poll the Siting Board conducted in Fairfield after that town had rejected the idea of continuing to study the possibility of volunteering, some residents complained that they had not known the disposal facility was a possibility until they received a letter from the mayor. Well, the purpose of that letter had been to tell them about it, and to describe the two or more years of public discussions and deliberations that would precede any decisions that would obligate the town.

In short, it seems that most people predisposed to actively oppose a proposal are never going to applaud the decision-making process, nor are they likely to appreciate any nuances added for their benefit. This is not to say that process is unimportant, but rather that it should be designed more for the benefit of the persuadable members of the community than as an attempt to anticipate all the objections likely to be raised by those whose minds are firmly made up. There came a point where we and others promoting voluntary siting strategies were tying ourselves in knots in vain attempts to impress people who were only looking for ammunition to add to their arsenal of opposition.

A similar conclusion can be made about the volumes of technical data generated by proponents of controversial developments. Most members of the public who become aware of them note only their girth and are not particularly impressed. The few who actually read the reports, other than the staff of regulatory agencies, are primarily looking for inconsistencies and errors to bolster the negative conclusions they have already reached.

The issue of openness in government permeated the Siting Board's work. The Board moved to a voluntary siting process out of belief that the path to success lay in involving the public in all decisions of any consequence concerning the location, design, and operation of the disposal facility. The process the Board adopted provided a degree of community participation that rivaled legendary New England town meetings, but expected more time and commitment from the public than most people felt they were able to give.

Although the Board felt its process embodied and exemplified the spirit of public participation, it kept tripping over the specifics of the state's Open Public Meetings Act, as well as the expectation by some that no conversation between two or more people in government should be off limits to public scrutiny.

In an article in *Governing* magazine called "The Endless Struggle Over Open Meetings," reporter Charles Mantesian writes:

> *A quarter-century after the first wave of open-meetings laws, it is clear that the basic premise is settled — in theory if not always in practice. At least publicly, just about everyone accepts the idea that openness is a worthy goal. Now, however, another question is coming to the fore: Just how much open government is enough — or practical?* [97]

This situation is by no means unique to New Jersey. A consultant hired to study a school system in Florida found he would be violating that state's Sunshine Law if he discussed any of his largely negative findings with two or more members of the school board. "The law creates a situation," he said, "where people with vested interests can derail proposals before they even get off the ground." [98]

Human interchange thrives on informality. To try to exclude such informality from efforts to address problems so intractable that we have asked government to handle them is designing the process to fail.

Former New Jersey Governor Tom Kean felt this was a major obstacle facing President Clinton's Race Initiative Advisory Board. In the midst of its deliberations, Kean, one of the seven members of the Commission, wondered whether the requirement that all meetings be held in public would render it useless:

> *Every time we meet, we find ourselves sitting around facing each other with an audience of 500 and the camera of CNN...It's very difficult for people who never met to meld as a commission and to have the disagreements that are necessary when everything is done before television cameras.* [99]

It is noteworthy for someone as prominent as New Jersey's former governor to advocate, in effect, that some government meetings be kept private. But as the president of Drew University, it is much easier for Tom Kean than it would be for a politician or government administrator currently in office. Paradoxically, enabling our elected and appointed representatives to sometimes operate with a little less openness may be one of the changes necessary to help make government more effective and thereby more trusted.

"Educating citizens about the issues their states and representatives face is worthwhile," writes Alan Rosenthal, but there is a far greater need for citizens to be educated as to how their political system works and what representative democracy entails." He continues:

> *This is not to defend the institutional status quo, but to be concerned about demands for wholesale governmental reform that result, in large part, from a lack of understanding of how government operates. Nor is this to challenge public skepticism, a reasonable amount of which is healthy in a democratic polity. But too much skepticism, cynicism, and distrust add up as a threat to the health of the polity.*[100]

"It is ethical to fight unjustified anxiety," writes Dietrich Schwarz in an assessment of "Ethical Issues In Radiation Protection."[101] If proponents of locally unwanted land uses succumb to the notion that only project opponents are permitted to plan, strategize, target, and even conspire, most siting struggles are going to be won by the opposition which, perfectly understandably and appropriately, will do most of their planning and strategizing in secret.

Land use struggles are often portrayed as David vs. Goliath battles, but the roles are miscast. Michael Gerrard, who as a lawyer has represented opposition groups, writes that the fight is "lopsided," and that "project opponents have much fuller quivers than do project proponents." This derives in part, he notes, from the fact that "far more laws are bunched toward the 'Must Avoid' end [on a stringency of regulation spectrum] than the 'Must' end."[102]

Peter Sandman notes, "Ironically, nearly everyone is impressed by the community's power of opposition — except the community, which sees itself as fighting a difficult, even desperate uphill battle to stop the siting juggernaut."[103] While the processes the New Jersey Siting Board stimulated in 12 municipalities never were able to survive for more than three months, the opponents in each community saw themselves as vastly overmatched underdogs.

New Jersey's voluntary siting plan was an attempt to grapple with the widespread prejudice about radiation that is nurtured by deeply rooted perceptions and consistent repetition in the popular culture. The task, while difficult, might have been possible if all the stars had been properly aligned. But they were not. Not only did the extent of the waste disposal crisis fall into doubt, but as this experiment occurred, public distrust of government and lack of civility in public discourse were skyrocketing.

"The NIMBY problem," concluded Michael O'Hare and Debra Sanderson after helping to draft and then monitor a hazardous waste siting law in Massachusetts, "is, at heart, symptomatic of the pessimistic expectations that citizens, industry and government all hold for each other and themselves; raising those expectations is not a task that can be accomplished by any legislated decision process."[104]

Pessimism and incivility are interrelated. Many people seem to accept the most negative view available of the magnitude of present and potential future problems, particularly in the areas of health and the environment. To some, anyone offering a different perspective is lying and being deceptive. If they feel this deceit is threatening their family's health, then polite and civil discourse often is an early casualty.

In addition, on many issues people seem to assume that differences of opinion are far wider than they really are. A study by psychologists Dacher Keltner of the University of California at Berkeley and Robert Robinson of Harvard University found that participants "thought there was twice to four times as much disagreement between their position and their opponents' position as there actually was." In what they called the "lone moderate" phenomenon, they concluded that "people sense they alone have got the facts right, that they're the balance between extreme opinions."

If "opposing partisans ... perceive their opponents as hostile, irrational, immoral and ideologically extreme," the researchers conclude, it "may be a self-fulfilling prophecy. If most people assume that ideological debates are matters of black and white and extremists standing in opposition to each other, then moderates and moderate solutions may not be heard."[105]

As Robert Putnam observed in his article, "Bowling Alone," finding a way for people to question their preliminary opinions and perhaps compromise is harder in groups that do not find other occasions to interact.[106] Reporter Dirk Johnson, in an article in *The New York Times* called, "Civility in Politics: Going, Going, Gone," noted that "the mobility of American society means that town meetings are more often a collection of strangers than a comfortable gathering of old friends scripted by Thornton Wilder."[107]

In varying degrees, all of these obstacles are well-known. Perhaps it was naive of us in New Jersey to think that one little agency might be able to overcome public fear of radiation, distrust of government, skepticism caused by seeing past "crises" defined away, lessened interest in civic engagement, and a lack of civility and openness to the opinions of others.

On the other hand, if the national Republican sweep in the 1994 elections had not carried David Beasley into office as South Carolina's governor, New Jersey's voluntary siting process might well have had a different outcome. The Barnwell facility would have remained closed to waste from New Jersey, and the disposal situation might have continued to be viewed as a crisis. Perhaps one of the 12 towns with which the Siting Board had close encounters would have found sufficient outside support and encouragement to stay with the process. Even if the 12th township had still dropped out in 1998, the Board probably would not have suspended the process at that point. Instead, it would have offered other municipalities the plan it had arrived at with Carneys Point.

It was in 1988 and 1989 — not too long ago really — that voters in Martinsville, Illinois voted twice to locate a disposal facility for low-level radioactive waste in their town. The local support continued through 1992 when the mayor and Board of Aldermen signed a binding agreement to be the host community. While the facility was never built, its demise was due to statewide regulatory and political factors and not to opposition in Martinsville.

Tom Kerr, who worked for the Illinois Department of Nuclear Safety at the time, attributes the unusual absence of overwhelming local opposition to three factors. One was that the private firm the state had hired to perform the site characterization was able to spend money immediately to respond to landowners' concerns. When a farmer claimed that the contractor had damaged his fence while inspecting a site, the contractor paid cash on the spot to settle the claim. A few such incidents became the talk of the town and gave the siting program a reputation for being a group of reasonable people you could work with and trust.[108]

The second factor was that Martinsville received grants from the beginning of the process. Before the referenda took place, the mayor was able to send all residents a letter telling them that they owed no money for one of the local taxes that year because the siting program had paid the bill.

The final factor, which may be the most important, was that the mayor of Martinsville, Truman Dean, believed in the disposal facility. He thought it would be good for the town, and he wanted it to be his legacy. His steadfast support and enthusiasm was inspiring to other local leaders and residents.

Leadership

The type of political leadership exhibited by Mayor Dean in Martinsville, Illinois is essential at the local, state, and federal levels, but how this is best manifested is subject to debate. The timing of public expressions of support by leaders can significantly affect their impact.

In each of the New Jersey towns where volunteering was seriously considered, a handful of people initially raised the issue. Others were needed to then endorse the process and give it momentum. Local officials needed to make clear that they would oppose any attempt to stop the process and end the dialogue before there was a chance to assemble the information needed for the town to make an informed decision.

The communities considering the disposal facility were municipalities in which the financial benefits would be a significant percentage of the annual budget. Almost by definition, the population of these areas was relatively small. As a result, while supporting consideration of the facility would take some political backbone, personal courage was the more important ingredient on the local level. Supporters would be subjected to virtually non-stop lobbying, often from their friends, neighbors, and family, and often expressed in incredulous, combative tones.

Politics is intensely personal at the local level and increasingly less personal as it moves toward higher levels and larger jurisdictions. Therefore, it was easier for county and state legislative representatives than it was for municipal officials to resist pressure to take stands against siting initiatives within their domains. The Siting Board encountered only two exceptions: Cumberland County's elected freeholders visibly and strenuously opposed Fairfield Township's interest in considering the disposal facility, and state Senator Robert Singer apparently singlehandedly ended the public discussion in Springfield before it could begin.

The appropriate role of New Jersey's governor in this issue is more complex. During New Jersey's siting process, several representatives of the waste generators advocated that Governor Whitman publicly endorse the siting process and encourage municipalities to consider volunteering to host the disposal facility. They thought she should increase the incentives, perhaps by directing other state agencies to give preferential treatment to towns that were looking into hosting the disposal facility, and that she personally could help increase public awareness of the problem.

While members of the Siting Board would have welcomed such actions, most felt that they were receiving the support they needed. Appointments to the Board and Advisory Committee, which had languished during the prior Florio administration, were acted upon, and the Board's general operations, from budgets to securing adequate office space, were routinely approved.

More active involvement and support from the governor would be most useful when the Siting Board and a town were in serious negotiations. That is when opponents would undoubtedly be urging her to stop any agreement by exercising her power to veto minutes of the Board's meetings.

The Board believed that the governor's routine approval of the minutes of its meetings and of its budgets implied tacit approval of its overall direction and activities; perhaps it did as long as the process did not seem likely to succeed. The Board's work was useful in case there was a waste storage emergency. The first time it appeared possible that the voluntary siting process might succeed, and the need for a decision reached the Statehouse, Governor Whitman, finding no immediate emergency, chose to end the process.

The siting experience could be seen to confirm the wisdom and approach of politicians who seek every possible way to avoid taking positions on tough issues. The free market, combined with unexpected election results in South Carolina, averted a crisis and saved the day. Whitman, like other governors across the country, took no action except to quietly withhold support for the siting process when it would have mattered, and for that neither she nor radioactive waste generators nor the state as a whole have suffered at all.

If waste was now piling up at 100 locations around New Jersey and drug companies were curtailing research in nuclear medicine out of fear that there would soon be no place to dispose of the waste, the governor would be guilty of shirking political leadership. If that had been true early in 1998, however, it is reasonable to assume that she would have made a different decision regardless of the position her Commissioner of Environmental Protection was advocating.

Since it has turned out that the state will get by without building a disposal facility, maybe benign neglect of the problem was appropriate. Governor Whitman saved the political capital such a stand would have cost and perhaps was able to apply it to other public problems. While no one could have predicted at the time that two years later she would be selected by President George W. Bush to head the U.S. Environmental Protection Agency, Whitman may have greater latitude to address issues related to waste disposal in her federal role than she would have had if she had become an outspoken supporter of building a facility in New Jersey. Moreover, in that case, Bush might have chosen someone else for the position.

While the Siting Board reluctantly ended the voluntary process because it lost Governor Whitman's support, it was at the federal level that political leadership on low-level radioactive waste was most lacking. To begin with, Congress, by passing the federal Low-Level Radioactive Waste Policy Act, handed the disposal problem to the states. This was not really an abdication of responsibility because at the time the nation's governors were asking for this assignment.

After delegating this important national problem, however, the federal government subsequently stood in the way of solutions the states suggested. This was most egregious during the Clinton administration, when the Department of the Interior blocked California's plan to build a disposal facility in Ward Valley. Here a state took on the responsibility assigned in a federal law and played by the rules Congress established, only to face a losing fight against a President and a Secretary of the Interior who as governors were instrumental in moving this mission away from Washington.

Improving Government

The one constraint in the private sector that is not faced by most government agencies is the need to show a profit. This is a significant advantage for governments, but often it is not sufficient to counter the many more disabling restrictions under which public agencies are forced to operate. Complying with the terms of those restrictions often inadvertently becomes an agency's primary mission, relegating to a lower priority the policy objectives of the laws it was created to administer.

The Siting Board, a small agency with a single clearly defined purpose, provides a good vantage point from which to view this conundrum. To begin with, the Board was able to offer speakers, conferences, publications, videotapes, and other services for free. This is far from unique, but it did contrast sharply with the movement in some parts of government toward user fees and other techniques to make programs self-supporting.

The Board itself worked with two sections of the federal government that demonstrated the difference between these two approaches. The Nuclear Regulatory Commission is required to charge everyone, including other government agencies, for any work it does. It operates like a law firm with no *pro bono* clients. The U.S. Department of Energy's National Low-Level Waste Program, on the other hand, offers all its services for free.

If New Jersey had progressed to build a disposal facility, the final major step would have been to apply for a license from the NRC. Since there was a federal law mandating that states address the problem of low-level radioactive waste disposal, it would have been reasonable to expect the NRC to help states to meet that requirement. Instead, any request the Siting Board or other states or regional compacts made for review of preliminary documents or for other advice was met by a discussion of how much time would be required of the NRC staff and how much they would then charge the state or other groups requesting the information. While NRC staff informally provided the Siting Board with some limited comments, they acknowledged they were doing so by skirting their rules.

The U.S. Department of Energy's Low-Level Waste Management Program, on the other hand, by being a readily available and unbureaucratic resource, saved the Siting Board considerable time and money. Upon request, they sent knowledgeable staff and consultants to New Jersey to meet with the Board and with interested community residents, and they prepared two major documents the Board needed. One was a book of responses to questions and issues that had been raised at some of the Siting Board's public meetings, and the other was a financial feasibility study for a disposal facility specifically for New Jersey.

In addition, several times each year the U.S. Department of Energy convened a task force of representatives from states pursuing or considering voluntary siting programs for a very useful exchange of ideas. All these forms of assistance were provided quickly and at no cost, and were of high quality.

Providing free information may seem like an obvious or unimportant governmental function. But at a time when government's role, structure, and relationship with its constituents are all under scrutiny, it is symbolic. It is one of the ways in which government can move subtly toward being — and being perceived as — a helpful friend rather than a seemingly alien force.

As a government agency, the Siting Board was unusual in that it had great flexibility in budgeting and hiring. The budget the Board developed in 1992 assumed the siting process would proceed as rapidly as the members could envision. It then assessed $12.7 million from the waste generators to cover the cost projected for that scenario. The process never came close to achieving this idealized vision, but the Board was able to structure its budget for success. Had a town stepped forward ready to have a full site assessment and to undertake an extensive public involvement process, the Board would not have had to ask the local leaders to put the request on hold until it could get adequate funds in the next budget cycle.

The Siting Board's ability to raise its operating funds as needed from waste generators was almost unique. Most agencies operate under the suspense and uncertainty inherent in one-year budgets. Multi-year funding is in use in few state budget processes. Vice President Gore's National Performance Review, among others, recommended two-year budgeting as a significant tool for helping reinvent government.[109]

The other way in which the Siting Board was different from most parts of state government was that its hiring process was based entirely on program need, employee merit, and available funds. Board members wrote descriptions of each job they needed to fill and then hired the best person they could find. When circumstances changed and the Board shifted to a much less active status, those of us on the staff were laid off.

Had the Legislature not stated in the Board's enabling statute that it could hire staff "and fix and pay their compensation without regard to the provisions of Title 11A of the New Jersey Statutes,"[110] the Board would have joined most other state agencies in following the detailed code of Civil Service requirements. Under this system, the Board would have had to mold its specific needs to fit into pre-existing job titles and descriptions, and then hire from a list of people who had taken a test for the particular title and scored near the top of the pack. Many talented, motivated individuals are included on the Civil Service lists, but they may or may not be a good fit with a specific job.

One result of this exemption for the Board was that it was able to choose staff who were all good communicators, a characteristic not evaluated by most Civil Service tests. Another was that everyone on the staff believed in the voluntary siting process. The Board did not have to include anyone on its staff, as many other agencies do, who disagreed with important aspects of its direction or policies. And, finally, the Board could dismiss people who didn't work out or were no longer needed.

Defenders of the Civil Service system will usually admit its inefficiencies, but extol the necessity of protecting individuals from discrimination and mistreatment by politicians and government executives. Most people agree that a person's political views should not be a factor in state government hiring, but what should happen if someone disagrees with the basic purpose of a particular agency? If the Legislature decrees that the state should, if necessary, build a disposal facility for low-level radioactive waste, how would the public interest be served by requiring the agency to hire someone who had done well on a Civil Service test but believed that building a disposal facility would only encourage the production of more waste? This is not an ignorant or foolish position nor one that should disqualify the person from most government jobs, but it is fundamentally at odds with the mission of the agency.

The other protection presumably offered by the Civil Service system is to avert arbitrary firing and provide job protection. If the Siting Board had operated under that system, when the staff was no longer needed, each of us would been moved to some other part of the state government, depending on our years of seniority.

Instead, we were laid off and had to seek other jobs. That the Siting Board had a finite life was never in doubt. As each of us was hired, we understood that this job could end relatively soon and, when it did in 1998, no one was surprised or felt mistreated. In return, while we worked at the Board, we all knew we were there because we were respected and valued, and not because someone had been stuck with us.

While the Board was liberated from hiring and budget constraints, its effectiveness was still limited by other governmental operating procedures. In addition to issues relating to open public meeting requirements, the processes of contracting for services and appointing Board and Advisory Committee members presented obstacles that seemed to serve little public purpose.

The problem with the contracting system is simple to summarize: It takes too long. With elected officials trying to keep down the number of state employees because of fear that it could become a campaign issue, and state hiring procedures so byzantine, issuing a contract for consultants to perform needed services often makes sense. Unfortunately, the time required for the competitive bidding process as currently designed discourages its use. The Siting Board's early experience may have been particularly stark. It began and completed its shift from a statewide screening process to voluntary siting during the months that its contract for a site characterization consultant was pending. As a result, the contract, awarded more than two years after the Request For Proposals was issued, did not mesh well with the Board's needs.

The problems with the appointment process for members of boards and committees are more serious for their effect on government and on public perception of government. At the same time, they should be more easily amenable to correction.

New Jersey's government has hundreds of positions for which the governor nominates candidates and the state Senate votes to confirm or reject them. A handful of these appointments, such as cabinet officers and Supreme Court justices, require great scrutiny. They are powerful posts second in importance only to the governor.

Most of the appointed positions, however, are more similar to the 11 seats on the Siting Board and 13 on the Advisory Committee. Appointment offers no pay and no glamour, and is attractive only to a small number of people with a specific interest in the subject matter and the time available to serve. Yet the state appointment process treats all vacancies as essentially equivalent. The irony is that filling the many lower profile slots that should require less attention usually takes many months longer than selecting a cabinet officer or Supreme Court justice.

One reason to create boards and commissions is to help people feel closer to their government. Yet the effect of New Jersey's appointment process is to leave most potential nominees feeling neglected, offended, and sometimes embarrassed. When they are approached about sitting on a board, they will typically consult with their employer and perhaps discuss it with their family. But then months and sometimes years pass before the actual gubernatorial nomination is announced, and more months go by before the Senate gets around to approving the nomination.

This protracted time frame is rarely because anyone objects to the nominee or has questions about the agency. But ensuring that government posts are filled promptly and that potential candidates are treated decently is a surprisingly low priority for the Governor's Office and the state Senate. In New Jersey, this has been true regardless of the individuals or parties in power. It is particularly peculiar because making appointments is considered an important power of the governor and would seem to be such an easy way to involve and reward supporters and other people the governor and her staff respect.

The appointment delay also takes a toll on the agency. The Siting Board was always able to obtain a quorum even when several seats were vacant, but other councils and commissions left for months or years to operate with several positions vacant have been prevented from making decisions because affirmative votes were required from a majority of their full boards.

Beyond that tangible problem are the more subtle messages that are given to the public by a government agency listing multiple positions as unfilled. Government appears sloppy and perhaps indifferent or arrogant as well.

No one likes government the way it is, but there are fundamental choices to be made about the direction in which it should change. Layers of checks and balances and restrictions can continue to be added in an attempt to protect the public from every real and imagined abuse government might commit. If one individual or agency misspends its budget, then require another sign-off for spending in all parts of state government. If one policy appears to be adopted without adequate public input or one meeting is held that appears to give unfair access to one side of a dispute, then add more public notice mandates for all.

An alternative perspective is to reexamine the tasks we give to government and ask what needs to change to enable them to be done well. Which requirements help ensure that the public knows what is being considered and has meaningful opportunities for input? Which procedures primarily serve to delay decisions and the resolution of issues? These questions are rarely easy to answer, but they still must be asked. The default response now is that more opportunities for participation and more steps in the decision-making process are always better. This reaction, however, is keeping us on a cycle of low expectations for government, followed by low performance and consequent inadequate attention to major public problems and opportunities.

Breaking that cycle will require recognition that effectiveness and public trust are interlocking goals. Public confidence is not likely to rise significantly unless public agencies are given power and discretion sufficient to perform their jobs effectively. Unfortunately, the wave of the future is in the opposite direction. As Rutgers Professor Alan Rosenthal notes, "proposals to strengthen the power of the governed by weakening the authority of the governors are in high fashion today."[111]

Confronting Risk, Radiation, and Distrust of Government

In June of 1998, just before the Siting Board went into hibernation and the staff started to go its separate ways, New Jersey hosted the final meeting of the Voluntary States Working Group. Over the previous four years, this informal group of 25 veterans from active or fledgling voluntary siting programs in Connecticut, Illinois, New York, Ohio, Massachusetts, Michigan, Pennsylvania, and New Jersey had met about a dozen times.

The meeting, held in the Lambertville Municipal Building, was a warm though bittersweet sharing of anecdotes and tentative conclusions. The previous weekend, I had seen the movie *The Truman Show,* which had recently opened to great attention and acclaim. Jim Carrey plays a man born into a stage set for a television program that is going to document his entire life without his knowledge. He thinks he is living in the real world, but he is really under a huge dome surrounded entirely by actors and props. During the meeting, I found myself wondering whether the movie was a metaphor for voluntary siting.

The siting process in New Jersey had received considerable attention from newspapers and fans of government around the state. Our search for just one willing town had been compared to Diogenes looking for one honest man, to the Maytag repairman waiting for the phone to ring, and, most often, to Sisyphus repeatedly pushing the rock partway up the hill.

Few observers expected that any local elected official or civic group would ever suggest that their municipality consider volunteering. Some suggested that we had misinterpreted our mandate: The Siting Board was only supposed to look busy so that the Southeast Compact Commission would continue to let New Jersey generators send waste to Barnwell. Others clearly rooted for the Board, but watched it more as an interesting laboratory experiment than as a process that was going to succeed.

Only those of us under the bubble — the members and staff of New Jersey's Siting Board and Advisory Committee, our visiting colleagues from other states, and a handful of others — thought it could be done. We cheerfully pictured the day when the ribbon would be cut to open New Jersey's Disposal Facility for Low-Level Radioactive Waste. We even joked about hiring the performance artist Christo to wrap the facility in beautiful cloths that would make it a tourist attraction.

New Jersey's voluntary siting experiment did not succeed in siting a disposal facility, but for reasons that are not part of mainstream political discussion or debate. The Board as the responsible agency had adequate funds, and the government officials involved were not incompetent, corrupt, or lazy. Neither battling between industry and environmental advocates nor the way in which campaigns are financed were factors, and the media — at least the print press — was not part of the problem.

The forces that defeated the voluntary siting process did not turn out to be individuals or organizations who could be engaged in debate, negotiation, and eventual compromise. That would have made the task easier. Instead, the enemies were our inability to deal with risk, and the pervasive distrust of government. In both areas, the huge gulfs between what is demanded, what is expected, and what is real or possible have helped create an increasingly cynical public climate. Cynicism flows inevitably when unrealistic expectations cannot be met.

A vicious cycle has been created by the response of much of the public and many legislators. Government agencies are given less authority and fewer resources, presumably until "they" get it right and show themselves to be worthy — that is, until government agencies present information that is more conclusive, and take actions that are more definitive and are admired both for their substance and for the process by which they were reached. Requirements for multiple avenues of public participation are such that government agencies —like modern candidates — are not permitted private lives. It should not be a surprise when government does not excel under all these constraints.

In an age in which we are bombarded with information about what has gone wrong and what can harm us, how do we absorb data and decide what is correct or acceptable? We want to avoid what is hazardous while coping with the fact that danger can be envisioned flowing from virtually any decision or activity. We need to decide what we believe is safe, particularly as technology advances and we can measure increasingly minute amounts of substances which, in much larger amounts, are unquestionably dangerous. Gaining greater scientific literacy will certainly help, but few people have the time, or the knowledge, to explore every conceivable risk.

The only workable option for us as a society is for us as individuals to place some trust in others. Our friends and acquaintances, chosen media personalities, and favorite non-profit groups will offer information and opinions that may sometimes change our individual behaviors. But when we think others must change or take some type of action, we need government.

New Jersey, Connecticut, and South Carolina solved their low-level radioactive waste disposal problems, and the rest of the nation eventually will probably do so as well.

The larger questions, however, are unresolved: Can our relationship with government change so that we trust public agencies and officials to help us know which of the many possible risks are really dangerous and worth worrying about? While government is far from infallible and, by nature, often slow-moving and inefficient, how can we still use it to make necessary decisions that can't be left to individuals? Finally, as we gain access to more and more information, can we learn how to choose the best option even when we know none are perfect?

Chronology,

Endnotes

&

Index

Chronology
Events Affecting Low-Level Radioactive Waste Disposal in New Jersey

(Federal activities and events from other states in italics)

1979   *Accident at Three Mile Island nuclear power plant in Pennsylvania.*

South Carolina Governor Richard Riley says his state's Barnwell disposal facility will not accept waste from Three Mile Island.

In response to leaking containers found at the nation's other two disposal facilities for low-level radioactive waste, the governors of Nevada and Washington each take action to restrict access to out-of-state waste.

1980   *Congress enacts the Low-Level Radioactive Waste Policy Act making states responsible for the waste generated within their borders. The law, based on recommendations of the National Governors Association, encourages the formation of interstate compacts.*

1983   New Jersey agrees to enter a regional compact as its Legislature passes the Northeast Interstate Low-Level Radioactive Waste Management Compact Act.

1986   *Congress enacts the Low-Level Radioactive Waste Policy Act Amendments of 1985, and also approves the Northeast Compact and six others. The Amendments Act requires Nevada, Washington, and South Carolina to continue accepting out-of-state waste, but only until December 31, 1992.*

Chernobyl nuclear reactor in the Ukraine, Soviet Union explodes.

1987   New Jersey Legislature enacts the Regional Low-Level Radioactive

Waste Disposal Facility Siting Act. The Northeast Compact Commission subsequently designates both New Jersey and Connecticut as hosts for disposal facilities.

1988 New Jersey Siting Board and Advisory Committee are constituted and hold first meetings.

1989 Siting Board hires staff and awards contracts to environmental and public relations consultants. Board issues preliminary siting criteria.

1990 Siting Board adopts final siting criteria and site identification methodology. Also issues "Guidelines for Designing A Multi-Purpose Local Assistance Program for Siting A Facility" prepared by Rutgers University's Center for Negotiation and Conflict Resolution.

1991 Legislature amends Siting Act to enable Board to assess and collect fees from waste generators instead of relying on annual state appropriations for its budget. Board also forms subcommittee to explore volunteer siting programs.

Connecticut proposes three sites for disposal facility selected through statewide screening process, but Legislature balks and instructs siting agency to try again.

1992 Subcommittee on the Volunteer Program recommends to Board that Advisory Committee undertake a public process to design a volunteer siting program. Board concurs and directs environmental consultant to stop statewide screening being conducted as part of site selection methodology. Advisory Committee, assisted by Department of the Public Advocate's Office of Dispute Settlement and New Jersey League of Women Voters, convenes focus groups and other public discussions on volunteer siting.

Governor of Nevada closes Beatty disposal facility as of December 31, 1992. Washington restricts use of Richland facility to waste generators from the eight states in the Northwest Compact. South Carolina agrees to continue unrestricted access to Barnwell until June 30, 1994, provided participating states demonstrate continued progress towards creating an alternate disposal option.

California receives needed licenses to build and operate regional low-

New Jersey Municipalities Involved in the Voluntary Siting Process

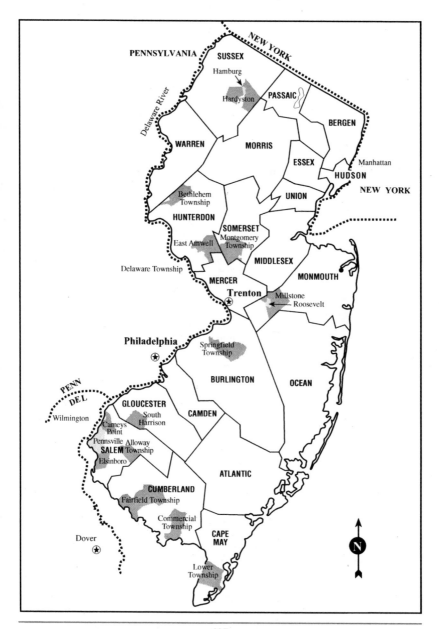

level radioactive disposal facility on federally owned land in Ward Valley. Only remaining step is transfer of land by the U.S. Department of the Interior.

Connecticut adopts volunteer siting program.

1993 Siting Board enters into contract with Rutgers University's Environmental and Occupational Health Services Institute to develop tools to help schools teach about radiation and low-level radioactive waste.

Texas Governor Ann Richards signs law creating a Texas-Maine-Vermont Regional Compact.

1994 Siting Board accepts proposed voluntary siting plan from the Advisory Committee. Distributes it for public comment and holds public meetings and hearings.

On July 1, disposal facility in Barnwell, South Carolina closes to low-level radioactive waste from outside the Southeast Compact region. New Jersey's 100 generators each begin storing waste on site.

1995 Siting Board adopts New Jersey's Voluntary Plan for Siting a Low-Level Radioactive Waste Disposal Facility.

Public discussions of volunteering to host the facility begin and end in Roosevelt (Monmouth County), Alloway (Salem County), Elsinboro (Salem County), and Hamburg and Hardyston (Sussex County).

Envirocare of Utah, a private company outside the regional compact system, receives approval to accept high-volume, low-radioactivity waste from all states.

South Carolina's new governor, David Beasley, reverses state's policy and reopens Barnwell facility to low-level radioactive waste from generators in all states except North Carolina. Action is challenged in court.

1996 Public discussions of volunteering to host the disposal facility begin and end in Springfield, (Burlington County), Lower (Cape May), Fairfield and Commercial (Cumberland).

Public discussion begins in Bethlehem (Hunterdon County).

"I'm sorry young man, but I'm afraid I can't bring you a low-level nuclear waste site."

South Carolina Supreme Court upholds reopening of Barnwell.

Pennsylvania, host state for the Appalachian Compact it had formed with Delaware, Maryland, and West Virginia, switches from a statewide screening siting process to a combined screening and volunteer approach.

1997 Public discussions of volunteering to host the disposal facility end in Bethlehem, and begin and end in Delaware (Hunterdon County), South Harrison (Gloucester County), and Carneys Point (Salem County).

Midwest Interstate Compact abandons plan to build a disposal facility in Ohio for Indiana, Iowa, Minnesota, Missouri, Ohio, and Wisconsin.

1998 New Jersey Siting Board ends the voluntary siting process.

Pennsylvania ends its voluntary siting process.

South Carolina Governor Beasley defeated, throwing future state policy on waste disposal into greater uncertainty.

1999 *California Governor Gray Davis abandons state's effort to build a regional disposal facility in Ward Valley for Southwestern Compact, which includes California, Arizona, North Dakota, and South Dakota.*

Texas Governor George W. Bush abandons his state's effort to build a disposal facility in Sierra Blanca for Texas, Maine, and Vermont.

Nebraska enacts law withdrawing state from the Central Interstate Compact, which had selected it as the host state in 1987. Compact included Arkansas, Kansas, Louisiana, and Oklahoma, as well as Nebraska.

North Carolina, host state for the Southeast Compact of Alabama, Florida, Georgia, Mississippi, South Carolina, Tennessee, and Virginia, abandons its planned facility due to budgetary constraints.

South Carolina's new governor, Jim Hodges, talks of having his state join a small compact, perhaps one with as few as two states.

2000 *South Carolina forms Atlantic Compact with New Jersey and Connecticut, and agrees to accept all low-level radioactive waste generated in the three states for the next 50 years. The other 35 states relying on Barnwell are told they must find other disposal options by 2008.*

Endnotes

Preface to the 2007 Edition

a. *The State*, Columbia, S.C., March 29, 2007.

b. See Energy*Solution's* web page at www.EnergySolutions.com.

c. "Trust in Government" at www.GallupPoll.com, March 1, 2007.

d. The American National Election Studies, *The ANES Guide to Public Opinion and Electoral Behavior*, University of Michigan, Center for Political Studies, Ann Arbor, Mich., at www.electionstudies.org.

e. See page 362.

Introduction

1. John Weingart, *New Jersey's Voluntary Siting Process for a Low-Level Radioactive Waste Disposal Facility: A History and Perspective*, National Low-Level Waste Management Program, U.S. Department of Energy, Idaho Falls, Idaho, July 2000.

Part 1. The Challenge

2. Richard W. Riley quoted in Michael B. Gerrard, *Whose Risk, Whose Backyard: Fear and Fairness in Toxic and Nuclear Waste Siting* (Cambridge, Mass.: The MIT Press, second printing, 1995), p. 153, citing article by Jon Jefferson, "Barnwell: The Radwaste Era Ends—But Not Quite Yet," Forum for Applied Research and Public Policy 88,89, 1993.

Chapter 1. Visualizing Success

3. Alan Karcher, author of *Multiple Municipal Madness* (New Brunswick: Rutgers University Press, 1998), telephone conversation, September 28, 1998.

4. Seymour Littman, conversation over breakfast in an out-of-town diner the former mayor chose in part to avoid being seen with a representative of the Siting Board, May 1995.

Chapter 2. Radiation and Risk

5. National Council on Radiation Protection and Measurements, "Sources And Magnitude Of Occupational And Public Exposures From Nuclear Medicine Procedures," Bethesda, Md., March 11, 1996.

6. National Low-Level Waste Management Program, U.S. Department of Energy, "Frequently Asked Questions about Commercial Low-Level Radioactive Waste Disposal" (www.inel.gov/national/faq.htm, 1999).

7. Marvin Goldman, "Cancer Risk of Low-Level Exposure," *Science*, Vol. 271, p. 1821, March 29, 1996.

8. Myron Pollycove, M.D., "The rise and fall of the linear no-threshold theory of radiation carcinogenesis," *Nuclear News*, June 1997, p. 34.

Chapter 3. Multi-State Compacts: Seeking Regional Solutions

9. This historical information is available from a number of sources including: United States General Accounting Office, "Report to the Chairman, Committee on Energy and Natural Resources, U.S. Senate: Low-Level Radioactive Wastes; States Are Not Developing Disposal Facilities," Washington, D.C., September 1999. Also, National Low-Level Waste Management Program, *Directions in Low-Level Radioactive Waste Management: A Brief History of Commercial Low-Level Radioactive Waste Disposal*, Idaho National Engineering Laboratory, August 1994. Also, Dennis Coates, Victoria Heid, and Michael Munger, *The Disposal of Low-Level Radioactive Waste In America: Gridlock in the States*, Center for the Study of American Business, St. Louis, Mo., December 1992.

10. Richard W. Riley, "Why South Carolina Said No," *Washington Post*, April 23, 1979, p. A23, cited by William F. Newberry, "The Rise and Fall of American Public Policy on Disposal of Low-Level Radioactive Waste," *South Carolina Environmental Law Journal*, The University of South Carolina School of Law, Winter 1993.

11. Richard W. Riley quoted by Fred Shapiro, "A Reporter At Large: Nuclear Waste," *The New Yorker*, October 19, 1981, p. 53, cited in Newberry, *ibid.*

12. U.S. Department of Energy, "Low-Level Radioactive Waste Policy Act Report: Response to Public Law 96-573 at 7," 1981, cited in Newberry, *ibid.*

13. In 1998, New Jersey adopted higher limits on some roads on a trial basis.

Chapter 4. First Steps: The New Jersey Siting Process Begins

14. New Jersey Low-Level Radioactive Waste Disposal Facility Siting Act (P.L. 1987, Chapter 333), 13:iE-50.

Chapter 5. Connecticut and Illinois: The Seeds of the Voluntary Siting Process

15. John Larson did, in fact, run for governor of Connecticut in 1994, but was defeated in the Democratic primary. In 1998, he was elected to the U.S. Congress from Connecticut's 1st District. He was reelected in 2000.

16. Egon French, R&R Ventures, Inc., "Presentation to NLLW Symposium on Understanding Community Decision Making," Philadelphia, Pa., March 4-5, 1998.

17. Paul E. Wyszkowski, letter, February 2, 1999.

18. In Vermont over the previous two years, the Low-Level Radioactive Waste Authority had fought a heated public battle with the state's Agency of Natural Resources over site selection methodology. The result was that the Agency allowed the Authority to conduct a volunteer siting program but insisted that they also conduct statewide screening. While it would be logical to assume that New Jersey was influenced by Vermont's experience, apparently it wasn't.

19. Much of the information about Martinsville is from "The Illinois Approach to Public Participation and Development of a Community Compensation and Benefit Package," an unpublished draft report by Tom Kerr, who directed the Illinois siting program during its deliberations with Martinsville.

20. Illinois Department of Nuclear Safety, and City of Martinsville, "Summary of the Martinsville Community Agreement" and "A Resolution Approving An Agreement Between The State of Illinois And The City Of Martinsville Concerning The Development And Operation Of A

Low-Level Radioactive Waste Disposal Facility And Granting And Affirming Approval Of Such Facility For Location Within The City Of Martinsville," Martinsville, Ill., June 3, 1992.

Chapter 6. 'One Tough Job: Peddling Radioactive Waste'

21.   New Jersey Low-Level Radioactive Waste Disposal Facility Siting Board, *Insight*, Trenton, N.J., Fall 1993.

22.   The League of Women Voters Education Fund, *The Nuclear Waste Primer: A Handbook for Citizens* (New York, N.Y.: Lyons and Burford, revised edition, 1993).

23.   Pew Research Center, Philadelphia, Pa., November/December 1996.

24.   Charles Peters, "Tilting At Windmills," *The Washington Monthly*, Washington, D.C., January 1995.

25.   John Chancellor, b. July 14, 1927, d. July 12, 1996.

Chapter 7. Culture

26.   Pete Seeger, *Gazette-Containing a Collection of Topical Songs, Old and New without direction as to content or pressure*, Folkways Records, Washington, D.C., 1958.

27.   Niels Bohr, *Atomic Physics and Human Knowledge*, (New York, N.Y.: Jon Wiley and Sons, 1958), p. 38.

28.   David R. Schwarz, "Science: From Hero to Villain in One Generation. What Next?" The Seminar Discussion Group Series of the Lyceum Club of the New York Academy of Sciences, New York, N.Y., April 9, 1996.

29.   Duck's Breath Mystery Theatre, reprinted in *Funny Times*, 1997.

30.   Vern Partlow, "Talking Atom," recorded by Pete Seeger, *op. cit.*

31.   *Atomic Café* soundtrack album, Rounder Records, Cambridge, Mass.

32.   Tom Lehrer, "The Wild West Is Where I Want To Be," *Songs and More Songs by Tom Lehrer*, Rhino Records, Los Angeles, Cal., written in 1953.

33.   Anonymous, "What Do We Do With It?" This song, written at the Fourth Annual N.J. Conference on Environmental Music in 1995, was sent to the Siting Board by Anna Sanders, who stressed that it was the work of an impromptu group.

34.   Tom Callinan, "The New Jersey Waste-Land," Cannu Music, Ltd., Clinton, Conn., 1995.

35.   Calvin Trillin, "Smear Window," *Time*, July 14, 1997.

Part 2. 1995 - The First Year of the Voluntary Process

Chapter 10 Selling Snake Oil at Rotary Clubs

36.   Robert D. Putnam, "Bowling Alone," *Journal of Democracy*, Cambridge, Mass., January 1995.

37.   See Rotary International web page at www.Rotary.org.

Chapter 13 Welcome to Allowayste

38.   See Ruritan web page at www.Ruritan.org.

39.   Lester Sutton, telephone conversation, September 17, 1998.

Chapter 16 Barnwell Returns

40. Nina Brook, reporting in *The State*, Columbia, S.C., June 30, 1995.

41. Harriet Keyserling, *Against The Tide—One Woman's Political Struggle*, University of South Carolina Press, 1998, p. 182.

42. "Sited states" was the term of art for the three states with operating, and therefore "sited," disposal facilities.

43. October 1, 1990 letter to Governor James Florio from Director Jerry Griepentrog of the Nevada Department of Human Resources, Chief Heyward G. Shealy of the Bureau of Radiation Health in the South Carolina Department of Health and Environmental Control, and Program Manager Roger Stanley of the Nuclear and Mixed Waste Management Program in the Washington State Department of Ecology.

44. January 28, 1991 letter to Governor Florio from Jerry Griepentrog of Nevada and Roger Stanley of Washington State.

45. January 2, 1992 letter to Governor Florio from Roger Stanley of Washington State.

46. Jeff Miller, *The State*, Columbia, S.C., May 29, 1992.

47. Letter to Richard Sullivan, Chair, Northeast Compact Commission, from Southeast Compact Commissioners Carlisle Roberts, Jr. and Heyward G. Shealy, July 1, 1992.

48. Harriet Keyserling, *op. cit.*, p. 183.

Part 3. 1996 - The Second Year of the Voluntary Process

Chapter 17 The Saturday Paper

49. *The New York Times*, December 9, 1995, p. B1.

Chapter 20 Environmental Justice and Other Lessons from Fairfield

50. For a listing of the 48 environmental justice complaints filed between September 1993 and April 1998, see *LLW Notes*, LLW Forum, c/o Afton Associates, Vol. 13, No. 3, Washington, D.C. (www.afton.com/llrwforum, *April 1998*.)

51. New Jersey Low-Level Radioactive Waste Disposal Facility Siting Board, *New Jersey's Voluntary Plan for Siting a Low-Level Radioactive Waste Disposal Facility*, Trenton, N.J., pp. 10, 14, and 16.

Chapter 22 The Poll

52. I now work at the Eagleton Institute and enjoy the perspective of having been a satisfied client of its Center for Public Interest Polling.

Chapter 23 More Planning for the Election

53. Vice President Al Gore, *The Gore Report On Reinventing Government* (New York, N.Y.: Times Books, 1993), p. 17.

54. By December 2000, the web page for the National Low-Level Waste Management Program read, "Due to federal funding cuts the National Low-Level Waste Management Program has closed its doors."

Part 4. "Working For The Government Is Cool"

Chapter 26 A Bureaucrat's Journal from April 28, 1997 to February 12, 1998

55.   The Washington Generals were the Harlem Globetrotters' most frequent opponent until they were disbanded in 1995. Now the Globetrotters generally play the New York Nationals.

56.   Harold Stassen was governor of Minnesota from 1938 to 1943. He sought the Republican nomination for President in 1948, but did not become a metaphor for a totally unrealistic pursuer of hopeless quests until he continued to run every four years thereafter. Although he announced his candidacy in 2000 when he was 93 years old, he mounted no visible campaign.

57.   *The West Wing*, the program introduced in the fall of 1999 on NBC-TV, is close to what I had in mind.

58.   By reinvigorating this open space commission, Governor Whitman began a process that led the Legislature to place a referendum on the 1998 ballot which New Jersey voters overwhelmingly approved, providing for $1 billion to be spent for open space preservation over a ten-year period.

59.   Cynthia Covie and Michael Hogan are now New Jersey Superior Court judges. Both were nominated by Governor Whitman and confirmed by the state Senate.

60.   This book is based largely on my recollection and notes I took at the time. When representatives of the New Jersey Department of Environmental Protection reviewed a draft of the report subsequently published by the National Low-Level Waste Management Program (See Endnote #1), they felt that my description of the January 30, 1998 meeting "left out some important points." They write: "According to Commissioner Shinn's recollection, the discussion began by noting the dramatic reduction in volume of LLRW that had taken place, approaching a 90% reduction from 10 years ago. Several facilities in other states seemed to be nearly ready to accept waste, including Texas, and the Envirocare facility in Utah was able to accept higher concentrations of radioactivity than previously. This provided additional disposal capacity to NJ generators than just access to Barnwell in South Carolina. Commissioner Shinn firmly believed that siting a facility in NJ was no longer achievable or necessary given the out-of-state options. The NJ LLRW Siting statute gives the Siting Board two main duties — to site a facility in NJ and to guarantee access to other facilities for NJ generators. Commissioner Shinn felt that the Siting Board should stop focussing on siting in NJ, and should concentrate on their responsibilities for guaranteeing access to out-of-state facilities. The Siting Board had over $8M that could be used for either option and the Commissioner argued against spending money on the siting effort in NJ due to its uncertainty, and the lack of necessity." (February 24, 1999 letter from Jill Lipoti, Assistant Director of DEP's Division of Environmental Safety, Health and Analytical Programs, to Kathleen A. Asbell, National Low-Level Waste Management Program.)

61.   New Jersey Low-Level Radioactive Waste Disposal Facility Siting Act (P.L. 1987, Chapter 333), 13:1-179, Section (a).

62.   *Ibid.*, Chapter 166, Section 5 (a), as amended 1991.

Part 5. False Starts, Potential Solutions

Chapter 28 The Problem Is Solved

63.   Keyserling, *op. cit.*, p. 86.

64. *Task Force Recommendations to the Governor and General Assembly*, South Carolina Nuclear Waste Task Force, Columbia, S.C., December 15, 1999, p. 1.

65. *Ibid.*, cover letter signed by Task Force Chair Butler Derrick.

66. Governor Jim Hodges quoted by the LLW Forum at www.afton.com/llwforum, May 26, 2000.

67. Representative Joel Lourie, as quoted by Joseph S. Stroud and Sammy Fretwell, *The State,* Columbia, S.C., and by Rachel Graves, *The Post and Courier*, Charleston, S.C., both May 25, 2000.

68. Paul Wyszkowski, personal communication, May 3, 2000.

Chapter 29 In-State Options

69. "Special Announcement For All Americans," Better World Technology, Newfoundland, N.J.

Chapter 30 Out-of-State Options

70. Governor George W. Bush, quoted in *LLW News Flash*, LLW Forum, c/o Afton Associates, Washington, D.C., October 23, 1998.

71. A photocopy of the February 21, 1996 memo from John Garamendi to Bruce Babbitt became public and was widely circulated among those following low-level radioactive waste issues. U.S. Senator Frank H. Murkowski, R-Alaska, entered it into the *Congressional Record* with the following introduction: "Mr. President, it is often useful to compare the public statements of Government officials with their private statements. Such a comparison can say a great deal about an official's true motives, not to mention his character..." After reading the memo, Murkowski continued, "Mr. President, here is the Deputy Secretary of Interior engaged in a PR campaign to portray the governor of California as a venal toady. For those in this Chamber who may not know the precise definition of a 'venal toady,' it means a deferential, fawning parasite open to bribery...Is this what Deputy Secretary Garamendi calls the high ground?" *Congressional Record*, October 21, 1997, p. S10907.

Chapter 31 High-Level Waste: The Federal Experience

72. U.S. General Accounting Office, *op. cit.*

73. Douglas County Republican Central Committee web site at www.douglasgop.org.

74. Senator Frank H. Murkowski, quoted in "Senate and Clinton Still Stalled on Nuclear Waste Disposal," *The New York Times*, February 11, 2000, p. A16.

Part 6. Fear and Distrust: The Modern Malady

Chapter 32 Environmentalists and Risk: Have We Met The Enemy?

75. Lou & Peter Berryman, "Why Can't Johnny Bowl?" This is one of their many funny songs from the CD *Cow Imagination*, Cornbelt Records, Madison, Wis. (members@aol.com/berrymanp, 1990).

76. NIRS Web Page at www.NIRS.org, December 1998.

Chapter 33 Fear of Risk – It Could Happen

77. Newberry, *op. cit.*, p. 58.

78. Terry Davies, *Center For Risk Management Newsletter*, Resources For The Future, Washington, D.C., Fall 1997.

79. James Walsh, for example, in his book *True Odds: How Risk Affects Your Everyday Life*, attributes at least 90 percent of premature deaths to these six causes. As quoted by Jim Shea in the *Hartford Courant*.

80. Jeremy Hall, *Real Lives, Half Lives: Tales From The Atomic Wasteland* (London, England: Penguin Books, 1996). In 1995, Hall, a British journalist and documentary film-maker, traveled the world and prepared profiles of a dozen people, including me, whose lives "have been exposed, in one way or another, to radioactivity." The book that resulted provides a fascinating range of experiences and perspectives. Hall managed to write sympathetically about each person regardless of their diverse points of view. The chapter based on Hall's interview with me is called "Fear of Frying."

81. The song, "NIMBY," is from *ARTISAN*, a British trio that sings *a cappella*. It is written by Brian Bedford of the group and is on their CD *Our Backyard*, Bedspring Music, Birdsedge, Huddersfield, England (artisan@artifact.demon.co.uk., 1996).

82. Jeff Wheelwright, "Atomic Overreaction," *Atlantic Monthly, April 1995*, p. 28.

83. Gina Kolata, "With an Eye on the Public, Scientists Choose Their Words," *The New York Times*, January 6, 1998, p. B12.

84. Albert Einstein, from the chapter "Attributed to Einstein" in *The Quotable Einstein* (Princeton, N.J.: Princeton University Press, 1996), p. 224.

85. American Council on Science and Health, *Facts Versus Fears: A Review Of The Greatest Unfounded Health Scares Of Recent Times* (third edition) by Adam J. Lieberman and Simona C. Kwon, New York, N.Y., June 1998, pp. 33-34.

86. Gerrard, *op. cit.*, p. 132.

87. Peter L. Spencer, "Asbestos risk or not?" *Consumers Research Magazine*, Washington, D.C., November 1993. Also M. Ross, "The schoolroom asbestos abatement program: a public policy debacle," *Environmental Geology*, New York, N.Y., 1995, pp. 182-188. Also Denise Scheberle, "Indecent Exposure," *Forum for Applied Research and Public Policy*, Knoxville, Tenn., Summer 1998.

Chapter 34 Fear of Government

88. Newberry, *op.cit.*, pp. 72-73.

89. Richard Sullivan, telephone conversation, July 1998.

90. Peter M. Sandman, "Siting Controversies: Some Principles, Paradoxes and Heresies," *CONSENSUS*, July 1992, p. 36, quoted in Newberry, *op. cit.*, p. 76.

91. Richard F. Paton, "The National Low-Level Radioactive Waste Act: Success or Failure?" *RADWASTE*, July 1997, p. 23.

92. Personal communication with members of the Kempton, Pa.-based musical group *LEFT FIELD* (www.infomonger.com, 1998).

Part 7. Politics Is More Difficult Than Physics

93. Albert Einstein, quoted in *The Quotable Einstein, op. cit.*, p. 141.

Chapter 35 Making Government Work In A NIMBY Age

94. Quoted by Allan R. Gold in "Counties Battle Radioactive Waste Sites," *The New York Times*, February 26, 1990, p. B1.

95. William D. Ruckelshaus, "From Conflict to Collaboration: Restoring Trust in Government," a speech delivered to the Institute for Environment and Natural Resources, University of Wyoming, May 1, 1997.

96. New Jersey Regional Low-Level Radioactive Waste Disposal Facility Siting Act, P.L. 1987, Chapter 333, Section 17.

97. Charles Mantesian, "The Endless Struggle Over Open Meetings," *Governing*, p. 49, December 1997.

98. *Ibid.*, p. 51.

99. Governor Tom Kean, quoted in "Board member wants progress," *The Star-Ledger*, December 3, 1997, p. 25.

100. Alan Rosenthal, *The Decline of Representative Democracy* (Washington, D.C.: Congressional Quarterly Press, 1998), p.340

101. Dietrich Schwarz, "Ethical Issues In Radiation Protection, Continued," *Health Physics*, Vol. 75, No. 2, August 1998, p. 186.

102. Gerrard, *op. cit.*, p. 64.

103. Peter Sandman quoted in Gerrard, *ibid.*, p. 106.

104. Michael O'Hare and Debra Sanderson quoted in Gerrard, *ibid.*, p. 130.

105. Cited by Richard Morin, "Unconventional Wisdom — Is This A Nation Of Extremists?" *Washington Post*, January 11, 1998, p. C-5.

106. Putnam, *op. cit.*

107. Dirk Johnson, "Civility in Politics: Going, Going, Gone," *The New York Times*, December 10, 1997, p. A20.

108. Tom Kerr, comments at Voluntary States Working Group Final Meeting, Lambertville, N.J., June 25, 1998.

109. Gore, *op. cit.*, p. 17.

110. New Jersey Low-Level Radioactive Waste Disposal Facility Siting Act, *op. cit.*, Section 4 f.

111. Rosenthal, *op. cit.*, p. 340.

Index

About the author

John Weingart served as Assistant Commissioner of the New Jersey Department of Environmental Protection and as Director of its Division of Coastal Resources before directing the state's Low-Level Radioactive Waste Disposal Facility Siting Board from 1994 to 1998. A 24-year veteran of state government, he served under Governors Brendan Byrne, Tom Kean, Jim Florio, and Christine Todd Whitman — two Democrats and two Republicans.

Weingart is Associate Director of the Eagleton Institute of Politics at Rutgers University. He lives in Sergeantsville, New Jersey with his wife, Deborah Spitalnik, and their daughter, Molly. On Sunday nights, he hosts *Music You Can't Hear On The Radio*, New Jersey's oldest folk music and bluegrass radio program, on WPRB at 103.3 FM and WPRB.com.

Weingart's previous book, *Reform of Undergraduate Education*, written with Arthur E. Levine, received the "Book of the Year" award from the American Council on Education.

He wrote *Waste Is A Terrible Thing To Mind* as a Senior Fellow at the Center For Analysis of Public Issues, Princeton.